GENETICS OF MAN

GENETICS OF MAN

F. Clarke Fraser, Ph.D., M.D.C.M., F.R.S.C.
(Acadia)

Professor of Medical Genetics, Department of Biology
Associate Professor, Department of Paediatrics
McGill University, Montreal
Director, Department of Medical Genetics
The Montreal Children's Hospital
Montreal, Canada

James J. Nora, M.D.

Professor of Pediatrics
University of Colorado
School of Medicine
Denver, Colorado

Lea & Febiger

Philadelphia

1975

Library of Congress Cataloging in Publication Data

Fraser, F Clarke, 1920-
Genetics of man.

Includes bibliographies and index.
1. Human genetics. 2. Medical genetics.
I. Nora, James J., joint author. II. Title.
[DNLM: 1. Genetics, Human. 2. Hereditary diseases. QZ50 F841g]
QH430.F7 573.2'1 75-4743
ISBN 0-8121-0484-6

Published in Great Britain by Henry Kimpton Publishers, London

PRINTED IN THE UNITED STATES OF AMERICA

We dedicate this book, with love, to the following important people, who will know whom we mean:

NORAH, NOEL, ALAN, SCOTT,

WENDY, PENELOPE, MARIANNE,

JAMES, AND ELIZABETH.

PREFACE

Human genetics has changed dramatically over the past twenty-five years, from a science devoted largely to the collection of pedigrees attempting to illustrate the mendelian inheritance of physical or pathological traits to a full blown branch of medicine. We felt that there was a need for a book, written for undergraduates, which would emphasize the social and medical implications of our burgeoning knowledge of human genetics and its practical applications. Much of the content is drawn from a course in human genetics given to junior and senior undergraduate students in a Bachelor of Science program. Much of it is drawn from our own experience in applying genetic knowledge to human problems.

Because we felt an even greater need for a book on genetics written for senior medical students and physicians we wrote *Medical Genetics: Principles and Practice* (Nora and Fraser, 1974) first. In *Genetics of Man* we have removed a good deal of the strictly clinical material, added chapters on "Race" and "Pharmacogenetics," provided a Glossary, amplified somewhat the chapters on basic genetics, and brought the material up to date. We have included enough basic genetics so that the book can be used as a text for an introductory course in genetics, with human emphasis, as well as for a course in specifically human genetics. The references cited are, of course, not exhaustive, but have been chosen either for their historical significance or as recent articles from which the reader can find her/his way into the literature.

Montreal, Canada — F. Clarke Fraser
Denver, Colorado — James J. Nora

ACKNOWLEDGEMENTS

Many people have helped us in many ways in the preparation of this book and we are grateful to all of them. In particular we would like to thank Dr. Marilyn Preus Fraser and Dr. Audrey Hart Nora for writing Chapters 15 and 18 and 19, respectively; Ms. M. Forster, Ms. F. Langton, and Ms. M. Peterson for secretarial help; and Ms. J. Ingram and Ms. R. Bayreuther who produced some of the illustrations. Several colleagues obliged us by reviewing specific chapters: Dr. A. Robinson and Dr. F. Hecht (cytogenetics), Dr. C. Scriver (biochemical genetics), Dr. K. Sittman (population genetics), and Dr. A. G. Steinberg (race).

CONTENTS

Chapter 1

GENETICS AND DISEASE

The idea that certain diseases, as well as normal physical and mental traits, were likely to "run in families" has been with us since long before the time of Hippocrates. The Mosaic law, for instance, contains rules about excusing from circumcision certain male relatives of "bleeders," which shows that the basic principles of sex-linked recessive inheritance were recognized at that time.

At the end of the nineteenth century inheritance was generally considered to be blending—hereditary factors for contrasting characters were thought to blend with one another when they occurred in the same individual. Mendel's chief contribution was to show that the hereditary factors did not blend when combined in one individual, but retained their specificity and could reappear unchanged in subsequent generations. When Mendel's laws, published in 1866, became generally recognized in 1900, examples of Mendelian inheritance were quickly found in man, one of the first being alkaptonuria, a rare metabolic disease shown by Sir Archibald Garrod, a physician, and the geneticist Bateson, to fit the expectation for a recessive trait. Blending inheritance was then reserved for "quantitative characters," those that did not lend themselves to classification into discrete categories, but rather to measurement. By 1910 it was clear that the genetic behavior of even quantitative characters could be accounted for in terms of Mendelian factors.[2] There followed a period where attempts were made to fit large numbers of human traits into the mold of Mendelian inheritance.

Many traits fitted beautifully; others had to be bent, by virtue of secondary assumptions such as reduced penetrance, to fit the expectations of Mendelian inheritance. Although some cases required an undue number of assumptions, enough of these efforts were successful that they set the stage for the eugenics movement.[2] Eugenics refers to the science of race improvement. If a major component of man's ills were due to Mendelian mutant genes, it was argued, could not many of man's health problems be eliminated by controlled breeding? Simply prevent the carriers of deleterious genes from breeding and one could remove a major source of man's woes. Soon a split arose between those who wished to promote our understanding of human heredity by the objective collection and critical analysis of data and those who wished to popularize the results and incorporate the conclusions into programs for social and political action. Some of the eugenicists were quite uncritical of the data; it was seriously claimed, for instance, that mental deficiency in general showed Mendelian recessive inheritance. Eugenic programs and some legislation were based on this assumption.

Unfortunately, the real situation is not so simple, and the dangers of compulsory selective breeding were demonstrated by the crimes committed in the name of eugenics by the Nazis in the 1930's and 1940's. In Germany the eugenics movement took the name of "*Rassenhygiene*," or race hygiene, and the direction in which it was leading was evident well before Hitler came to power. Its integration into public policy was a dramatic demonstration of the problems resulting from the premature and ill-considered application of genetic advances to social issues. The Nazis, in 1933, promulgated a law for sterilization of the "unfit," often based on highly erroneous genetic information, and for the exclusion of Jews from the new state, ostensibly to prevent "contamination" of the Aryan "race." These laws were supported by some outstanding German geneticists. In 1941 von Verscheuer said: "Decisive for the history of a people is what the political leader recognizes as essential in the results of science and puts into effect. The history of our science is most intimately connected with German history of the most recent past. The leader of the German state is the first statesman who has wrought the results of genetics and race hygiene into a directing principle of public policy." And the Minister of the Interior, responsible for enforcing the laws, said: "The fate of race-hygiene, of the Third Reich, and the German people will in the future be indissolubly bound together."[2] It was only a step from here to mass genocide, in which millions of Jews, Gypsies, and other "undesirables" were exterminated, officially on the grounds of "racial hygiene." The results of this disaster were to give eugenics a bad name from which it has yet to recover.[4]

Nevertheless, rapid advances have been made in human, and particularly medical, genetics, due largely to three factors. First, advances in molecular genetics made it possible to identify the effects of mutant human genes in precise biochemical terms, and hence to work out rational methods of precise diagnosis, treatment, and carrier detection.[6] Second, the advent of the atomic age aroused general interest, not to say alarm, about the possible genetic damage that could be done by atomic fallout, and this raised questions about man's present mutation rates and genetic load. (The geneticists had long since been expressing concern about the harmful genetic effects of man-made radiation, but it took the atomic bomb to blast the problem into public awareness.) The realization that we were woefully ignorant of these matters led to a great increase of activity in the field of human population genetics,[5] resulting in a better understanding of how mutations, selection, and genetic drift affect the frequencies of genes in human populations. There came also a better recognition of the multifactorial diseases—those familial diseases resulting from the interaction of many genetic factors, individually largely indefinable, and a variety of environmental factors.[1] Third, the development of improved methods of observing human chromosomes led to the recognition of a whole new category of diseases, chromosomal diseases, resulting from excesses or deficiencies in the chromosomal material.[3] This, perhaps even more than the other two factors mentioned above, aroused the interest of the medical practitioner and led to such a demand for services in clinical genetics that a department of medical genetics where cases can be referred for diagnostic help and for genetic counseling is an accepted feature of most modern large medical centers today.

The aim of this book is to present the fundamentals of genetics, particularly as they relate to man, and to show how our increasing knowledge of these fundamentals is improving our understanding of human ills from the point of view of biochemical and cytogenetic diagnosis, our understanding of multifactorial diseases, the management of hereditary diseases and genetic counseling, and man's genetic load, now and in the future.

SUMMARY

Human genetics has developed from fairly vague ideas about the familial nature of disease, and the study of pedigrees that fit, or did not fit, the Mendelian laws, into an important branch of medicine. Recent rapid progress has resulted largely from (a) interest in gene frequencies and mutation rates resulting from the advent of the atomic age, (b) advances in biochemistry leading to an understanding of gene structure and function in molecular terms, and (c) improved techniques of visualizing human chromosomes leading to the recognition of chromosomal diseases.

REFERENCES

1. Carter, C. O.: Multifactorial inheritance revisited, p. 227–232. *In* F. C. Fraser and V. McKusick (eds.), Congenital Malformations. New York, Excerpta Medica, 1970.
2. Dunn, L. C.: Cross currents in the history of human genetics. Am. J. Hum. Genet. 14:1, 1962.
3. Hamerton, J. L.: Human Cytogenetics. Vols. I and II. New York, Academic Press, 1971.
4. Ludmerer, K. M.: Genetics and American Society. Baltimore, The Johns Hopkins Press, 1972.
5. Morton, N. E.: The future of human population genetics, p. 103–124. *In* A. G. Steinberg and A. G. Bearn (eds.), Progress in Medical Genetics. Vol. 8. New York, Grune & Stratton, 1972.
6. Stanbury, J. B., Wyngaarden, J. B., and Fredrickson, D. S. (eds.): The Metabolic Basis of Inherited Disease. 3rd ed. New York, McGraw-Hill Book Co. Inc., 1972.

Chapter 2

CHROMOSOMAL BASIS OF HEREDITY

The word *chromosome* was introduced in 1888 by Waldeyer.[16] Several investigators, working in the late nineteenth and early twentieth centuries, appreciated before the rediscovery of Mendel that the chromosomes could be the carriers of heredity.

However, it was the rediscovery of Mendel that provided the catalyst for the reaction that synthesized the discoveries of cytology and genetics into the discipline of *cytogenetics*. It became apparent to the cytologists that the behavior of the hereditary characters of Mendel was reflected by the behavior of the chromosomes in *meiosis*. Sutton and Boveri independently proposed the chromosomal hypothesis of inheritance.

The remarkable contributions to knowledge of the chromosomal basis of heredity made over the next decades were, of necessity, derived from studies in lower animals, particularly the Drosophila. As early as 1910, T. H. Morgan was able to locate a specific *gene locus* on a specific chromosome of Drosophila melanogaster.

Recognizing that the hereditary material was carried by the chromosomes did not, of course, define the nature of the unit of inheritance, for which Johannsen coined the term "gene." This line of investigation will be developed in Chapter 5. It is sufficient for this discussion to state that the chromosome consists of the hereditary material *deoxyribonucleic acid* (*DNA*), embedded in a protein matrix (largely histone in man) to form a deoxyribonucleoprotein fiber, *chromatin*. The units of inheritance, the genes, are segments of DNA. The number of genes distributed throughout the 46 chromosomes of the human cell has been estimated to be of the order of 100,000. An excellent review of the structure and function of chromosomes has been presented by Comings.[4]

The chromosomal constitution of each individual is derived equally from mother and father, 23 chromosomes being contributed by each parent in the form of a *gamete* (ovum or sperm). The cell formed by the fertilization of the ovum by the sperm is the *zygote*. Each of the 23 paternal chromosomes in the sperm has a homologue in the ovum. Thus the end result of the fusion of two germ cells (gametes), each with a *haploid* number of chromosomes, is a *diploid* cell having 23 homologous pairs of chromosomes.

The human is an unsatisfactory subject for genetics research in many ways, and especially in the area of cytogenetics. It was not until 1956 that the diploid number of human chromosomes was demonstrated to be 46.[15] For 33 years prior to this, students of medicine and biology had been taught that the human diploid complement was 48.[10] The reason for this discrepancy was not carelessness on the part of the cytogeneticists, but inadequacy of the existing cytogenetic techniques.

CHROMOSOMES

Chromosomes (*chromos* = color; *soma* = body) are not individually distinguishable except during cell division, at which time they may be seen under the light microscope as rodlike bodies that stain with basic dyes. Twenty-two pairs are identical in males and females and are designated *autosomes.* The homologous chromosomes in each pair of autosomes are usually indistinguishable. The chromosomes in the remaining pair are called the sex chromosomes. In the female the two sex chromosomes are identical and are referred to as X chromosomes. In the male there is one X chromosome and a distinctly different Y chromosome.

Figure 2-1 is a photomicrograph of the chromosomes of a single human peripheral-blood leukocyte as they appear under the light microscope in metaphase, the stage of cell division during which they are most readily studied. In Figure 2-2 chromosomes from a normal male have been individually cut out of the photomicrograph and arranged on the basis of their size and the position of the centromere. This convention was established at a meeting of human cytogeneticists in Denver in 1960 and is thus known as the Denver classification. At a similar meeting in London in 1963, it was

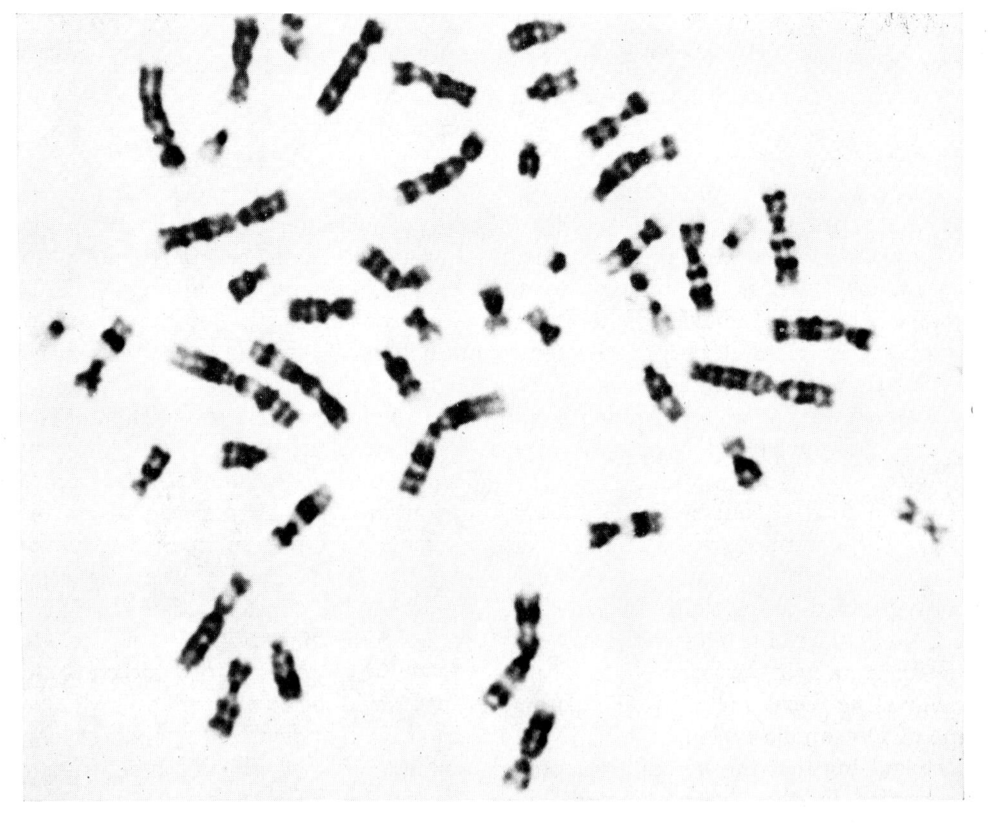

Figure 2-1. Photomicrograph of human chromosomes in metaphase showing G-banding by the trypsin technique. (Courtesy of A. Robinson and D. Peakman.)

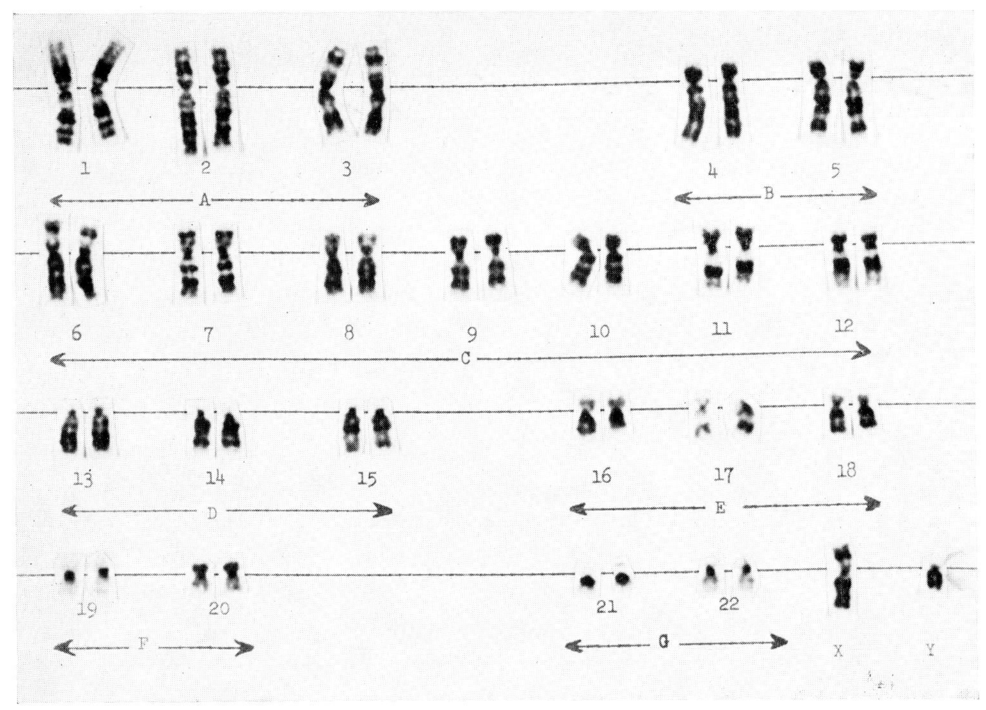

Figure 2-2. Chromosomes from a normal human male have been individually cut out of the photomicrograph shown in Figure 2-1 and are displayed in a karyotype. (Courtesy of A. Robinson and D. Peakman.)

agreed to use letter designations for the various groups as shown in Figure 2-2. This array of chromosomes in a form suitable for analysis is called a *karyotype*. Further modifications were added at yet another conference in Chicago in 1966,[3] to code for numerical and structural aberrations. Most recently, in Paris, in 1971, some slight alterations were made in the Chicago nomenclature.[12] Figure 2-3 illustrates some of the terminology for the position of the centromere and for structural alterations. If a centromere is located in the middle of a chromosome, the chromosome is called *metacentric* (or mediocentric). If the centromere is off-center, the chromosome is *submetacentric;* and if the centromere is located near one end of the chromosome, the term *acrocentric* is used. The short arm of a chromosome is designated by letter "p" (for "petit") and the long arm by "q." Tables 2-1 and 2-2 summarize the morphologic features and nomenclature proposed at the Denver, London, Chicago, and Paris conferences.

The absence or addition of a whole chromosome is indicated by a minus sign or a plus sign *before* the symbol for that

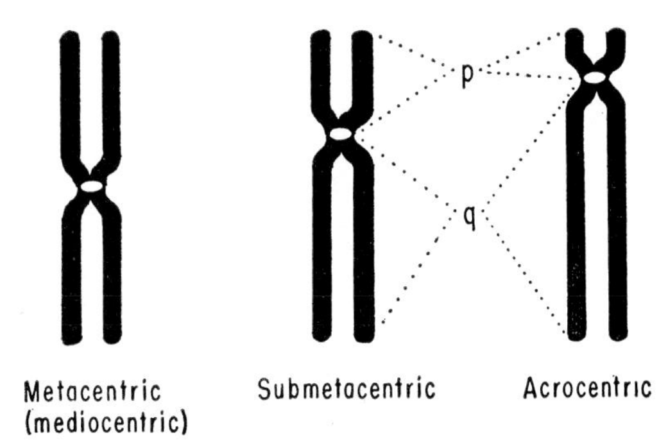

Figure 2-3. Chromosome nomenclature. The centromere of a metacentric (mediocentric) chromosome lies in the middle; the centromere is off-center in a submetacentric chromosome and is near the end of an acrocentric chromosome. The short arm of a chromosome is designated "p" and the large arm "q."

TABLE 2-1. Description of Human Mitotic Chromosomes (Adapted from the 1960 Denver and 1963 London Conferences)

Group 1-3 (A)	Large chromosomes with approximately median centromeres (metacentric, mediocentric) readily distinguished from each other by size and centromere position. In No. 1, a secondary constriction may be observed in the proximal region of the long arm.
Group 4-5 (B)	Large chromosomes with submetacentric centromeres. Chromosome 4 is slightly longer.
Group 6-12 and the X chromosome (C)	Medium-sized submetacentric chromosomes. Chromosomes 6, 7, 8, 11, and X are more metacentric than 9, 10, and 12. The X chromosome most resembles No. 6. In individuals having two X chromosomes, one of these characteristically incorporates tritiated thymidine later than any other chromosome.
Group 13-15 (D)	Medium-sized acrocentric chromosomes with satellites on the short arms.
Group 16-18 (E)	Rather short submetacentric chromosomes. No. 16 is comparatively more metacentric and may have a secondary constriction in the proximal part of the long arm.
Group 19-20 (F)	Short metacentric chromosomes.
Group 21-22 and the Y chromosome (G)	Very short acrocentric chromosomes. Nos. 21 and 22 may have satellites.

TABLE 2-2. Karyotype Nomenclature Adapted from the Chicago Conference[3] and the Paris Conference[12]

Karyotype	*Condition*
46,XX	46 chromosomes, (normal female)
46,XY	46 chromosomes, (normal male)
45,X	45 chromosomes, 1 X chromosome (Turner syndrome)
47,XXY	47 chromosomes, 2 X chromosomes and 1 Y chromosome (Klinefelter syndrome)
47,XYY	47 chromosomes, 2 Y chromosomes (XYY syndrome)
49,XXXXY	49 chromosomes, 4 X chromosomes and 1 Y chromosome (XXXXY syndrome)
47,XY,+21	47 chromosomes, male (21 trisomy, Down syndrome)
46,XY,−D,+t(Dq21q)	46 chromosomes, male, chromosome absent from D group, additional translocation chromosome made up of a long arm of a 21 chromosome and long arm of a D group chromosome (unbalanced D/21 translocation Down syndrome)
46,XY,i(21q)	46 chromosomes, male, isochromosome of long arm of No. 21 chromosome (Down syndrome)
46,XY/47,XY,+21	Double cell line mosaic (normal/21 trisomy), male (mosaic Down syndrome)
47,XX,+18	47 chromosomes, female, additional No. 18 chromosome (18 trisomy)
46,XX,18q−	46 chromosomes, female, deletion of long arm of a No. 18 chromosome (18q− syndrome)
46,XX,5p−	46 chromosomes, female, deletion of short arm of a No. 5 chromosome (cri-du-chat syndrome)
46,XX,r(D)	46 chromosomes, female, group D ring chromosome

chromosome. Thus, the nomenclature for a male having an additional chromosome 21 would be 47,XY,+21. A plus or minus sign placed *after* a symbol means an increase or decrease in length. A short-arm deletion of chromosome 5 is termed 5p−, and the nomenclature for a female patient having the cri-du-chat syndrome is 46,XX,5p−, the total chromosome number, followed by the sex chromosomal constitution, followed by the autosomal abnormality. An increase in the length of the long arm of a chromosome 18 would be termed, in a female patient, 46,XX,18q+.

MORPHOLOGY OF THE HUMAN KARYOTYPE

On morphologic grounds alone some of the human chromosomes within each group can be confidently distinguished, but many cannot. *Autoradiography* (in which the location of radioactively labeled molecules is demonstrated photographically) has helped to differentiate between certain morphologically similar chromosomes. It is the only method for identifying the late-replicating X chromosome. It can be shown by adding tritiated thymidine to the cell culture and then visualizing the labeled chromosomes by autoradiography that one X chromosome replicates later than all other chromosomes (when more than one X chromosome is present). This chromosome is heavily labeled in cells where there is virtually no label on the other chromosomes, showing that it is still replicating at a time when the other chromosomes have completed DNA synthesis. Chromosome 18 appears to replicate later than the morphologically similar chromosome 17, chromosome 4 later than 5, and chromosome 13 later than the rest of Group D.

Some regions of a chromosome, referred to as *heterochromatin*, are irregularly variable in their staining properties, often staining more darkly, but sometimes more lightly than the rest of the chromosome (*euchromatin*). The exact nature and function of the heterochromatin is still unclear.[3] It is thought to be genetically less active than euchromatin. It is found most consistently near the centromere, and near the ends of the chromatid arms, though blocks may occur in other regions. Some of it, at least, is associated with the so-called redundant DNA, which contains many identical base sequences that are rich in thy-

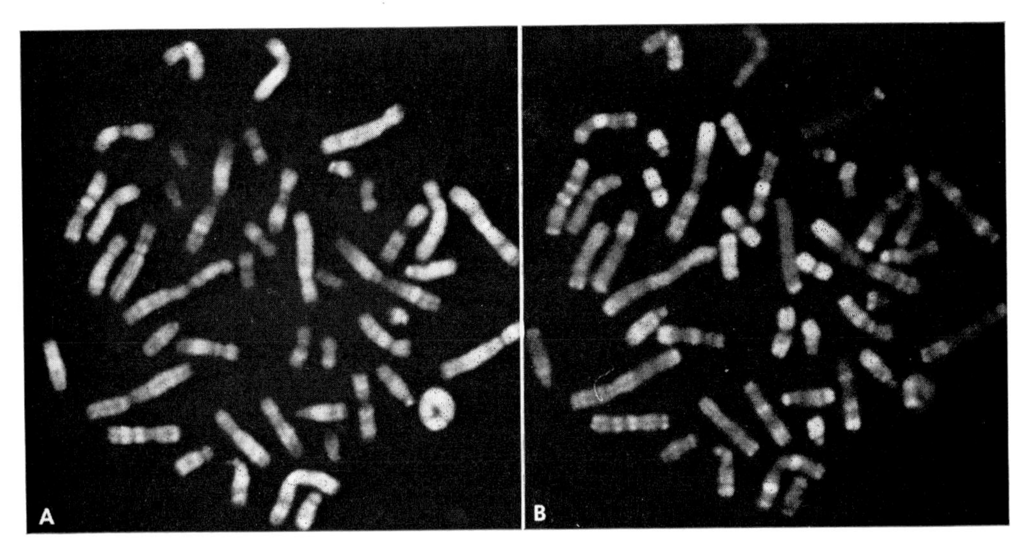

Figure 2-4. The same chromosome preparation sequentially stained by (*A*) quinacrine mustard (Q-banding) and (*B*) acridine orange (R-banding). That the fluorescence in R-banding is the reverse of Q-banding (and G-banding) may be observed most readily in the chromosomes located at 11 o'clock. (Courtesy of H. E. Wyandt and H. Lebowitz.)

mine and adenine. "Constitutive" heterochromatin is consistently heterochromatic; "facultative" heterochromatin is variable, as in the case of the X chromosome, in which the inactive, heterochromatic X is the maternal one in some cells and the paternal one in others.

A recent innovation by Caspersson and colleagues[2] has provided a powerful new tool for discriminating the chromosomes: quinacrine mustard (QM) fluorescence ("Q-staining methods"). Quinacrine binds preferentially to certain regions of metaphase chromosomes to produce characteristic banding patterns ("Q-bands") (see Fig. 2-4). These banding patterns are sufficiently reproducible to enable the identification of each chromosome when these findings are added to the usual information, such as centromere index and morphologic features.

By inspection alone, good fluorescent preparations reveal banding patterns adequate for distinguishing the chromosomes. A further refinement, the photometric recording of patterns, has been used by Caspersson in the development of the Q-staining methods. This refinement requires expensive equipment and is not used by most laboratories. Quinacrine fluorescence clearly differentiates between the G + Y group chromosomes and, taken with other cytogenetic findings, provides distinctions between the eight chromosome pairs of the C + X group not possible with earlier techniques.

Techniques that may be more easily employed by many laboratories are the "G-staining methods." In one such method, the Giemsa stain, conventionally used at pH 6, is used at pH 9, revealing banding patterns ("G-bands") that are, with few exceptions, similar to those obtained with quinacrine fluorescence.[11] Various methods of pretreatment are recommended by various laboratories.[8]

A variation on the "G-staining methods" is the reverse-staining Giemsa method ("R-staining methods"), which gives patterns ("R-bands") opposite in staining intensity to the "G-bands" (Fig. 2-4). Methods that demonstrate constitutive heterochromatin are the "C-staining methods" and the chromatin stained, the "C-bands."

At a number of laboratories throughout the world, research in banding techniques is being carried out, and many methods in addition to the ones described in this chapter are being developed.[8] The selection of Q-, G-, R-, and C-methods is based on the Paris Conference.[12]

The nomenclature of banding is illustrated in Figure 2-5 by a diagrammatic representation of chromosome 1. The convention of referring to a morphologic landmark by chromosome number, arm, region, and band (e.g., 1p36) is becoming more commonly observed in the literature.

To summarize the methods now being employed to identify chromosomes:

1. Measurements and morphology by Giemsa 6 or other standard stains
   a. arm ratio (long arm : short arm)
   b. centromere index (short arm : total chromosome)
   c. morphologic features (satellites, secondary constrictions, etc.)
2. Autoradiography
3. Q-staining methods; Q-bands. Quinacrine mustard (or dehydrochloride) fluorescence
4. G-staining methods; G-bands. Giemsa staining with altered pH (i.e., Giemsa 6 at pH 9)
5. R-staining methods; R-bands. Reverse-staining Giemsa methods (e.g., with temperature treatment)
6. C-staining methods; C-bands. Methods that demonstrate constitutive heterochromatin.

MITOSIS

Mitosis, or somatic cell division, is the means by which the cells of the body duplicate themselves for growth and maintenance of tissues. Cells that are not dividing are said to be in interphase. This is the

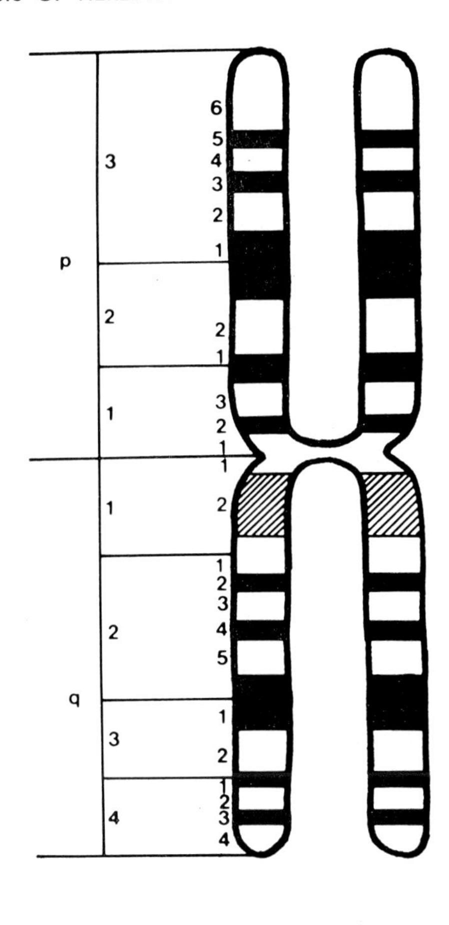

Figure 2-5. Diagrammatic representation of chromosome bands in chromosome No. 1 as observed with Q-, G-, and R-staining methods (centromere representative of Q-staining method only). Regions and bands are numbered consecutively from the centromere outwards. The two regions next to the centromere are labeled "1." A band used as a landmark is considered as belonging entirely to the region distal to the landmark and is accorded the band number of "1" in that region. To illustrate, the hatched band near the centromere of the long arm of chromosome No. 1 would be designated 1q12 (chromosome No. 1, long arm, region 1, band 2). Another example, the most distal band as seen on the short arm of our diagram of chromosome No. 1, would be labeled 1p36 (chromosome No. 1, short arm, region 3, band 6). Redrawn and modified from *Paris Conference* (1971).[12]

normal condition in which cells are engaged in their designated functions. Also during interphase the genetic material duplicates itself, so that before cell division actually takes place, a double DNA content is present, which is then divided between the two daughter cells during mitosis. The chromosomes, which are metabolically active and greatly elongated during interphase, are not visible at this stage. They appear as formless granularity (Fig. 2-6).

In mitosis, four stages are recognized: *prophase, metaphase, anaphase* and *telophase*. It is during this process that a precise sequence of events occurs that results in the production of two daughter cells, each having the exact chromosome complement and genetic material of the parent cell. The cytoplasmic material divides in half while the nucleus undergoes this series of changes.

Prophase

This stage begins when the chromosomes, which have not been distinguishable in interphase, condense and become visible

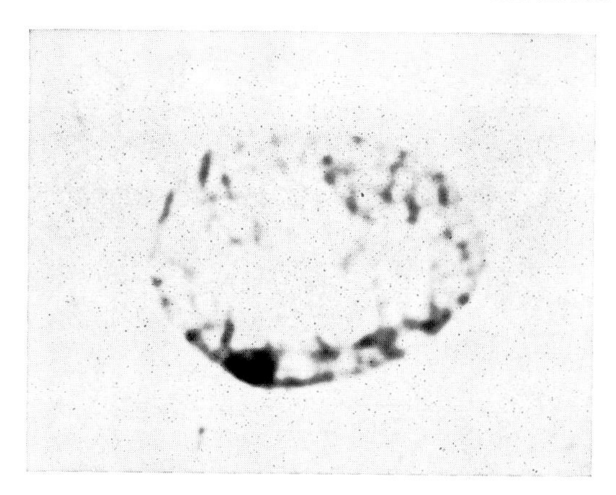

Figure 2-6. Nucleus of buccal mucosal cell in human female. Observe the formless granularity of chromosomes during interphase except for the darkly staining mass (Barr body) adjacent to the nuclear membrane at 4 o'clock. (See Chapter 4 for discussion of the significance of the Barr body.)

by light microscopy. The DNA content has already doubled, and each chromosome is visualized as two parallel strands, *chromatids,* joined together in one place, the *centromere.* The nuclear membrane begins to disappear, and two small bodies, the *centrioles,* start to migrate to opposite poles from a position immediately external to the nuclear membrane (Fig. 2-7*A*).

Metaphase

This is the stage during which the individual chromosomes are most clearly visualized. The *karyotype,* the display of human chromosomes for analysis, is taken from metaphase plates. As may be seen in Figure 2-7*B* the nuclear membrane disappears and the chromosomes line up along the equator and are connected at their centromeres to the *spindle,* which consists of protein fibers radiating from the polar *centrioles.*

Anaphase

The separation of the two chromatids from each other signals the beginning of anaphase (Fig. 2-7*C*). The centromere divides longitudinally into two, and the two daughter centromeres move toward opposite poles, dragging their chromatids with them. Thus *each pole of the dividing cell will receive a set of chromosomes identical to that of the original nucleus.*

Telophase

The *daughter chromosomes,* which are now single-strand chromatids, arrive at the poles of the cells as the cytoplasm begins to divide in the area of the equatorial plane (Fig. 2-7*D*). The two daughter cells go on to separate as the chromosomes become less densely staining until they are indistinguishable. When the separation is complete, two new daughter cells are recognized in interphase.

MEIOSIS

There are two critical events that occur in *meiosis* (reduction division) but do not occur in *mitosis* (equational division): (1) Each chromosome pairs with its homologue, so that there are 23 pairs at the metaphase plate; (2) the homologues separate to opposite poles so that the two daughter cells each have a set of 23. That is, the complement has been reduced from diploid to haploid. A mitotic division follows so that each haploid gamete is one of four descendants of the original meiotic cell. This is the essence of meiosis. The morphological details will now be described.

First Meiotic Division

Prophase. This stage of division may be seen as five clearly defined substages. The important events in meiotic prophase that do not occur in mitotic prophase are that the chromosomes pair (*synapse*) and that the strands of the paired chromosomes may break and recombine so that a piece of one homologous chromosome may be exchanged for a comparable piece of its homologue. This is known as "crossing over."

1. ***Leptonema.*** The chromosomes first become visible and appear to be single threads, although the DNA has already reduplicated (Fig. 2-8*A*).

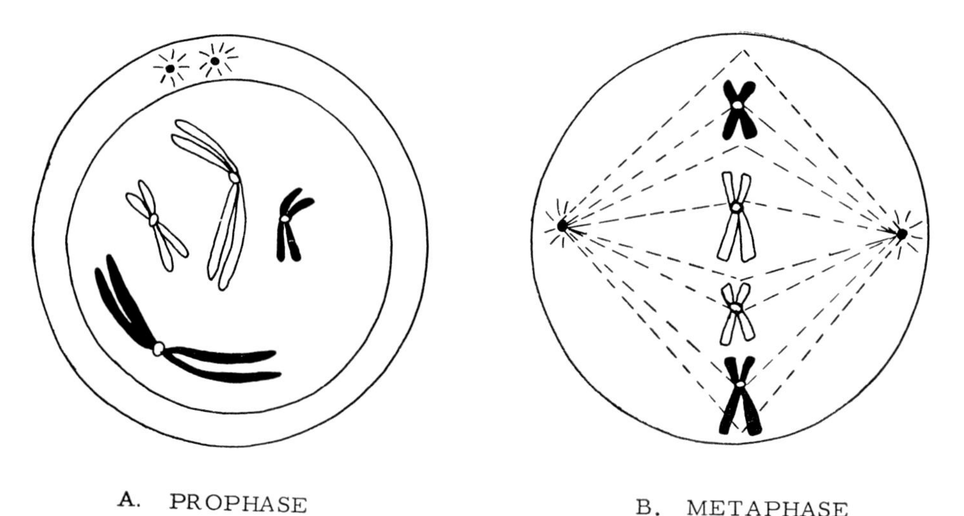

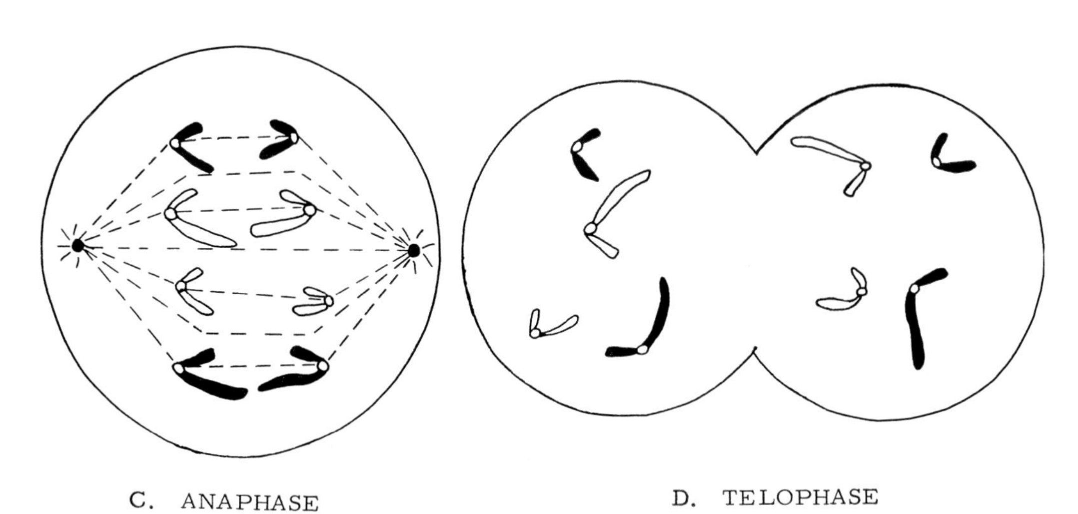

Figure 2-7. Mitosis. Two of 23 pairs of chromosomes are shown passing through the four stages from double-stranded in prophase to two sets of single-stranded chromosomes at cell division. See text.

2. ***Zygonema.*** Each chromosome now pairs with its counterpart in such a way that each part of one chromosome is associated with the identical part of its homologue. These *synapsed chromosomes,* or *bivalents, do not form in mitosis* (Fig. 2-8*B*).

3. ***Pachynema.*** Each chromosome is now visible as a double strand (Fig. 2-8*C*).

4. ***Diplonema.*** The two members of the bivalent now begin to move apart except where crossing-over has occurred, and the exchange of strands results in X-like formations, known as *chiasmata,* which hold the homologues together. Figure 2-8*D* illustrates one such chiasma in which the material from one chromatid (in black) has been exchanged with the homologous seg-

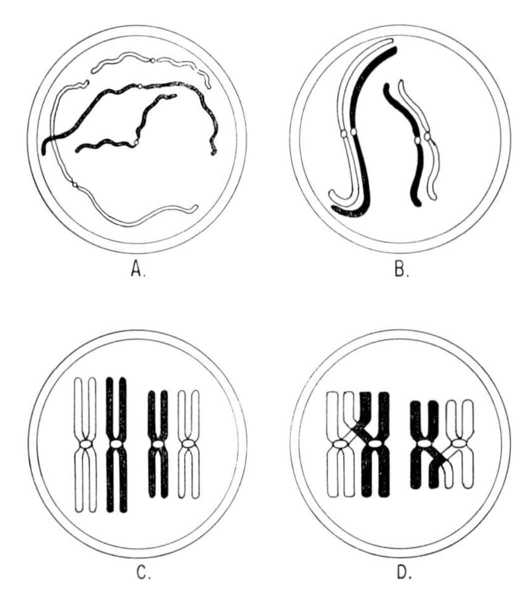

Figure 2-8. Prophase of the first meiotic division. Two of 23 pairs of chromosomes are shown in the stages of prophase: *A*, leptonema; *B*, zygonema; *C*, pachynema; *D*, diplonema. Diakinesis is not illustrated. Note that there is pairing of homologous chromosomes and crossing over. See text.

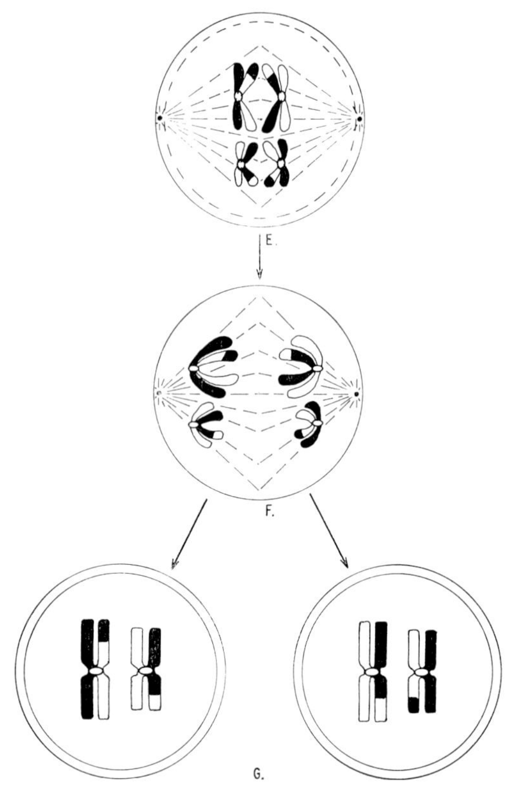

Figure 2-9. Continuation of first meiotic division: *E*, metaphase; *F*, anaphase; *G*, telophase. Note that double-stranded chromosomes go into the daughter cells (each gets a set of 23 double-stranded chromosomes instead of 46 single-stranded chromosomes as in mitosis). See text.

ment of material from a chromatid of its homologous chromosome (in white). Exchanges of genetic material by crossing-over add almost infinite variety to the ultimate genetic makeup of a given individual.

5. ***Diakinesis.*** The final stage of prophase is characterized by the chromosomes becoming more condensed and darkly staining. It is not illustrated in the accompanying figures.

Metaphase. This stage is the same in meiosis as in mitosis except that the homologues are paired, as bivalents. The nuclear membrane disappears and the chromosomes line up in an equatorial plane connected at their centromeres by protein fibers radiating from the centrioles (Fig. 2-9*E*).

Anaphase. The chromosomes, each still consisting of two chromatids joined at a centromere, now separate from their homologues and *23 double-stranded* chromosomes go into each of two daughter cells (Fig. 2-9*F*). Note the difference from mitotic anaphase in which the two chromatids of each chromosome separate at the centromere and *46 single-stranded* chromosomes go into each cell. In first meiotic anaphase we find the chromosomal basis for two of Mendel's laws of inheritance: *segregation* and *independent assortment.* The separation of the homologous chromosomes, with either the maternal or paternal member of the pair going (after further divisions) to a given gamete, is the basis for the segregation of genes. The decision as to which pole gets the maternal and which the paternal homologue is independent for each pair. This is the physical basis for the independent assortment of genes.

Telophase. Telophase in meiosis I is comparable to mitotic telophase except that there are 23 double-stranded daughter chromosomes that congregate at the poles of the cells rather than 46 single-stranded chromosomes (Fig. 2-9G).

Second Meiotic Division

An interphase in which the DNA is replicated does not occur between the first and second meiotic divisions. In fact, no interphase at all may separate the two meiotic divisions. As may be appreciated in Figure 2-10*H–J* the stages metaphase, anaphase and telophase are essentially the same as those found in somatic cell mitosis except for the fact that *only 23 chromosomes are involved.* At metaphase, the 23 chromosomes, each consisting of two chromatids joined by a centromere, line up on the equator; at anaphase, the centromeres divide longitudinally and migrate to the poles, each drawing its single strand along with it; at telophase, the chromatids arrive at the poles and the cytoplasm begins to divide. The cytoplasm divides evenly in cells destined to become sperms (Fig. 2-9G) so that two equal spermatocytes are present after first meiotic division and four equal spermatids after second meiotic division. In the female, however, the cytoplasm is unevenly divided. After meiotic division the major share of the cytoplasm goes to one cell, the secondary oocyte, and a polar body containing the full 23 chromosomes but negligible cytoplasm is discarded. After second meiotic division the same unequal distribution of cytoplasm occurs, yielding an ovum and another polar body.

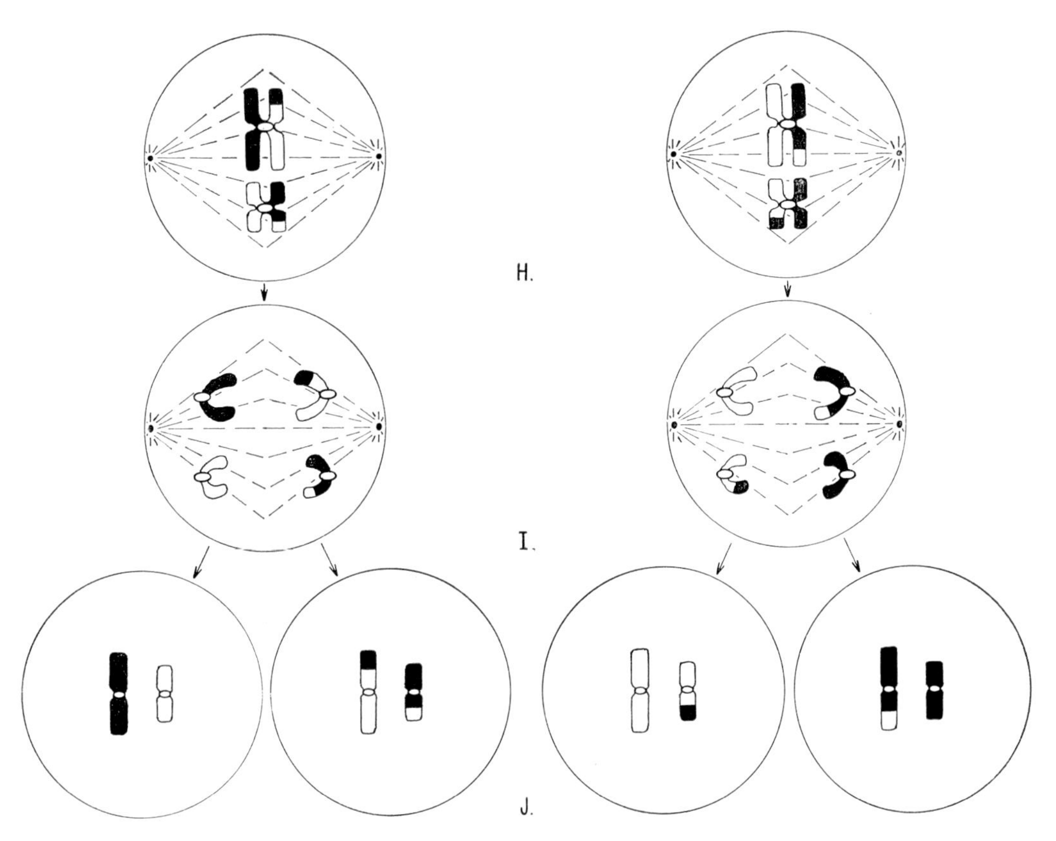

Figure 2-10. Second meiotic division: *H*, metaphase; *I*, anaphase; *J*, telophase. The 23 double-stranded chromosomes now separate into 23 single-stranded chromosomes.

GAMETOGENESIS AND FERTILIZATION

The end-products of the events of meiosis are *gametes,* and fusion of the maternal and paternal gametes during *fertilization* produces the first cell of the new individual, the *zygote*. Within this single cell resides all the genetic information required for growth and differentiation into a complex, multicellular human organism.

Spermatogenesis

This is the process through which the early male germ cells (*spermatogonia*) undergo a series of changes, including the previously described first and second meiotic division, and subsequent differentiation into mature sperm. Spermatogenesis occurs in the seminiferous tubules of the testes, and takes place only in the mature male. The entire process, from spermatogonium through primary and secondary spermatocytes and spermatid to mature sperm, requires 74–75 days.[6] The first and second meiotic divisions occupy approximately half of this period. About 200 million sperms are normally present in an ejaculate, only one of which (if any, of course) will participate in fertilization of the egg.

Oogenesis

This process, by which the early female germ cells (*oogonia*) differentiate into ova, may take anywhere from 12 to 45 years, depending on when, during the reproductive life of the female, the mature ovum is expelled. When the fetus is about 3 months old, the oogonia begin to differentiate into primary oocytes. At the time of birth of a female infant, it is thought that every oocyte she will ever possess is already present, although this concept has been challenged. These primary oocytes are already in first meiotic prophase and remain in a sort of temporary suspension (dictyotene) until sexual maturity. In the sexually mature female, a Graafian follicle progresses to maturity each month and extrudes an oocyte, which, having completed first meiotic division, continues through second meiotic division in transit through the Fallopian tube.

Fertilization

Fertilization usually occurs in the lateral portion of the Fallopian tube (Fig. 2-11) when one of the many sperm that surround the secondary oocyte penetrates it. It is usually not until after fertilization that the second meiotic division of the ovum is completed. During fertilization the tail of the sperm rapidly disappears as the head is embedded in the ovum. Soon all that remains of the sperm is the pronucleus containing the 23 chromosomes. The sperm pronucleus makes its way into the pronucleus of the ovum, the nuclear membranes disappear, and fusion occurs, producing a zygote that now has a single nucleus containing 46 chromosomes. The chromosomes now embark on a series of mitotic divisions. At interphase there is replication of DNA; the usual mitotic sequence is followed through to the division into two cells, four cells, eight cells, and so on through the blastula, gastrula, and embryo stages until a mature individual develops, who is capable of reproduction and initiating a similar series of events.

Chromosomal Aberrations

What are the possible errors in meiosis, and in early mitosis in the zygote, that can lead to chromosomal anomalies? First, some definition of terms is in order. The number of chromosomes in the gamete is 23, and this is the *haploid* (n) number. Forty-six is the *diploid* (2n) number. Any exact multiple of the haploid number is called *euploid.* Therefore, haploid (n) and diploid (2n) numbers of chromosomes, which are multiples of 23, are euploid. However, if a patient has a chromosome number that is not an exact multiple of 23, such as in Down syndrome with 47 chromosomes, the number is called *aneuploid. Polyploidy* is a euploid condition, other than diploid, in which an exact multiple of the haploid state is present; 69 (3n)

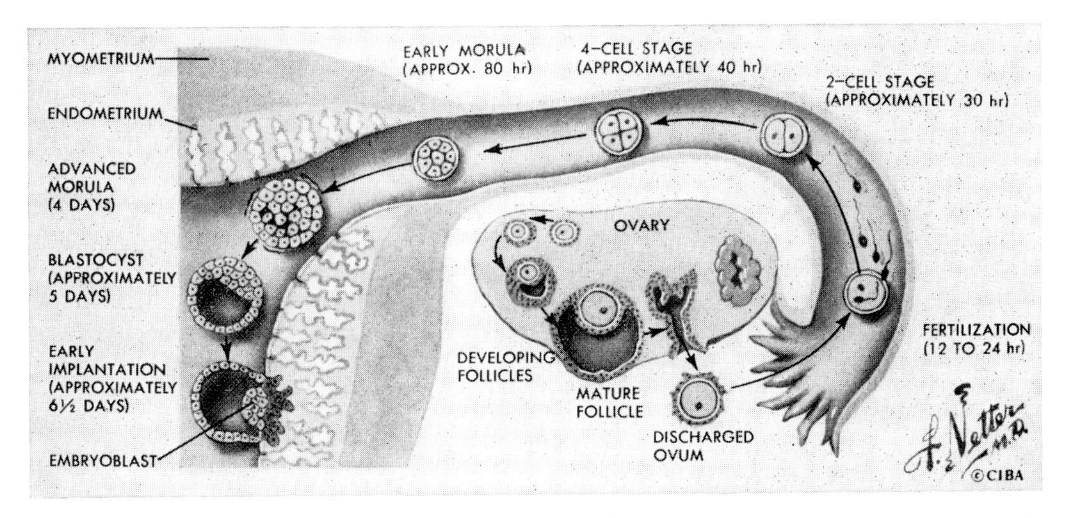

Figure 2-11. Fertilization and early meiotic cell divisions to implantation. (Copyright © 1969 Ciba Pharmaceutical Company, reproduced with permission from The Ciba Collection of Medical Illustrations by Frank H. Netter, M.D.)

chromosomes is the triploid number, 92 (4n) the tetraploid, and so on through higher multiples. It is rarely encountered in humans except in abortuses, certain differentiated states (such as bladder epithelium), and tumors.

An aneuploid state in which a third homologous chromosome is present in addition to the normal autosomal pair (as in Down syndrome) is called *trisomy*. Down syndrome is the eponym for 21 trisomy, the presence of chromosome 21 in triplicate rather than in duplicate. Absence of a chromosome of a pair is called *monosomy* for that pair. Although monosomy for the X chromosome in Turner syndrome is relatively common, monosomy for an autosome is almost nonexistent. As will be discussed in Chapter 4, whether there are 1, 2, 3, or more X chromosomes, only one X chromosome is fully active, so that monosomy for an X chromosome involves very little loss of active genetic material. However, although the human can survive the amount of developmental confusion produced by the presence of some autosomes in triplicate, he cannot withstand the absence of an entire autosome.

If only a piece of a chromosome is present in triplicate, as is found in translocation, this is termed partial trisomy. If a piece of a chromosome is missing, the terms partial *monosomy* or *deletion* may be used. In the following sections on numerical aberrations and structural alterations, further essential terms will be defined.

Numerical Aberrations. *Aneuploidy* is a manifestation of a mistake that may occur at first or second meiotic divisions or during a mitotic cell division. There are three main categories of mistakes: First, the chromosomes may fail to pair in meiotic prophase, and may therefore not segregate to opposite poles at anaphase. Second, one of the chromosomes may be late in reaching the pole of the spindle (anaphase lag) and be excluded from the newly formed nucleus. Third, the homologous pair may fail to disjoin at the first meiotic division, or the double-stranded chromosome may fail to separate into chromatids at second meiotic division or a mitotic division (*nondisjunction*). The term nondisjunction is often loosely used to refer to the results of all three kinds of mistake.

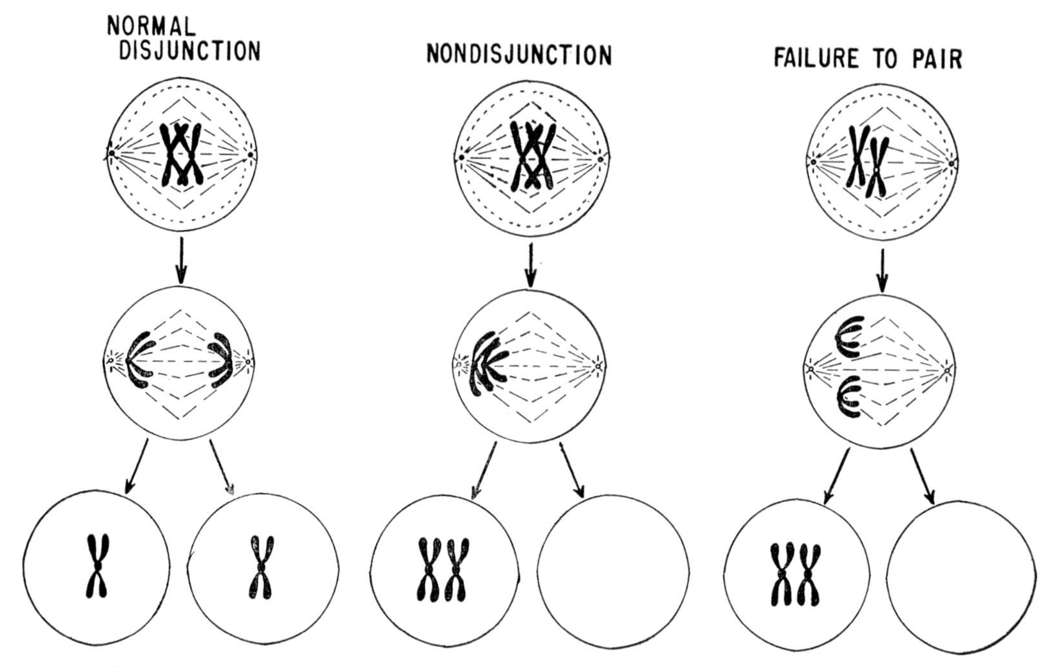

Figure 2-12. Mistakes in first *meiotic division* leading to aneuploidy.

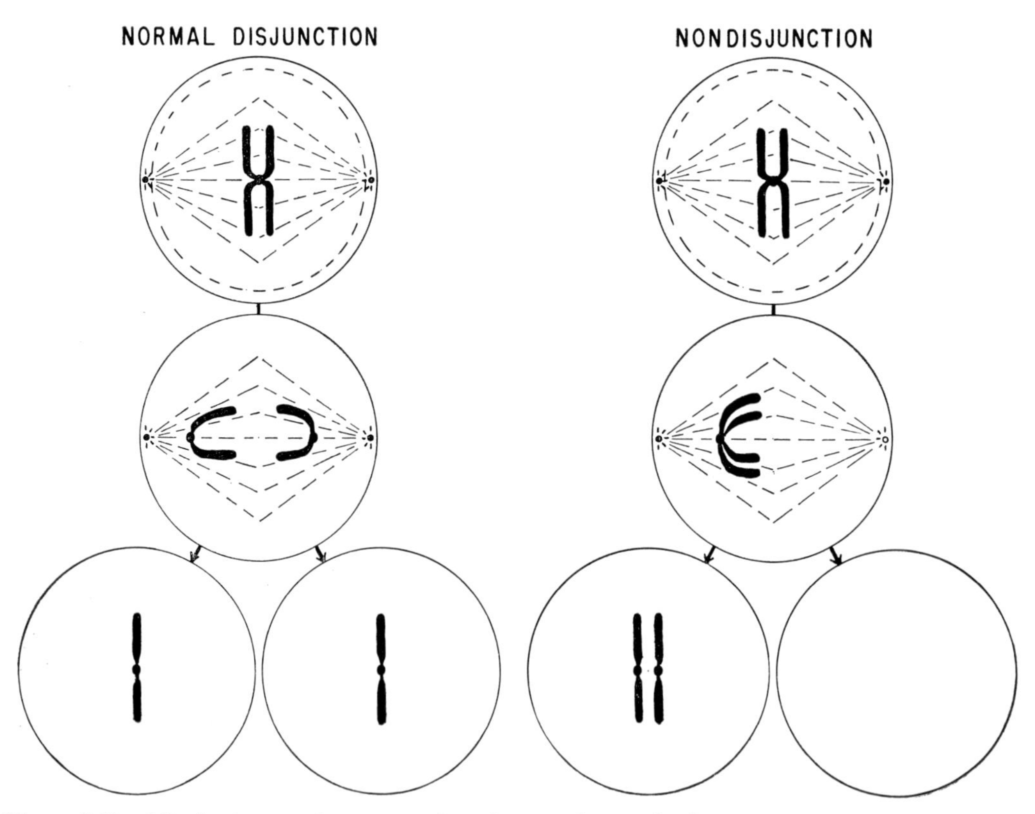

Figure 2-13. Mistake in *second meiotic* or first cleavage division leading to aneuploidy.

If the pair fails to disjoin in first meiotic anaphase, as visualized in Figure 2-12, both chromosomes of a homologous pair may end up in one daughter cell (24 chromosomes) and neither member in the other cell (22 chromosomes). But the same result may occur if the homologous chromosomes fail to pair and are randomly assorted between the two daughter cells. Experimental evidence suggests that failure to pair is the more frequent error in first meiotic division.

In the second meiotic division, the double-stranded chromosome may fail to divide at the centromere, and both chromosomes will go to the same pole (Fig. 2-13). Again the result is one cell having 24 chromosomes and the other cell only 22. Fusion of such a gamete with one having the normal haploid number (23) results in aneuploidy—either trisomy (47) or monosomy (45) for the pair concerned.

Use of specific markers, either genetic or cytologic, is beginning to provide information on where the errors occur that give rise to aneuploidy. Clearly, for instance, an XYY male must arise from a YY sperm, and this could only arise from an error in the second meiotic division of spermatogenesis. Use of cytogenetic variations has shown that the error resulting in trisomy 21 arose (in five cases that were informative) at the first meiotic division of oogenesis. X-linked markers have shown that nondisjunction can occur in both oogenesis and spermatogenesis. Consider, for instance, a case of an XXY individual (Klinefelter syndrome; see Chapter 4) who is color-blind, and whose parents both have normal color vision. The gene for color-blindness is sex-linked re-

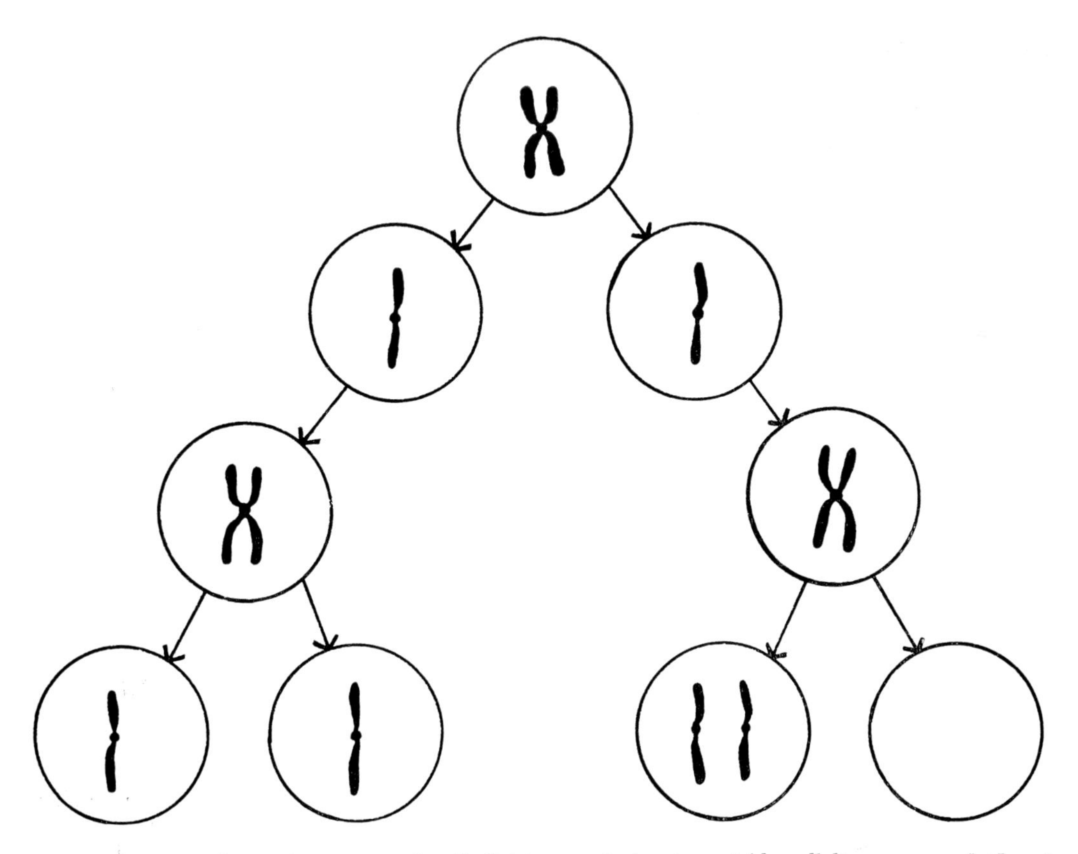

Figure 2-14. Nondisjunction at second cell division producing two viable cell lines, one euploid and aneuploid (mosaicism).

cessive (see Chapter 5), and the son did not receive an X chromosome from the father since he is color-blind. Therefore, the mother must have carried the gene on one X chromosome, and the son must have resulted from fertilization of an XX egg by a Y-bearing sperm. Since both the X chromosomes must have carried the mutant gene, nondisjunction must have occurred during the second meiotic division of oogenesis.

What happens at the second meiotic division may also occur during the first cell division of the zygote (Fig. 2-13), producing a viable cell line with an aneuploid number and a usually nonviable cell line. If nondisjunction occurs at the time of second or later cell division, then a different result is observed (Fig. 2-14). Two viable cell lines may be formed, one with a normal diploid number and one with an aneuploid number. This is called *mosaicism* or *mixoploidy*. Mosaicism may result from nondisjunction at the third or fourth or subsequent cell divisions, and the percentage of aneuploid cells compared with diploid cells depends on when the nondisjunction occurs. The capacity for establishing new cell lines leading to mosaicism is rapidly lost. The predominance of one cell line over another appears to affect the phenotypic features and severity of manifestation of the disorder. For example, a patient who has a cell line of 21 trisomic cells constituting 50% of his cell population is more severely affected than an individual having only 10% 21 trisomic cells.

Structural Aberrations. Chromosome rearrangements may occur within chromosomes or between chromosomes, and take place at the G_1 period of the cell cycle (Fig. 2-15), before the chromosome reduplicates into chromatids. Rearrangements may also occur in the S and G_2 periods involving chromatids (which will not be discussed). Misdivision of the centromere leading to isochromosome formation is perhaps the easiest structural aberration to visualize, so this will be discussed first.

Isochromosomes. In Figure 2-16 the results of normal longitudinal division of the centromere are compared with what happens when division occurs at right angles

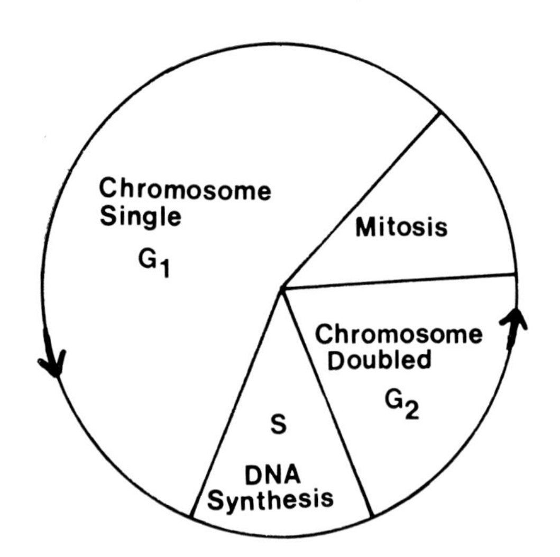

Figure 2-15. Periods of the cell cycle: G_1—after division, the chromosome is single and does not synthesize DNA; S—the period of DNA synthesis; G_2—the chromosome is now doubled; **mitosis** occupies a relatively small portion of the cell cycle.

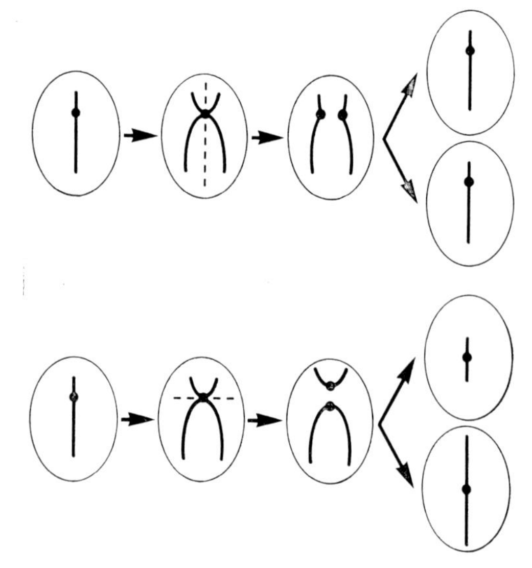

Figure 2-16. Isochromosome formation resulting from division of the chromosome in the plane of the short axis (below) rather than normal division in the long axis (above).

to the long axis. The chromosomes produced by this abnormal division are one chromosome having the two long arms of the original chromosome, but no short arms, and the other chromosome consisting of the two short arms with no long arm. Each of these, called isochromosomes, constitutes a simultaneous duplication and deletion. If the isochromosome of the long arm of a chromosome is present in a diploid cell, the genes of the long arm are present in triplicate ("trisomy" of the long arm) and the genes of the short arm are only represented once (monosomy of the short arm). As an example, in the nomenclature of the Chicago and Paris conferences, a female patient having an isochromosome of the long arm of the X chromosome would be 46,XX,i(Xq), with "q" being the abbreviation for long arm and "i" for isochromosome.

Rearrangements within Chromosomes. Several rearrangements, some of which lead to loss of chromosomal material or *deletion*, are shown in Figure 2-17. If a chromosome breaks, only that portion retaining the centromere is able to orient on the spindle and be maintained through successive cell divisions. The break may be at an end, or two breaks may occur within an arm so that a piece is removed from the middle. Where there is a break, the broken fragments are "sticky" and have a tendency to reunite. However, if there is no chromosomal material nearby to reunite with, the amputated end "heals" but sustains a loss of genetic material. If breaks occur at the ends of both arms, the two "sticky" ends may curl back and unite with each other, forming a ring chromosome, which, in many cases, also produces loss of genetic material. There are further complexities in ring formation which exceed the scope of this presentation. Finally, if two breaks take place within an arm and they are not properly reunited, the broken piece may fall out, and the "sticky" proximal and distal ends reunite, causing a loss of gene loci—an in-

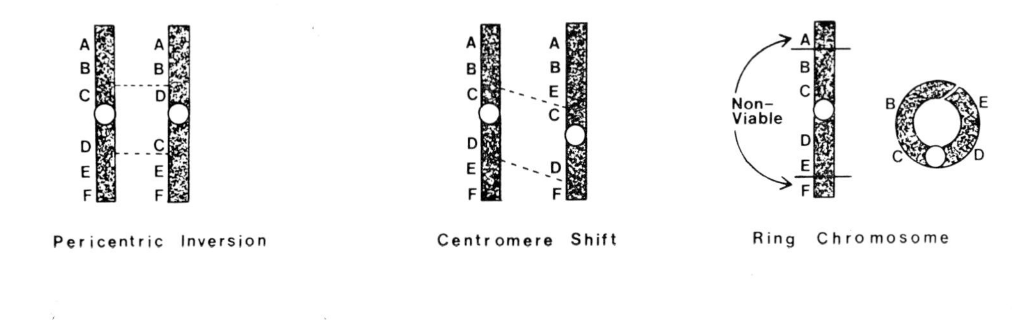

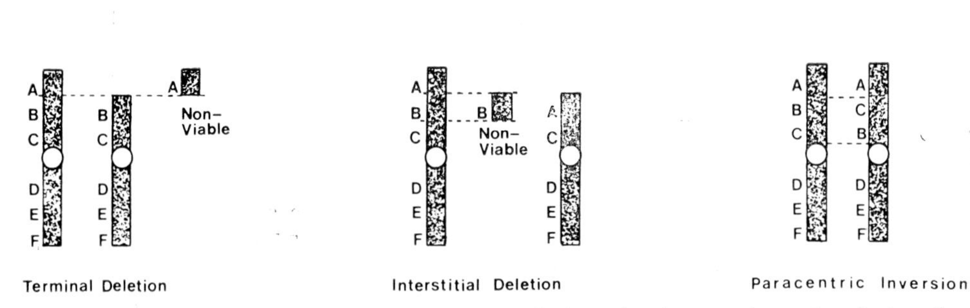

Figure 2-17. Rearrangements within chromosomes result from breakage and reunion during the G_1 phase of the cell cycle, before chromosome reduplication. Several examples are shown here.

terstitial deletion. As illustrated earlier, the Chicago Conference nomenclature for a female having the deletion syndrome cri-du-chat would be 46,XX,5p– if the deletion of the short arm of chromosome 5 is "open." The same syndrome may occur if the deletion is in the form of a ring chromosome, and the nomenclature would be 46,XX,r(5). Paracentric and pericentric inversions and centromere shifts, although illustrated in Figure 2-17, are relatively rare events in human beings.

Rearrangements between Chromosomes. A *translocation* refers to a rearrangement in which a piece of one chromosome is transferred to another chromosome. In a *reciprocal* translocation two chromosomes are broken and the broken pieces switch places (Fig. 2-18*A*). A *Robertsonian* translocation is one in which there is fusion of two acrocentric chromosomes at their centromeres, the two short arms being lost. An example is the translocation of the long arm of a chromosome 21 to a chromosome 14 (Fig. 2-18*B*). In a patient having 14/21 translocation mongolism, an *unbalanced* translocation, the affected individual has 45 chromosomes (a homologous pair of chromosome 21; a normal chromosome 14 and a structurally abnormal chromosome 14, which has an addition to the short arm consisting of the translocated long arm of chromosome 21). This may be referred to as *partial* trisomy, although, strictly speaking, trisomic means "three bodies" and there is no extra chromosome. A person with a 14/21 *balanced* translocation would be a translocation carrier and would have a chromosome count of 45 (one normal chromosome 14; one normal chromosome 21; and one abnormal chromosome made up on the long arms of chromosomes 14 and 21). This patient would have almost all of the genetic material of a pair of chromosome 14 and a pair of chromosome 21 and would thus be phenotypically normal.

The Chicago Conference nomenclature for a male patient with 14/21 translocation Down syndrome would be 46,XY,–D,+t(14q21q). The patient has 46 chromosomes, 1 X and 1 Y chromosome, and is missing a normal D group chromosome (–D). This chromosome has been replaced by a chromosome made up of the long arm of a D group chromosome, now known to be 14, and a G group chromosome identified as 21[+t(14q21q)].

Non-reciprocal or insertion translocations are rare events requiring three-break rearrangements, and have so far been described in mouse but not in man (Fig. 2-18*C*).

Clinical Disorders Produced by Chromosomal Aberrations. Ongoing studies reveal that one infant in every 200 newborns has a recognizable chromosomal abnormality.[7] An investigation of therapeutic abortions between 5 and 20 weeks has shown that between 1.5% and 2.9% of specimens have gross chromosomal anomalies,[1,9] suggesting that the majority of conceptuses that have chromosomal aberrations do not survive to term. In support of this inference is evidence that 28% of first trimester spontaneous abortions have chromosomal anomalies. A commonly used figure for the percentage of all pregnancies that end in recognizable spontaneous abortion is 15%.[17] It has been suggested that as many as 40% of fertilized ova are abnormal and probably result in early abortion, often unrecognized.

The question that is uppermost in the minds of the parents of a patient with a chromosomal anomaly has still not been approached: What caused the chromosomal aberration? Tracing a patient's disease to a chromosomal aberration is not an insignificant accomplishment. It provides a firm diagnosis, which is the basis for subsequent clinical management, prognosis, and recurrence risk data for genetic counseling.

But why a specific chromosomal anomaly occurs in a specific patient can seldom be answered. The vast majority of aberrations are sporadic, although on rare occasions a chromosomal disorder is directly inherited (as in an unbalanced 14/21

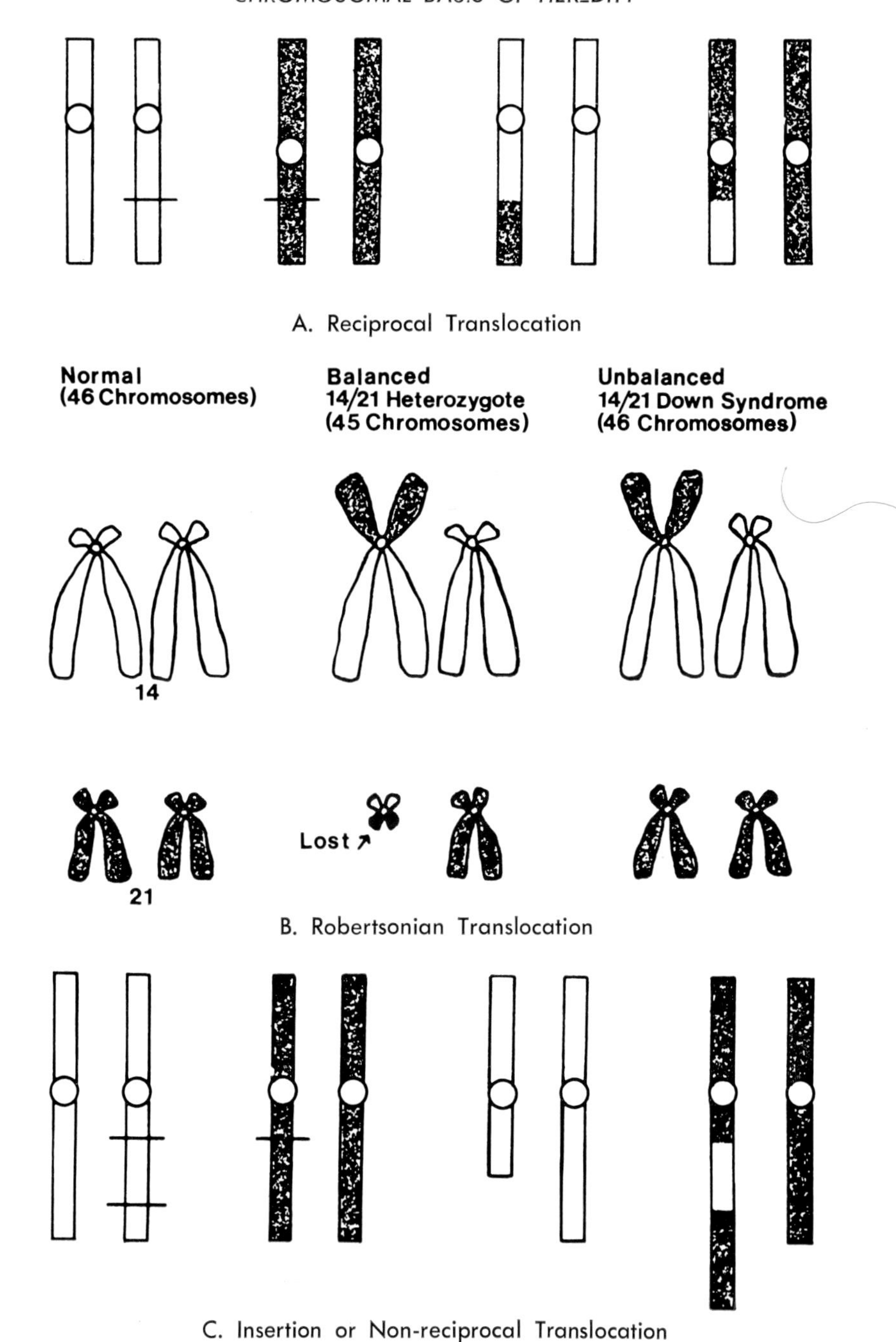

Figure 2-18. *A,* Reciprocal translocation involves a mutual exchange of chromosome segments between two non-homologous chromosomes and results in a balanced chromosome complement, as shown by two hypothetical pairs of chromosomes. *B,* Robertsonian ("centric-fusion") rearrangements involve acrocentric chromosomes. The D/G (in this case 14/21) "centric-fusion" is a common example. Note that in balanced 14/21 heterozygote, there are only 45 chromosomes. Only a small centric fragment is lost, leaving this individual with sufficient gene loci to be indistinguishable from normal. The breakage and reunion most likely occurred during G_1 as shown in Figures 2-17 and 2-18. *C,* Insertions or non-reciprocal translocations are rare events involving three breaks and are illustrated here in the G_1 phase with two pairs of hypothetical chromosomes.

translocation transmitted from a balanced 14/21 translocation heterozygote).

There are some general underlying causes of these sporadic events.[18] Late maternal age has clearly been implicated in mongolism and probably influences other aneuploidies. There also seems to be a slight familial predisposition to nondisjunction, with more than one "sporadic" aneuploidy occurring in siblings somewhat more often than expected by chance. A delay between ovulation and conception may predispose to nondisjunction—the "over-ripe egg" hypothesis. Environmental agents such as radiation, chemicals, and viruses have been demonstrated to cause chromosomal damage in experimental models. These environmental insults have not been so much associated with nondisjunction as with chromosomal breakage. Whether parental irradiation increases the probability of nondisjunction is still a matter of controversy. A particularly intriguing recent line of investigation has implicated an immunologic mechanism in nondisjunction through the discovery of antithyroid antibodies in mothers of infants with 21 trisomy.[5]

In Chapters 3 and 4, the commonly occurring autosomal and sex chromosomal anomalies will be reviewed, emphasizing the main clinical features, as well as problems in counseling, management, and ultimate prognosis.

SUMMARY

The chromosome carries the units of inheritance (genes), which are segments of DNA in a protein matrix (histone in the human). The normal (euploid) chromosome number in the human is 46. Techniques for the study of human chromosomes have been developing rapidly since 1956 and now permit identification of every one of the 23 pairs of chromosomes by their morphologic features, banding patterns shown by special staining techniques and, to a lesser extent, autoradiography.

Mitosis or somatic cell division is the means by which body cells reduplicate for growth or maintenance of tissue. The DNA replicates in the interphase between cell divisions. During prophase, the first of four stages of mitosis, chromosomes condense and become visible under light microscopy and are seen as two parallel strands (chromatids) joined in one place by a centromere. At metaphase the chromosomes are best visualized; they line up along an equator and are attached at their centromeres to a spindle. The centromeres divide at anaphase, and the 46 single stranded chromatids migrate to opposite poles. Cytoplasmic division with the formation of two daughter cells takes place in telophase.

Meiosis, or reduction division, precedes the formation of germ cells and differs from mitosis in that (1) there is pairing of homologous chromosomes (with crossing-over and exchange of genetic material) and (2) there are two successive divisions of nuclear material resulting in cells (gametes) having 23 chromosomes.

Deviations from normal chromosomal number and structure produce significant abnormalities of development. Approximately one in 200 individuals has a recognizable chromosomal aberration. An abnormal number that is not an exact multiple of the haploid number of 23 is called aneuploidy. The errors of meiosis or mitosis resulting in aneuploidy are generally covered by the term nondisjunction, although other mechanisms such as failure to pair and anaphase lag may be more common than nondisjunction. Structural anomalies of clinical importance include isochromosomes, deletions, and translocations.

REFERENCES

1. Carr, D. H.: Genetic basis of abortion. Ann. Rev. Genet. 5:65, 1971.
2. Caspersson, T., and Zech, L. (eds.): Chromosome identification. Nobel Symposia on Medicine and Natural Sciences. New York, Academic Press, 1973.
3. Chicago Conference: Standardization in Human Cytogenetics. Birth Defects II:2, 1966. New York, The National Foundation.
4. Comings, D. E.: The structure and function of chromatin, p. 237. *In* Harris, K. and

Hirschhorn, K., Advances in Human Genetics. Vol. 3. New York, Plenum Press, 1972.
5. Fialkow, P. J., Thuline, H. C., Hecht, F., et al.: Familial predisposition to thyroid disease in Down's syndrome: Controlled immuno-clinical studies. Am. J. Hum. Genet. 23:67, 1971.
6. Heller, C. G., and Clermont, Y.: Spermatogenesis in man: An estimate of its duration. Science 140:184, 1963.
7. Lubs, H. A. and Ruddle, F. H.: Chromosomal abnormalities in the human population: Estimation of rates based on New Haven newborn study. Science 169:495, 1970.
8. Miller, O. J., Miller, D. A., and Warburton, D.: New staining techniques and the human chromosomes, p. 1. *In* A. G. Steinberg and A. G. Bearn (eds.), Progress in Medical Genetics, Vol. 9. New York, Grune & Stratton, 1973.
9. Nishimura, H.: Frequency of malformation in abortions, p. 275. *In* F. C. Fraser and V. A. McKusick (eds.), Congenital Malformations: Proceedings of the Third International Conference, 1970. New York, Excerpta Medica, 1970.
10. Painter, T. S.: Studies in mammalian spermatogenesis. J. Zool. 37:291, 1923.
11. Patil, S. R., Mersick, S., and Lubs, H. A.: Identification of each human chromosome with a modified Giemsa stain. Science 173:821, 1971.
12. Paris Conference (1971): Standardization in Human Cytogenetics. Birth Defects VIII:7, 1972. New York, The National Foundation.
13. Polani, P. E.: Chromosome anomalies and abortions. Develop. Med. Child. Neurol. 8:67, 1966.
14. Rodman, T. C.: Human chromosome banding by Feulgen stain aids in localizing classes of chromatin. Science 184:171, 1974.
15. Tjio, J. H., and Levan, A.: The chromosome number of man. Hereditas 42:1, 1956.
16. Wang, H. C., and Federoff, S.: Banding in human chromosomes treated with trypsin. Nature New Biol. 232:52, 1971.
17. Warburton, D., and Fraser, F. C.: Spontaneous abortion risk in man: Data from reproductive histories collected in a medical genetics unit. Am. J. Hum. Genet. 16:1, 1964.
18. Warkany, J.: Congenital Malformations. Chicago, Year Book Medical Publishers, 1971.

Chapter 3

AUTOSOMAL ANOMALIES

Before abnormalities of human chromosomes could be recognized, there had to be satisfactory preparations and knowledge of normal human chromosomes. Tjio and Levan provided the techniques for preparation and the definition of the normal number in 1956,[16] and Lejeune and associates followed, in 1959, with the first description of a chromosomal abnormality in the clinical disorder called Down syndrome or mongolism.[9] In the same month, Ford and associates described a patient with both Down and Klinefelter syndromes: 47,XY,+G,+Y.[8]

Since then, a large number of chromosomal anomalies have been described in the world literature. There are three well-established autosomal trisomy conditions and six other aberrations of autosomes that produce recognizable patterns of anomalies that appear to justify clinical description. The selection of these nine autosomal abnormalities is, to an extent, arbitrary and reflects the experience of the authors as well as the (probably correlated) frequency with which these syndromes appear in the literature. There are, of course, reports of many numerical and structural aberrations of chromosomes that are not included in this chapter or are only briefly mentioned. Some general observations should precede the description of specific syndromes.

PARTIAL TRISOMY AND MOSAICISM

Presence of an autosome in triplicate produces a somewhat variable but quite characteristic syndrome. When the excess of genetic material does not involve a whole chromosome, as in an unbalanced Robertsonian translocation (partial trisomy) or mosaicism, the syndrome may be complete or incomplete. In the case of Down syndrome, for instance, an unbalanced D/21 translocation results in the phenotype pattern of 21 trisomy (see p. 29), even though the short arm of the 21 chromosome is not present in triplicate as it is in regular 21 trisomy. On the other hand, translocations resulting in excess of a smaller portion of a chromosome may produce a more variable phenotype. Patients having mosaicism consisting of a cell line with 21 trisomy and a euploid line may have anything from the full syndrome to virtually no abnormalities, presumably depending on the proportions of aneuploid cells and in which tissues they occur.

TRISOMY PHENOTYPES WITHOUT APPARENT CHROMOSOMAL ABERRATION

The diagnostic pattern of anomalies accepted for the clinical diagnosis of any one of the chromosomal syndromes may be found in patients having no demonstrable chromosomal abnormality. This is true not only for the less well-established deletion syndromes but also for well-known trisomy syndromes. The classic manifestations of Down syndrome and 13 trisomy have been found in patients with normal karyotypes. Possible explanations are that such patients represent submicroscopic translocations or duplications of a segment of chromosome responsible for the phenotype of the syndrome. Undetected mosaicism may also be suggested, based on the possibility that the number of cells or the tissue studied in such patients may have been inadequate to reveal the aneuploid line responsible for the clinical abnormalities or that the abnormal cell line has disappeared before testing.

DOUBLE ANEUPLOIDY AND FAMILIAL NON-DISJUNCTION

One of the earliest reports of a chromosomal anomaly presented a patient with two separate trisomic aberrations, 21 trisomy and XXY.[8] Since then, a large number of other examples of double aneuploidy have entered the literature. These may represent the effects of an error in distribution of one chromosome at cell division affecting the distribution of another chromosome, or they may reflect some general predisposition to nondisjunction. In support of the latter concept are reports of families having one type of aneuploidy (e.g., Turner syndrome) in one offspring and a different aneuploidy (e.g., 13 trisomy) in another offspring. It is appreciated by the genetic counselor that, although the chance of recurrence of a sporadic trisomic condition within a sibship is remote, a family already having a member with aneuploidy is at greater risk than a family in the general population. A commonly used estimate is that the risk of recurrence of sporadic aneuploidy in a sibship is about double the risk in the general population. Recent data from amniocentesis material show that in Down syndrome the recurrence of "sporadic" trisomy may be as high as 2–3%.[12] This is higher than one would expect from the previous rule of thumb, for unknown reasons. Perhaps some trisomic individuals detected at amniocentesis would have aborted before term.

ISOCHROMOSOMES

Although examples of isochromosome formation are not uncommon for the X chromosomes, these structural aberrations are rarely encountered in the autosomes. Apparent isochromosomes of the long arm of a D group autosome (presumably 13) have been observed in patients having phenotypic stigmata of the 13 trisomy syndrome. Isochromosomes attributed to the G group have also been found in patients with features of Down syndrome.

RING CHROMOSOMES

The physician is interested in what happens to a patient possessing a ring autosome. The clinical features of such a patient *usually* reflect deletion because most often there is a net loss (partial monosomy) of genetic material. Thus, patients with a ring D chromosome [r(D)] have stigmata consistent with a D deletion syndrome (13q−). Reports of patients with ring D who have features more consistent with 13 trisomy illustrate the possibility of duplication through other mechanisms, which will not be amplified here.

21 TRISOMY (DOWN SYNDROME, MONGOLISM, G_1 TRISOMY)

This is the most common autosomal abnormality. An accepted frequency for the occurrence of mongolism in the general population is 1 : 660.[1] The clinical disorder was first recognized by Down in 1866, and before techniques were available to demonstrate the cytogenetic abnormality, Waar-

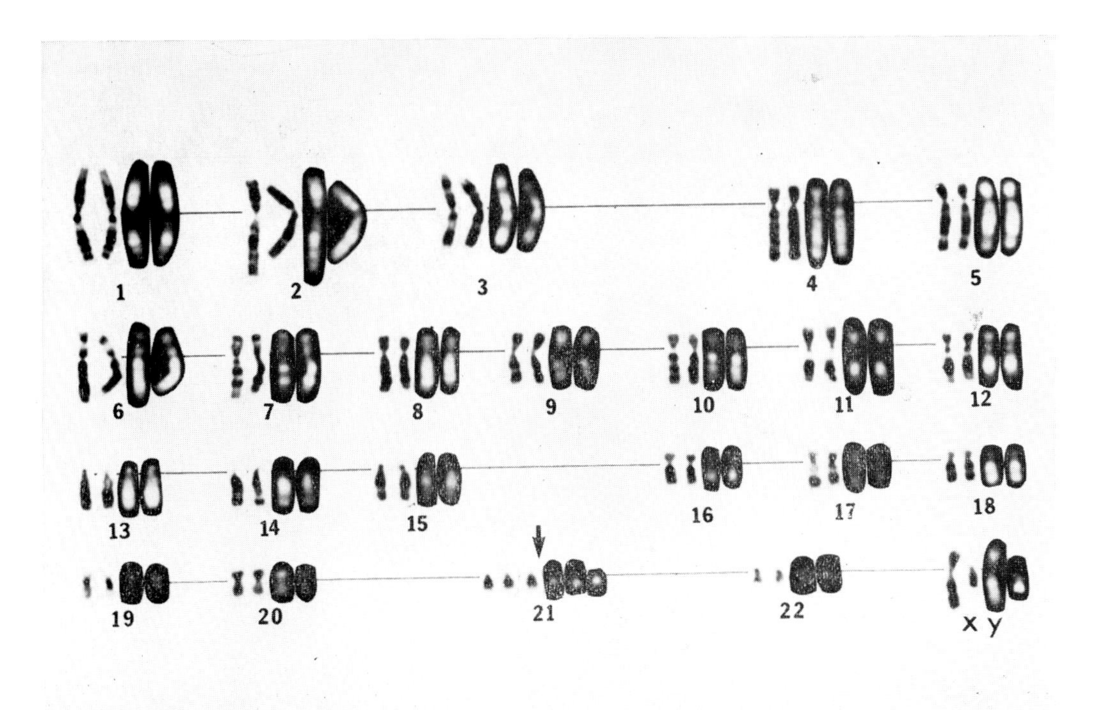

Figure 3-1. Karyotype of male infant with Down syndrome—stained by G (on the left) and Q (on the right) staining methods.

denburg, in 1932, suggested that a chromosomal anomaly could be responsible for the familial features of the syndrome, i.e., largely sporadic occurrence but a high concordance rate in monozygotic twins.

After Lejeune demonstrated that the underlying chromosomal anomaly was a trisomy of a small acrocentric chromosome, he designated chromosome 21 as the autosome present in triplicate. By cytologic and autoradiographic evidence it was not possible to distinguish chromosome 21 from chromosome 22, so many preferred the term G or G_1 trisomy to 21 trisomy. Now, of course, with the newer banding techniques, the G group chromosomes are readily distinguishable, and although the chromosome trisomic in Down syndrome turns out to be the shorter of the two G chromosomes, it is still referred to as number 21, for historical reasons.

The majority of patients who have Down syndrome have a complete 21 trisomy (Fig. 3-1). However, a significant number have translocation or mosaicism as the underlying aberration (see p. 31).

COMPLETE 21 TRISOMY

The syndrome occurs in males and females with equal frequency. The length of survival may be measured in weeks or in decades, and something approaching a normal life expectancy may be predicted for a well-cared-for patient who does not have a congenital heart defect and survives infancy.

In infancy the patient is observed to have poor muscle tone and may appear to remain "floppy" for months or even years. Growth is poor. Those patients who have congenital heart lesions and frequent pneumonia do not thrive. However, patients without these complications also fail to achieve the height of their sibs.

Psychomotor development is the major area of concern, and here there is a fairly

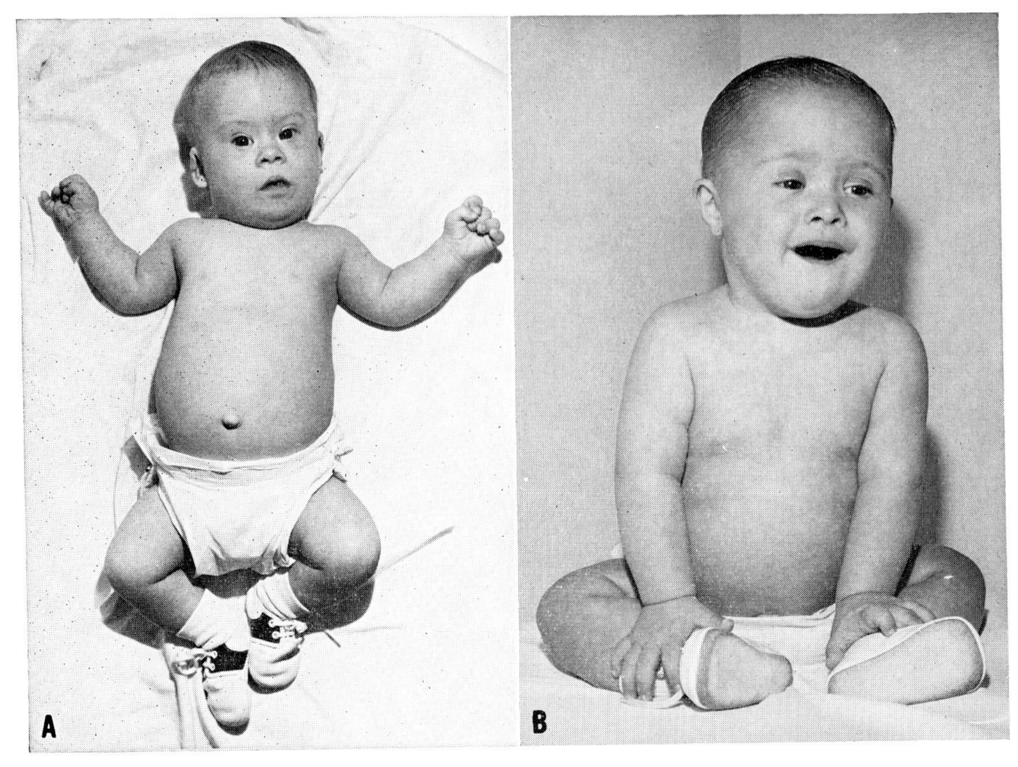

Figure 3-2. Photographs of unrelated one-year-old infants with 21 trisomy.

broad spectrum of achievement. Although most patients are in the IQ range of 25-50, an occasional patient is educable and able to learn to read and write.

The facial appearance in Down syndrome is so typical that patients with 21 trisomy, whatever their racial origin, tend to have facial features more like other patients with 21 trisomy than like those of their own sibs (Fig. 3-2). The palpebral fissures tend to slant upward laterally, and there is an epicanthic fold, which together contribute to the "mongoloid" appearance. The iris may have a ring of round, grayish "Brushfield spots" (poorly developed areas of the iris). The back of the head is somewhat flat; the nasal bridge is also flattened; the maxilla is small and the palate narrow, which makes the oral cavity inadequate to accommodate the tongue, which frequently protrudes and often has a fissured appearance. It should be emphasized that none of the above physical features is always present, and each may be present in occasional normal individuals. It is the unusual concentration of features in one individual that is characteristic.

The adult male mongoloid is sterile. Adult female patients have been reported to reproduce, and about 50% of their offspring have had Down syndrome, as expected.

About half the patients with this aberration have congenital cardiac malformations, which are responsible for much of the morbidity and early and late mortality in this syndrome.

The fingers are short, and the fifth finger is incurved. There is a gap between the first and second toes with a furrow extending down the plantar surface from this gap.

That certain dermatoglyphic patterns are characteristic of Down syndrome has been appreciated for over 30 years.[5] A bilateral simian line is found in the palms of about 30% of mongoloids and a bilateral distal axial triradius more frequently (82%). Ulnar loops on all ten fingers, rare in the general population, are found in one third of these patients. When present, a hallucal arch is the most useful discriminating feature, as it occurs in 72% of cases and less than 0.5% of controls. (For further details, see Chapter 15.)

For many decades it has been recognized that there is a relationship between maternal age and the risk of having an offspring with Down syndrome. Although the overall frequency of Down syndrome in the general population is about 1 : 660, the risk increases precipitously from 1 : 1500 for mothers under 30 years of age to 1 : 50 for mothers over 45 years old.[2] Since cases resulting from translocations do not show the maternal age effect, they are relatively more frequent in younger mothers. Thus, in the absence of chromosome studies, the recurrence risk is greatly increased in the young mothers. A 50-fold increase in risk has been estimated for mothers under 25 years of age. (This is still not very great, in absolute terms.)

The role of maternal thyroid disease and antithyroid antibodies in younger mothers of mongoloids has also been emphasized, and we have had examples of this situation in our clinic. This immunologic finding and the finding of the "Australia antigen" in 30% of patients with Down syndrome provide an area of continuing interest and speculation, although either finding may be the result of a virus, more frequent in individuals living in institutions.

TRANSLOCATION

At any maternal age the most frequent cause of Down syndrome is complete 21 trisomy. However, because there is such a clear relationship between increased age and increased frequency of offspring with Down syndrome, the affected infant of a young mother may be suspected of having an inherited translocation. For mothers under 30 years of age, translocation accounts for 9% of all patients with mongolism; one fourth of these are inherited from a parent and three fourths are sporadic translocations. Thus, in the younger age groups there is only 1 chance in 11 that the mongoloid patient has a translocation Down syndrome; and even if he has a translocation, the chances are only 1 in 4 that it is inherited. Thus, *the overall chance that a baby with an unbalanced translocation Down syndrome, born to a woman under 30 years of age, inherited this aberration is only 1 in 44.*[20]

The translocations that occur most commonly involve D group chromosomes. Autoradiographic evidence suggested that the D chromosomes participating in the translocation do not belong to pair number 13; however, evidence from QM fluorescence suggests that D/G translocations involve 13 and 14, but not, as yet, 15. Translocation within the G group and isochromosome formation must also be considered. Patients with mongolism resulting from an unbalanced translocation usually have the full features of 21 trisomy. However, an occasional patient will have a less complete expression of the syndrome, which may be related to the quantity of extra genetic material.

The Chicago Conference abbreviation for a male who has a D/G unbalanced translocation is 46,XY,—D,+t(DqGq). The patient has 46 chromosomes, but one of the D chromosomes is missing (−D) and is replaced by a chromosome made up of the long arm of a D chromosome and the long arm of a G chromosome (+t(Dq Gq)). Although the chances are only 1 in 4 of this being an inherited translocation, let us assume that the mother is a balanced carrier. The nomenclature for her chromosome complement would be 45,XX,–D, —G,+t(DqGq). She would have 45 chromosomes, a missing D chromosome (−D) and a missing G chromosome (−G); the long arms of these are united in a single chromosome (+t(DqGq)) that contains essentially all of the genetic information of the D and the G chromosomes.

The next observations would not have been predicted. It had been expected on the basis of random segregation that 1 in 3 viable offspring of either parent who was a balanced D/G translocation carrier would have translocation Down syndrome (1 in 3 would be a carrier and 1 in 3 would be normal). The empiric data from different series reveal that between 6% and 20% of offspring of a mother who is a balanced translocation carrier have Down syndrome, and only 2% of offspring of a father with a balanced translocation are mongoloid. Comparable recurrence risk figures have also been suggested for the G/G translocation (7% for female carriers, 3% for male carriers), unless it involves a translocation between homologues.

Therefore, if a patient has an inherited translocation, subsequent sibs have a relatively high recurrence risk, but a lower risk than had been previously thought. Now, with amniocentesis and antenatal diagnosis of chromosomal aberrations being more common, therapeutic abortion can be recommended if an unbalanced translocation is demonstrated. One caution: the rarely encountered isochromosome 21 or 21/21 translocation, if transmitted by a balanced carrier, will produce mongolism in all viable offspring. Isochromosome 21 and 21/21 translocation are now distinguishable from 21/22 translocation by banding techniques.

MOSAICISM

Patients having a normal cell line and a 21-trisomy cell line have the widest range of ultimate intellectual and physical attainment. Estimates that mosaicism is found in only 1.5% of patients with Down syndrome appear to be too low. There may be an appreciable number of patients having minimal findings to suggest Down syndrome who are 21-trisomy mosaics. It has not been uncommon in our experience for a pediatrician to ask for a chromosomal evaluation of a baby or child who may have few stigmata (or perhaps no clear stigmata) of mongolism and to find that the patient is a G-trisomy mosaic (presumably 21-trisomy mosaic). These are patients we would have predicted not to be particularly good bets for chromosomal anomalies.

Patients with 21-trisomy mosaicism may have the classical appearance of Down syndrome or may look essentially normal, presumably depending on the preponderance of the abnormal cell line. The realization of this is important to the genetic counselor, who should maintain a fairly high index of suspicion regarding the intellectually dull-normal young mother of a patient with 21 trisomy. If a parent has a mosaicism, the risk of 21 trisomy in future offspring is significantly increased. In fact, one may speculate that unrecognized mosaicism may account partially, if not completely, for the occasional high-risk family attributed to a "predisposition to nondisjunction."

TRISOMY 18 (E TRISOMY, TRISOMY 17, TRISOMY 16–18, EDWARDS SYNDROME)

In the same issue of *The Lancet* in 1960, two teams of investigators described the phenotypic abnormalities produced by a trisomy for a chromosome in the E group. Edwards and his colleagues described in detail the findings in a female child and designated the extra chromosome as a number 17.[6] Patau, Smith, and co-workers presented a preliminary description of this trisomic condition in two patients in a report that also included the first recognition of a patient with trisomy in the D group.[13] In their full presentation of the clinical findings of the E group trisomy, David Smith and associates considered the extra chromosome to be homologous with pair 18, and it has since been accepted that this is the appropriate assignment of the extra chromosome on the basis of centromere indices and autoradiographic and, more recently, fluorescence patterns[15] (Fig. 3-3). This appears to be the second most com-

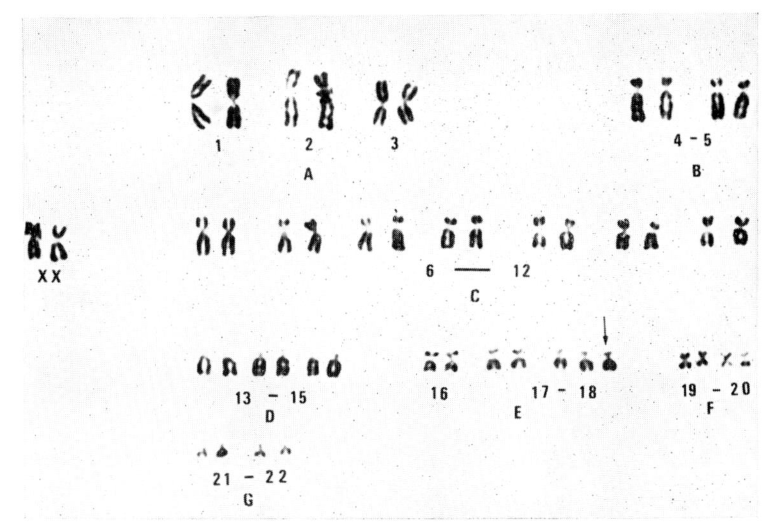

Figure 3-3. Conventional karyotype of 18 trisomy.

mon of the autosomal trisomy syndromes, with a frequency of 1:3,500 live births in a North American population.

A preponderance of patients have been reported as being female, although this may reflect the longer-term survivors. Growth and developmental retardation and early death, usually before 6 months, are typical of the course, but some patients may survive for years. The affected individual is usually hypertonic; the third and fourth fingers are clenched tightly against the palm and the second and fifth fingers overlap them. The back of the head is prominent, the ears low-set and malformed, and the chin small (Fig. 3-4). A short sternum, small pelvis with limited hip abduction and rocker-bottom feet are common. All patients in the literature, with the exception of one, have had congenital heart lesions. The dermatoglyphics in trisomy 18 provide useful diagnostic evidence (see Chapter 15).

As with Down syndrome, a patient having the stigmata of trisomy 18 most often has a complete trisomy but may have a partial trisomy or mosaicism consisting of a normal cell line and a +18 trisomic line. The phenotypic expression may be complete or incomplete in these patients, presumably related to the quantity of genetic information appearing in triplicate. Double aneuploidy has also been reported in patients with trisomy 18 who have had an additional trisomy of the X chromosome, chromosome 21, or chromosome 13. It should also be noted that in this syndrome, as in other chromosomal disorders, classic stigmata of the trisomy (except for the dermatoglyphics) may be found in patients who do not have a *demonstrable* chromosomal anomaly.

Unbalanced translocation is rare in this syndrome. The parents of a patient with trisomy 18 should be reassured that this has been a sporadic event which occurs to one infant in 3,500 and that the likelihood of a recurrence in their family is negligible.

TRISOMY 13 (D_1 TRISOMY, 13–15 TRISOMY, PATAU SYNDROME)

Although the pattern of anomalies found in trisomy 13 has been traced back through 300 years of the world literature, the first report that these abnormalities were associated with a chromosomal aberration was presented by Patau, Smith and colleagues in 1960.[13] A detailed clinical description by Smith[15] followed. To indicate that this was the first clinically recognizable disorder in-

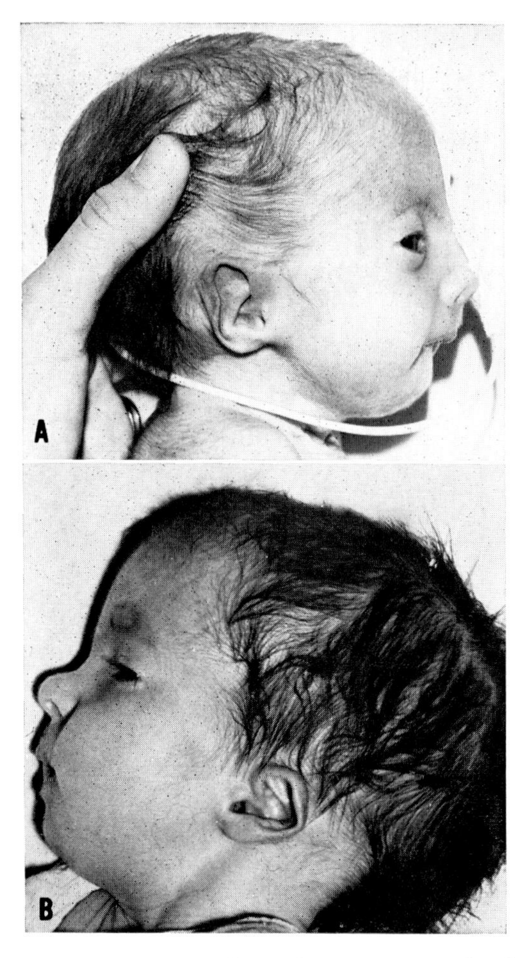

Figure 3-4. Two infants illustrating craniofacial characteristics of 18 trisomy (prominent occiput, low-set malformed ears and small chin).

volving trisomy of a D group chromosome, the term D_1 trisomy was used. Subsequent autoradiographic and fluorescence studies have provided evidence that the trisomic chromosome in this syndrome is homologous with pair 13 (Fig. 3-5). Of the three major autosomal trisomy syndromes, this is the least common, occurring in approximately 1 in 7,000 live births.

A patient with this syndrome has the appearance of being more severely malformed than patients having the other two trisomic syndromes (Fig. 3-6). Most 13 trisomic patients have cleft lip and palate and eye abnormalities that range from colobomata of the iris through microphthalmia to complete absence of the eye. Seventy-five per cent of the patients have defects of the midface and forebrain. The head is small and the forehead sloping. Those patients not having cleft lip and palate have a characteristic midfacial appearance. The ears are low-set and malformed and the chin small. Polydactyly is common and the hands are often held clenched as in 18 trisomy, with the second and fifth fingers overlapping the third and fourth. Cardiovascular defects are common. The dermatoglyphics are diagnostically useful (see Chapter 15).

The majority of patients having the stigmata of 13 trisomy have a complete trisomy. Occasionally, a patient who has the

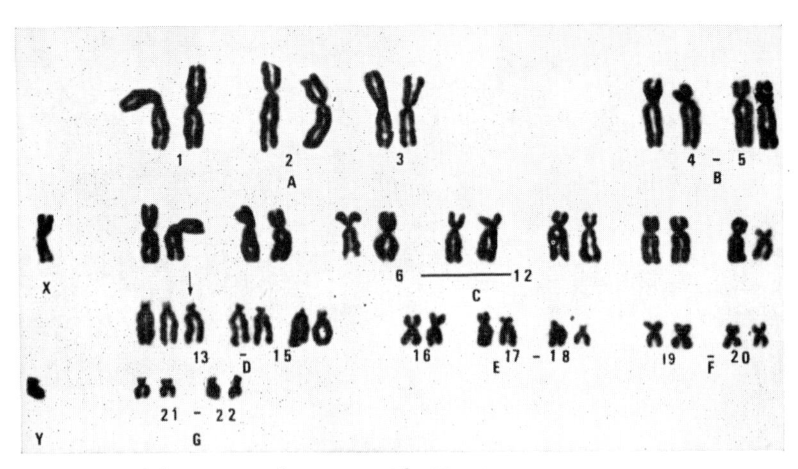

Figure 3-5. Conventional karyotype of patient with 13 trisomy.

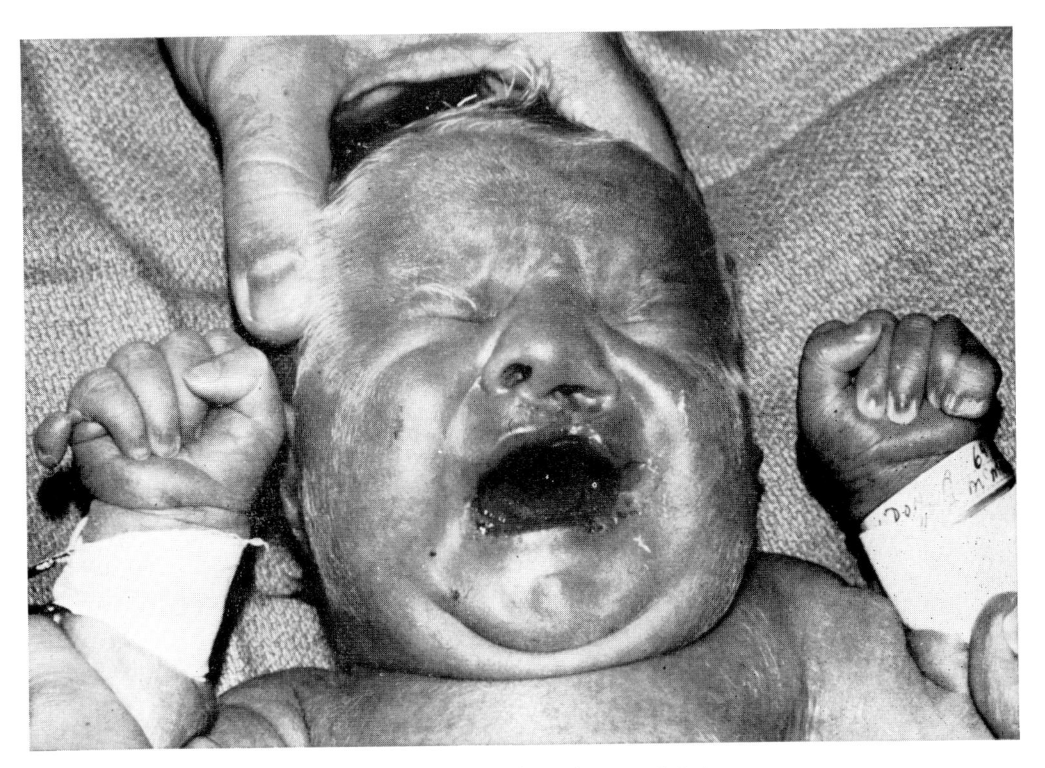

Figure 3-6. Note cleft lip and palate, polydactyly and microphthalmia.

phenotypic features of 13 trisomy has no detectable chromosomal anomaly. Balanced and unbalanced familial translocations involving a D group chromosome (presumably 13) are among the most common chromosomal aberrations. The frequency in the North American population of balanced D group translocations may be as high as 1:1,000. Clinical features of patients with unbalanced D group translocations usually reflect the full expression of the 13-trisomy syndrome. Mosaicism (normal/13 trisomy) also occurs.

Parents of a patient with complete 13 trisomy may be counseled that the incidence of 13 trisomy is about 1:7,000 in the general population. The risk of recurrence of nondisjunction in their family, although higher than in the general population, should be very small (of the order of 1:3,500). If the patient has an unbalanced translocation, chromosomal evaluation of the parents is required. Often the unbalanced translocation proves to be a sporadic event, but if a balanced translocation is found in a parent, the frequency of affected offspring of such balanced carriers is high enough to justify antenatal chromosomal studies of subsequent pregnancies by amniocentesis.

OTHER STRUCTURAL AND NUMERICAL ABERRATIONS OF AUTOSOMES

Group B

Cri-du-Chat (5p—) Syndrome. This clinical condition was first attributed to a chromosomal aberration in 1963 by Lejeune and co-workers.[10] On morphologic grounds the chromosomal anomaly was interpreted as being a deletion of the short arm of chromosome 5. Autoradiographic evidence has since added support to the hypothesis that the deletion involving the earlier-replicating B-group chromosome (5) was associated with the clinical features of

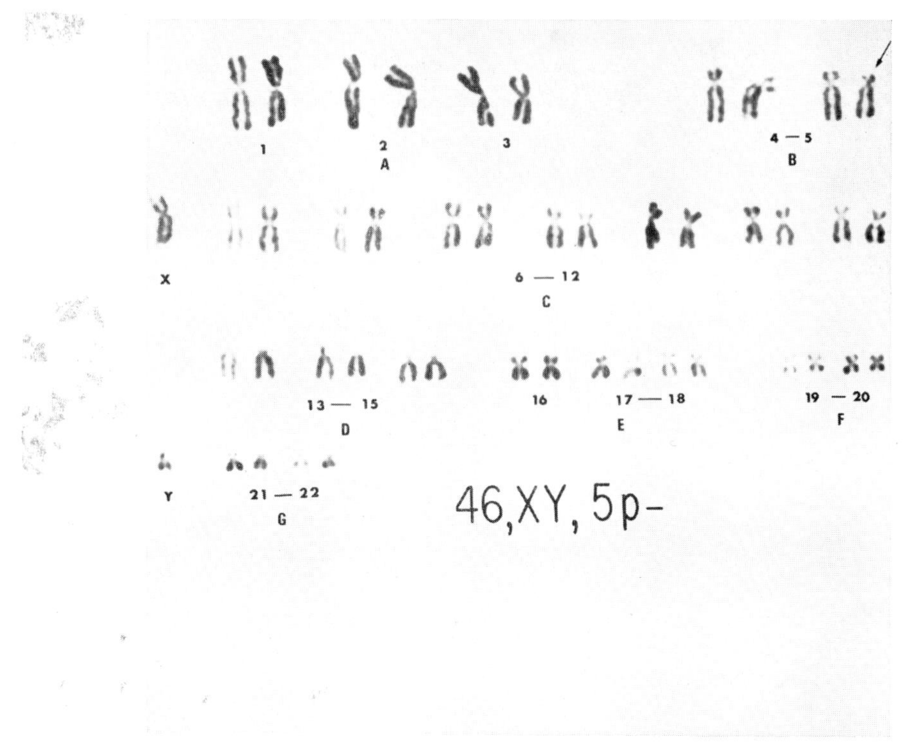

Figure 3-7. Conventional karyotype of patient with the cri-du-chat syndrome.

this syndrome.[17] The fluorescence patterns are very different and provide the clearest distinction between the B-group chromosomes.[3] The incidence of this lesion in the general population has been estimated as being about 1:50,000.

The striking clinical manifestation of this syndrome, which apparently affects girls more commonly than boys, is the cry, a mewing, plaintive sound like that of a kitten in distress. The physical features are not as diagnostic in this disorder as in the major autosomal trisomies (Fig. 3-8). However, the appearance of the face is often round; the expression may be paradoxically alert; and the head is small. The eyes are widely spaced and the chin is frequently receding. A congenital heart defect is present in about half of the patients. These patients are, of course, severely retarded. Parents may be reassured that this is a rare sporadic event that is not likely to recur. However, familial translocation and ring chromosome cases have been recognized.

4p— Syndrome. A number of patients have been observed who have a deletion of a B group chromosome but who do not have the features of the cri-du-chat syndrome. When it has been investigated, this deletion appears to be in the short arm of the later-replicating B chromosome and thus has been called a 4p— anomaly.[19] A further advance on the autoradiographic distinction among B group chromosomes is provided by the banding methods. In the older literature, the cri-du-chat syndrome is attributed to a 4p— deletion, reflecting the earlier difficulty in distinguishing among B group chromosomes.

Males and females appear to be affected equally in this condition, which is characterized by severe mental retardation and a midline facial defect consisting of wide-set eyes and a broad nose, root and bridge. A defect in the iris (coloboma), droopy eye-

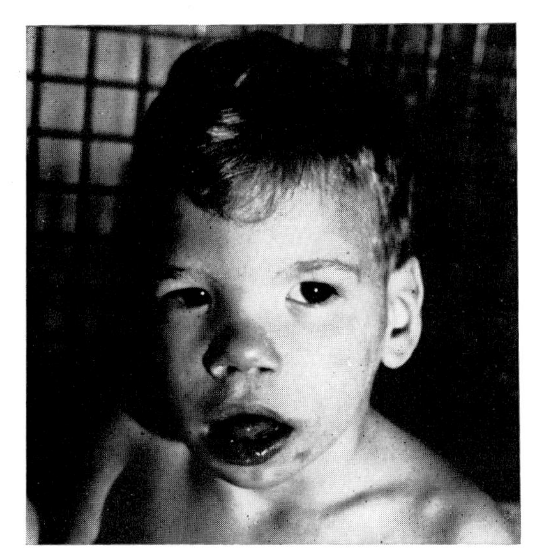

Figure 3-8. Patient with the cri-du-chat syndrome. The eyes are widely spaced with an inner canthic fold, but there is nothing pathognomonic about the facial features of this syndrome.

lids (ptosis), low-set ears with preauricular dimples or tags, carp-mouth, cleft lip and/or palate, and midline scalp defects complete the abnormalities of the head and face. A cat-cry is probably not present. The dermatoglyphics of a patient with the 4p— syndrome do not show any unusual features, other than dysplastic dermal ridges. This chromosomal anomaly has not been reported to recur in families. Patients with the phenotype but without any demonstrable chromosome anomaly have been recognized.

Group C

Several cases of trisomy of a group C chromosome have been reported, but relatively few in which the extra chromosome has been positively identified by banding techniques. The information is still too scanty to permit clinical delineation of the syndromes produced.[7]

Group D

Deletions of the long arm of a D group chromosome, presumably number 13 (13q—) have been repeatedly described. Short-arm deletions (13p—) and ring chromosomes [r(D)] have also been reported.

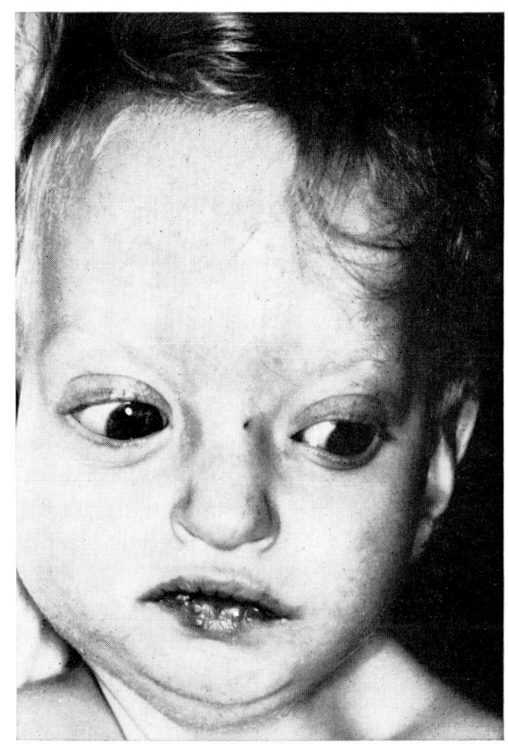

Figure 3-9. Craniofacial appearance of patient with 4p— syndrome.

Group E

Structural abnormalities, including deletions of the long arm (18q—), short arm (18p—), and rings [r(18)] have been reported for group E chromosomes. The pattern of anomalies is more uniform in patients with deletions of the long arm than in those with deletions of the short arm, so only the 18q— syndrome will be discussed in detail. Some 18p— patients have features that are difficult to distinguish from Turner stigmata, while others have varying patterns of abnormalities that are not consistent.

18q— Syndrome. Patients with a long-arm deletion of chromosome 18 are retarded and may survive periods ranging from months to many years.[18] In infancy they are hypotonic and have a low-pitched voice. Soon it becomes apparent that the middle of the face is retracted and that the mouth is broad and down-turning (carp-mouth). With further growth, the jaw becomes jutting, which accentuates the retraction of the middle of the face. The head is small and the eyes widely spaced. Epicanthic folds and nystagmus (rapid eye movements) are common. The ears are frequently large, and in contrast, the ear canal may be small or sometimes not patent. Long, tapering fingers and dimples of the knuckles, elbows, and knees are occasionally found. Patients with this syndrome usually represent sporadic events, although there is at least one report of occurrence of this syndrome in sibs (Fig. 3-10).

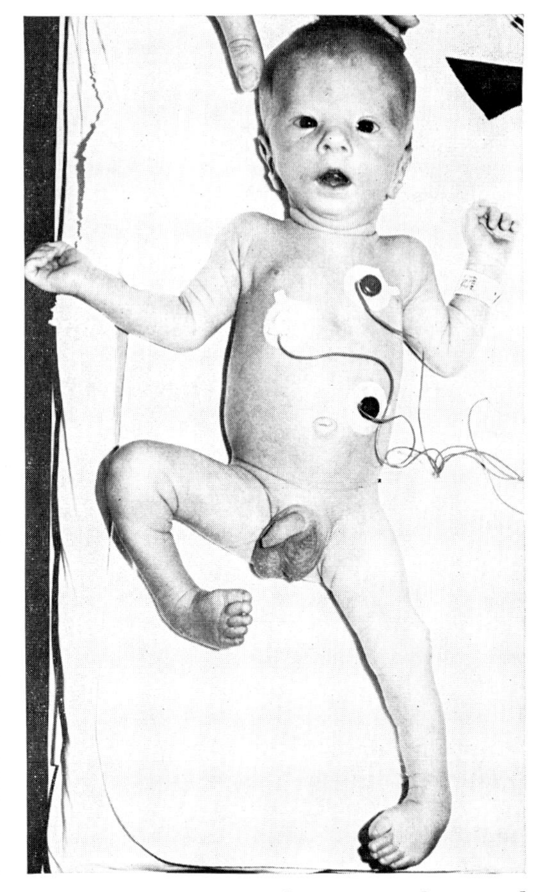

Figure 3-10. Patient with 18q— syndrome and characteristic "carp-mouth."

Group G

Several structural and numerical aberrations of group G chromosomes have been described other than the anomalies producing Down syndrome and its variants and 22 trisomy. Abnormalities reflecting loss of chromosomal material include long-arm deletions (21q—), ring chromosomes [r(21)], and, in at least three reports, a complete absence of a G group chromosome (45–G).

It was stated in Chapter 2 that the human being cannot easily survive the loss of an entire autosome. Small additions of chromosomal material and even smaller deletions of chromosomes are compatible with survival, although responsible for maldevelopment. As will be discussed in Chapter 4, numerical and structural alterations of the X chromosome represent a different situation, because only one X chromosome is normally "completely active" in an individual no matter how many X chromosomes are present.

Therefore, if one were to try to predict the autosome whose loss is most compatible with survival, the prediction would be an autosome with the least amount of genetic information (i.e., the smallest). And this has proved to be the case. Monosomy for a G group chromosome has been reported in three patients, as has mosaicism involving a normal cell line and a 21-monosomy cell line (46,XX/45,XX,—21).

On the assumption that the monosomy involved the chromosome that, in triplicate, produces Down syndrome, Lejeune[11] proposed that monosomy leads to the "contre-type" of polysomy. Although *some* features of a patient with a monosomy may be the opposite of a patient with a polysomy, there is no justification based on present knowledge of differentiation and development for the concept of "contre-type" or "anti-mongolism," except as clinical descriptions of physical findings.

Thus, certain findings in complete or partial monosomy of chromosome 21 may be the opposite of findings in complete or partial trisomy. The patient with 21 monosomy (complete, partial, or mosaic) is generally hypertonic rather than hypotonic, has an "anti-mongoloid" slant to the eyes, an elevated nasal bridge, and an elongated skull. Micrognathia, hypospadias, and pyloric stenosis are also found. The dermatoglyphic patterns have not been consistent; however, a proximal (normal) axial triradius has been repeatedly observed. These disorders are rare sporadic events.

One final association with partial deletion of a G group chromosome is leukemia with the Philadelphia chromosome. The leukocytes of a large percentage of patients who have chronic granulocytic leukemia have a deletion of the long arm of a G group chromosome (Gq—), the Philadelphia chromosome. When these patients are in remission, the G chromosomes appear to be normal, but the deletion reappears when the patient is in relapse. Recently, Caspersson and his associates[4] have established by quinacrine fluorescence that the Philadelphia chromosome is number 22. Recent reports indicate that the missing piece of the Philadelphia chromosome is translocated to (strangely enough) chromosome 9.

SUMMARY

There are three autosomal trisomy syndromes of clinical significance: 21 trisomy (Down syndrome), 18 trisomy, and 13 trisomy. Of these three, 21 trisomy is the most common, is associated with survival beyond infancy, and commands considerable involvement of the medical team. There are many syndromes of structural aberration that are less frequent than trisomy syndromes, but that require skill in recognition in order to provide optimal management and counseling.

REFERENCES

1. Carter, C. O., and MacCarthy, D.: Incidence of mongolism and its diagnosis in the newborn. Brit. J. Soc. Med. 5:83, 1951.
2. Carter, C. O., and Evans, K. A.: Risk of parents who have had one child with Down's syndrome (mongolism) having another child similarly affected. Lancet ii:785, 1961.
3. Caspersson, T., Lindsten, J., and Zech, L.: Identification of the abnormal B group chromosome in the "cri du chat" syndrome by QM fluorescence. Exp. Cell Res. 61:475, 1970.
4. Caspersson, T., Hulten, M., Lindsten, J., et al.: Identification of the Philadelphia chromosome as a number 22 by quinacrine mustard fluorescence analysis. Exp. Cell Res. 63:238, 1970.
5. Cummins, H.: Dermatoglyphic stigma in mongoloid imbeciles. Anat. Rec. 73:407, 1939.
6. Edwards, J. H., Harnden, D. G., Cameron, A. H., et al.: A new trisomic syndrome. Lancet i:787, 1960.
7. Feingold, M., and Atkins, L.: A case of trisomy 9. J. Med. Genet. 10:184, 1973.
8. Ford, C. E., Jones, K. W., Miller, O. J., et al.: The chromosomes in a patient showing both mongolism and the Klinefelter syndrome. Lancet i:709. 1959.
9. Lejeune, J., Gautier, M., and Turpin, R.: Etude des chromosomes somatiques de neuf enfants mongoliens. C. R. Acad. Sci. 248:1721, 1959.
10. Lejeune, J., Lafourcade, J., Berger, R., et al.: Trois cas de délétion partielle du bras court d'un chromosome 5. C. R. Acad. Sci. 257:3098, 1963.
11. Lejeune, J.: Types et contre-types. J. Génét. Hum. 15 (suppl.):20, 1966.
12. Nadler, H. L.: Indications for amniocentesis in the early prenatal detection of genetic disorders. Birth Defects VII:5, 1971.
13. Patau, K., Smith, D. W., Therman, E., et al.: Multiple congenital anomalies caused by an extra autosome. Lancet i:790, 1960.
14. Smith, D. W., Patau, K., Therman, E., et al.: A new autosomal trisomy syndrome: multiple congenital anomalies caused by extra chromosome. J. Pediat. 57:338, 1960.
15. Smith, D. W., Patau, K., Therman, E. et al.: The D_1 trisomy syndrome. J. Pediat. 62:326, 1963.
16. Tjio, J. H., and Levan, A.: The chromosome number of man. Hereditas 42:1, 1956.
17. Warburton, D., Miller, D. A., Miller, O. J., et al.: Distinction between chromosome 4 and chromosome 5 by replication pattern and length of long and short arm. Am. J. Hum. Genet. 19:399, 1967.
18. Wertelecki, W., and Gerald, P. S.: Clinical and chromosomal studies in the 18q— syndrome. J. Pediat. 78:44, 1971.
19. Wolf, U., Reinwein, H., Porsch, R., et al.: Defizienz an den kurzen Armen eines Chromosoms Nr. 4. Humangenetik 1:397, 1965.
20. Wright, S. W., Day, R. W., Muller, H., et al.: The frequency of trisomy and translocation in Down's syndrome. J. Pediat. 70:420, 1967.

Chapter 4

SEX-CHROMOSOMAL ANOMALIES

In 1959, the year the first autosomal anomaly was described, two sex-chromosomal anomalies were reported. Jacobs and Strong[8] reported a patient who had the clinical features of Klinefelter syndrome and had an XXY (47,XXY) sex-chromosomal constitution, and Ford and co-workers[6] demonstrated monosomy for a C group chromosome in a patient having the stigmata of Turner syndrome and inferred that the missing chromosome was an X.

Two earlier events are important in the history of the study of human sex chromosomes. The first is the investigations by Painter[13] of the presence of a Y chromosome in the human male. Strangely enough, it was not Painter's conclusion that the human chromosome complement was 48 that provoked a challenge, but rather the question of the male-determining role of the Y chromosome. Other early investigators believed that it was the number of X chromosomes that determined the sex of the individual. The sex-determining function of the Y chromosome was finally established by the previously mentioned studies of Jacobs and of Ford and their associates. The second historical event was the recognition of nuclear sex and development of the technique for the study of sex chromatin.

NUCLEAR SEX

Barr and Bertram[1] observed that there was a distinguishing difference in the interphase cells of males and females. The observation, first made in the nerve cells of cats, was that the female possessed a dense mass of chromatin in the nuclei of a significant percentage of cells that was not present in the male.

An extension of this investigation through several species of mammals to the human provided a simple yet powerful clinical tool. In the human, a light scraping of cells from the inside of the cheek (buccal mucosa) is spread on a slide, fixed, and stained to reveal the presence or absence of *sex chromatin*, or *Barr body*, which distinguishes the female from the male—or, more precisely, determines if an individual has more than one X chromosome.

A normal female, possessing an XX chromosomal constitution, will have a mass of densely staining chromatin pressed against the inner surface of the nuclear membrane in 20% to 60% of her buccal mucosal cells (Fig. 4-1). A normal XY male, having only one X chromosome, will

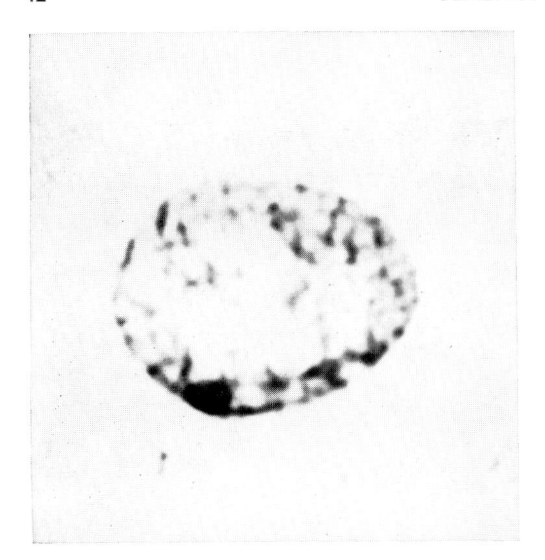

Figure 4-1. Barr body. Note that this densely staining mass lies against the nuclear membrane.

TABLE 4-1. Number of Barr and Fluorescent Y Bodies in Selected Conditions

Conditions	*Barr Bodies*	*Fluorescent Y*
Normal male (XY)	0	1
XO Turner syndrome	0	0
XYY	0	2
Normal female (XX)	1	0
Klinefelter syndrome (XXY)	1	1
XXYY	1	2
XXX	2	0
XXXXY	3	1

not have sex chromatin, but 1–2% of his cells may contain darkly staining masses that could be mistaken for Barr bodies.

Of less clinical importance is the finding of a "drumstick" in the nuclei of polymorphonuclear leukocytes. The "drumstick" is a projection from the nucleus that occurs in only about 5% of mature polymorphonuclear cells in the female and does not occur in the male. It is often difficult to distinguish from other nuclear structures and is therefore of limited diagnostic value.

The buccal smear for Barr bodies is a useful screening procedure for patients in whom an X chromosomal anomaly is suspected. The rule is that the number of Barr bodies is one less than the number of X chromosomes. Thus, a normal male has no Barr bodies, a normal female or an XXY male has one, an XXX female has two, and so on. There is no relationship of Barr bodies to the number of Y chromosomes. Table 4-1 lists the normal male and female chromosome constitutions and several common sex chromosomal anomalies, together with the number of Barr bodies one would expect to find in a major proportion of buccal mucosal cells. Some cells of a chromatin-positive individual will have no visible Barr bodies, and other cells may have fewer than the expected number of chromatin masses.

What precisely is the Barr body or sex chromatin mass? Ohno and Hauschka[12] observed in female cells at prophase a darkly staining (heteropyknotic) chromosome about the size of a Barr body, which they assumed to be an X chromosome. In the normal male no such heteropyknotic chromosome is seen. Autoradiographic studies using tritiated thymidine, which is incorporated into the DNA of replicating chromosomes, demonstrate that one chromosome completes replication later than all the others. This chromosome is located at the periphery of the nucleus where the Barr body is found. Individuals such as the normal XY male and the patient with the XO Turner syndrome do not have late-replicating X chromosomes and do not have Barr bodies. As will be discussed in the section on the single-active-X hypothesis, this late-replicating X chromosome is considered to be essentially inactive.

The phenotypic sex of the individual is determined (with few exceptions) by the presence or absence of a Y chromosome. Although a patient who has the XO Turner syndrome (45,X) has only one X chromosome, as does the normal male, she does not possess a Y chromosome, and is thus phenotypically female. A patient with Klinefelter syndrome (47,XXY) has two X's, like the normal female, but also has a Y chromosome and is therefore a phenotypic male. The absence of Barr bodies (normal for male) in the buccal smear of the female patient with the XO Turner syndrome and the presence of Barr bodies (normal for female) in the male patient with Klinefelter syndrome merely reflect the number of X chromosomes possessed by the patient. They do not imply that the individual with XO Turner syndrome is not a female or that the Klinefelter patient is not a male.

In addition to numerical aberrations of X chromosomes producing numerical aberrations of Barr bodies, structural anomalies of X chromosomes may produce structural differences in Barr bodies. An example of this is the patient who has an X isochromosome involving the long arm of an X chromosome [46,X,i(Xq)]. The patient has 46 chromosomes, including a normal X chromosome and an abnormal X chromosome. The abnormal X chromosome, the isochromosome, is consistently "inactivated." The patient has a loss of chromosomal material from the short arms of the X chromosomes and therefore has many phenotypic features of the XO Turner. (This concept will be developed later.) However, because the patient has two X chromosomes, there will be a positive buccal smear—i.e., a Barr body will be present, but it will be larger than normal, presumably because it consists of more chromosomal material (two long arms). Conversely, a patient with a deletion of part of an X chromosome may have a Barr body smaller than normal.

The work of Caspersson and colleagues[3] has shown that fluorescent alkylating agents such as quinacrine hydrochloride bind most avidly to the Y chromosome in metaphase preparations. Pearson[14] and coworkers found that, in interphase cells obtained by buccal smear, a brightly fluorescing body may be found in normal males; two fluorescent bodies are found in patients with the XYY constitution. The number of fluorescent Y bodies has already become a standard technique for identifying the number of Y chromosomes possessed by an individual. This provides a useful adjunct to the Barr body analysis in determining nuclear sex from cells obtained at amniocentesis, from Wharton jelly (material from the umbilical cord), or from buccal smears. The two procedures should be used together to confirm each other and to provide information as to the number of X chromosomes (Barr bodies plus 1) and the number of Y chromosomes (number of fluorescent Y bodies). Findings in selected X and Y chromosomal constitutions are summarized in Table 4-1.

Two cautions should be advanced. First, in addition to the brightly fluorescing Y chromosome, chromosomes 3 and 13 also have bright segments when treated with quinacrine mustard. However, as a rule,

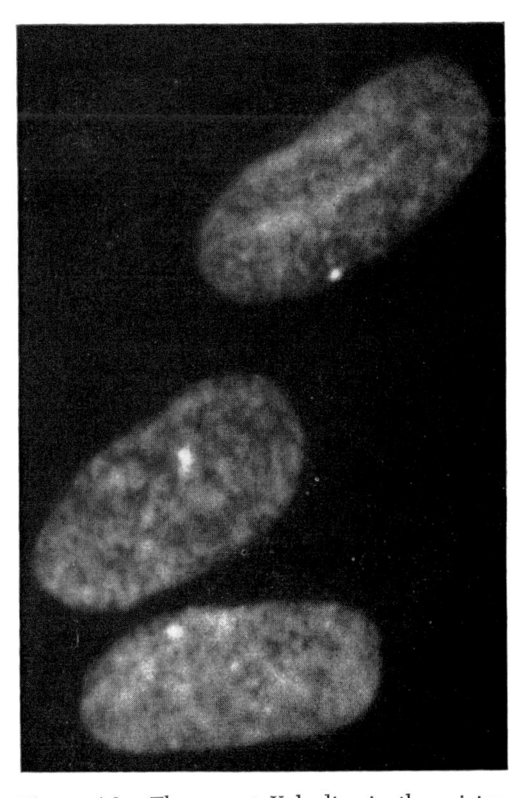

Figure 4-2. Fluorescent Y bodies in three interphase nuclei. Observe that the Y bodies are smaller than the Barr body in Figure 4-1 and that they are more often not against the nuclear membrane.

there is little difficulty in identifying the fluorescent Y body in cells treated with quinacrine hydrochloride (Fig. 4-2). The second caution is that quinacrine hydrochloride fluorescence of white blood cells is disappointing in its reliability—at least in some hands. The percentage of cells with fluorescent Y bodies decreases from 80–90% in fibroblasts and Wharton jelly cells to 70% in buccal mucosal cells to less than 50% in white blood cells—and in the white cells, the fluorescent Y is often difficult to distinguish.

SINGLE-ACTIVE-X HYPOTHESIS (LYON HYPOTHESIS)

In 1961 Mary Lyon[10] and, independently, several other workers[16] advanced a hypothesis that accounted very neatly for several hitherto unexplained observations. First, there was the phenomenon of "dosage compensation," the fact that genes on the X chromosome produce the same amounts of their corresponding proteins in females as in males, instead of twice as much, as one might expect from the fact that females have twice as many X chromosomes. Second, there was the cytologic observation that one of the X chromosomes in females was heterochromatic and corresponded to the sex chromatin in female somatic cells. Third, one X chromosome in dividing female somatic cells completed its DNA replication later than any other chromosome. The clue to the Lyon hypothesis came from a mutant mouse. A sex-linked gene causing a light coat color in hemizygous males caused variegated patches of light and dark fur in heterozygous females. (The same phenomenon occurs in the familiar tortoiseshell cat.) Lyon suggested that in the dark patches the X chromosome carrying the mutant gene was inactive, and in the light patches the X chromosome carrying the normal allele was inactive, so the mouse was in effect a mosaic of mutant and nonmutant tissue. This idea was then generalized, as follows:

1. In a somatic cell only one X chromosome is genetically active. The other (or all others in cases where there are more than two) is genetically inactive, replicates its DNA late in the cell cycle, and forms the sex chromatin. Thus a female is a genetic mosaic for any X-linked genes for which she is heterozygous.
2. Whether the maternal or paternal X is inactivated is a decision that occurs early in embryogeny (at the morula stage, about 16 days after conception) and that is decided *independently* for each cell of the embryo.
3. All cells descended from the one in which inactivation of the X first occurred have the same X inactivated.

Many studies support the single-active-X hypothesis. Cytologic evidence came from studies of the mule (the offspring of a horse and a donkey). The X chromosomes of the horse and donkey are morphologically different. In a female mule it could be shown by radioautography that the horse X chromosome was late-replicating in some cells and the donkey X was late-replicating in others.

One of the most definitive proofs came from the study of women heterozygous for electrophoretic variants of the enzyme G6PD, the gene for which is X-linked. *Tissues* from heterozygous females show a mixture of the two variants, showing that both genes are active. However, *clones* of fibroblasts (lines derived from a single cell) show either one variant or the other, but not both, proving that one X is inactivated in each cell and that it is the same X in all descendants of a particular cell.[4]

The mechanism that determines that one X chromosome will be inactivated is still unknown. One intriguing theory[2] argues from the fact that in marsupials the paternal X is always inactivated, and proposes that a gene (the "sensitive site") on the X chromosome in marsupials produces a single molecule of an "informational entity" that, when it combines with an adjacent X chromosome site, inactivates that X chromosome. Since there is only one molecule, only one X is inactivated, and it will be the one that produces the molecule. In the testis the sensitive site is "imprinted" in such a way that the paternal X site is inactive in the next generation. In mammals, the sensitive site has been translocated to an autosome, but it is still imprinted during male meiosis, so that the female receives only one active sensitive site, and the informational molecule will activate either X chromosome at random.

Besides accounting for a number of hitherto puzzling observations on mammalian X chromosomes the recognition of "Lyonization" has led to several practical applications. In X-linked genetic diseases where the effects of the gene are expressed at the cellular level, the carrier female should manifest the disease in some of her cells, thus allowing heterozygote detection for purposes of counseling. For instance, in choroideremia an X-linked gene causes degeneration of the cells of the retina with progressive loss of vision in hemizygous males. The heterozygous females can be diagnosed by a patchy "pepper and salt" appearance of the retina, although their vision is unimpaired. Similarly, X-linked anhidrotic ectodermal dysplasia causes absent sweat glands and defective teeth in males, and female heterozygotes show patches of skin free of sweat glands. If the gene affects an organ or tissue that is derived from relatively few cells at the morula stage, when X inactivation occurs, it may happen, just by chance, that inactivation affects the same X chromosome in all the precursor cells. For instance, if there were 10 precursor cells, then there would be 1 chance in 2^{10}, or a 1/1024 chance that they would all be normal. That may be why an occasional female heterozygous for hemophilia or X-linked muscular dystrophy, for instance, is affected with the disease and others cannot be distinguished from noncarriers by biochemical means. An application of the Lyon hypothesis in supporting the somatic mutation theory of cancer is referred to in Chapter 19.

If one X chromosome is completely inactive, then why should patients with the XO Turner syndrome or XXY Klinefelter syndrome have any abnormalities? And why should increasing numbers of X chromosomes as found in patients with XXX, XXXX, XXXY, and XXXXY be accompanied by progressively greater abnormality? The individual with XXY Klinefelter syndrome may graduate from high school or even go to college, but the patient with the XXXXY syndrome has an IQ in the 20 to 50 range and is not educable.

Russell[16] has suggested, from studies on the mouse, that portions of the "inactivated" X chromosome may not be inactivated. The inactive X chromosome may contain loci that are required to be present in duplicate if normal development and function is to take place. Thus Ferguson-Smith[5] has proposed that there are loci on the Y chromosome homologous with certain loci on the X chromosome, and that the normal XX female and normal XY male have these loci in duplicate. The XO Turner syndrome would be deficient in these loci, and the patient with XXY Klinefelter syndrome would have these loci in triplicate. Some such modification of the single-active-X hypothesis seems to be required to comply with the clinical observations.

SEX CHROMOSOMES

The X Chromosome

The X chromosome ranks between numbers 7 and 8 in total length and short-arm length, making it one of the larger chromosomes. The banding pattern by fluorescence is distinctive, with a band in the short arm separated by a broad paracentric dark area from an intense band in the long arm. It would be expected that a large chromosome should contain a large number of genes, and this appears to be the case. Almost 100 genes have already been assigned to the X chromosome. To reiterate, the normal human female possesses two X chromosomes, one of which is "inactive" in each cell. The normal male has only one X chromosome, and this chromosome is active in every cell. The X chromosome in the human assumes a passive role in the determination of sex. In the absence of a Y chromosome, the sex of an individual is female no matter what the number of X chromosomes may be.

The Y Chromosome

The Y chromosome, by contrast, is one of the smaller chromosomes and is most similar to the G group chromosomes in length and morphology. However, it is readily distinguished from chromosomes 21 and 22 by the most intense fluorescence found in any chromosome (located at the distal long arm). The Y chromosome takes the active role in sex determination, and its presence produces the male phenotype regardless of the number of X chromosomes an individual may possess. (Rare exceptions such as testicular feminization will be discussed later.) Other than determining "maleness," no significant role for the Y chromosome has been determined. A few decades ago several so-called Y-linked genes were described, but this number has been reduced to one inconsequential gene —the gene for hairy ears.

XO TURNER SYNDROME (TURNER SYNDROME, STATUS BONNEVIE-ULLRICH, GONADAL DYSGENESIS)

Most of the phenotypic features of what is now called Turner syndrome were described in 1930 by Ullrich,[19] who reported a combination of anomalies in an 8-year-old girl that included webbing of the neck, cubitus valgus, congenital lymphangiectatic edema, prominent ears, ptosis, small mandible, dystrophy of the nails, and hypoplastic nipples. In 1938, Turner observed webbing of the neck and cubitus valgus together with sexual infantilism in young women.[18] These three findings have become the cardinal signs of Turner syndrome.

In 1959 Ford and colleagues examined the chromosomes of a 14-year-old girl with clinical evidence of Turner syndrome, found that she had only 45 chromosomes, and suggested that the missing chromosome appeared to be an X.[6] From this point on, then, patients with gonadal dysgenesis and Turner stigmata who have a 45,X chromosomal constitution have been said to have Turner syndrome (Fig. 4-3).

The incidence of XO Turner syndrome in the newborn population is approximately 1 in 5,000, or about 1 in 2,500 females as estimated by chromatin-negative buccal smears. However, the frequency with which the XO Turner syndrome is encountered in spontaneous abortions (as high as 7.5%) suggests that perhaps as few as 1 in 50 conceptions with the XO Turner syndrome result in live births.[15]

A distinct maternal age factor has not been demonstrated in this syndrome. Normal life expectancy is the rule, but this may be affected by associated lesions, such as coarctation of the aorta and renal disease.

The most constant feature of the XO Turner syndrome is shortness of stature. Eventual height attainment rarely exceeds 60 inches. It appears to be correlated with the height of the parents. A patient with the XO Turner syndrome whose parents are tall is likely to reach 59 or 60 inches in height, whereas a patient whose parents are short may only grow to 53 or 54 inches. Intellectual development also appears to be related to "mid-parent" intelligence. In general, the patient with the XO Turner syndrome will be somewhat less gifted intellectually than her sibs, as well as significantly shorter. However, it is not unusual for these patients to finish college and to earn graduate degrees.

The appearance of the face is distinctive. A narrow maxilla, small chin, "shark" mouth, low-set, malformed or prominent ears, epicanthic folds, and drooping eye-

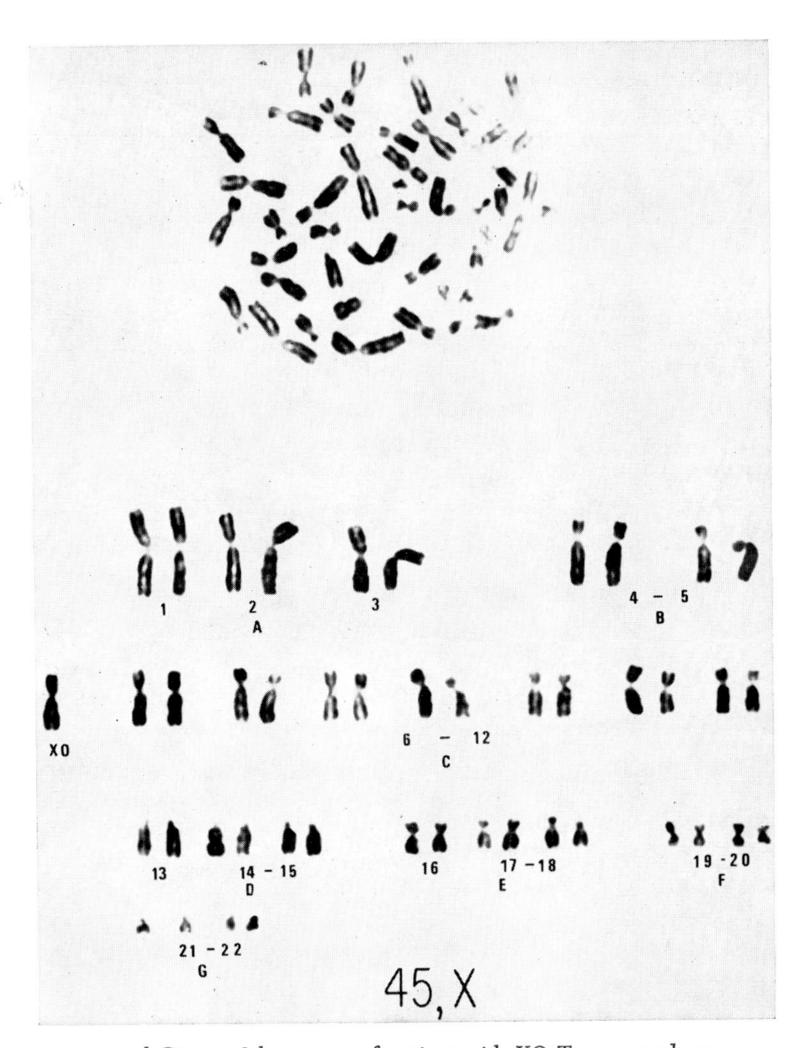

Figure 4-3. Conventional Giemsa 6 karyotype of patient with XO Turner syndrome.

lids (ptosis) comprise the typical facies. Only about 50% of patients have webbing of the neck, which is considered to be the characteristic anomaly of the syndrome (Fig. 4-4). Excessive looseness of the skin of the neck may be observed in infancy (Fig. 4-5), which may not necessarily produce obvious webbing in childhood. A low posterior hairline and pigmented nevi are frequently found.

A shield-shaped chest and widely spaced hypoplastic nipples are common features. Approximately 35% of these patients have cardiovascular disease.

Wilkins and Fleischmann found in 1944 that patients with Turner syndrome had, in place of normal ovaries, streaks of ovarian stroma without follicles.[20]

X CHROMOSOME MOSAICS

A patient bearing Turner stigmata may have more than one cell line, one of which is XO. A variety of mosaics have been reported, including XO/XX, XO/XXX, XO/XX/XXX, and XO/Xi(Xq). These patients may have the same stigmata as seen in XO Turner syndrome, although, in general, the expression of the phenotype is

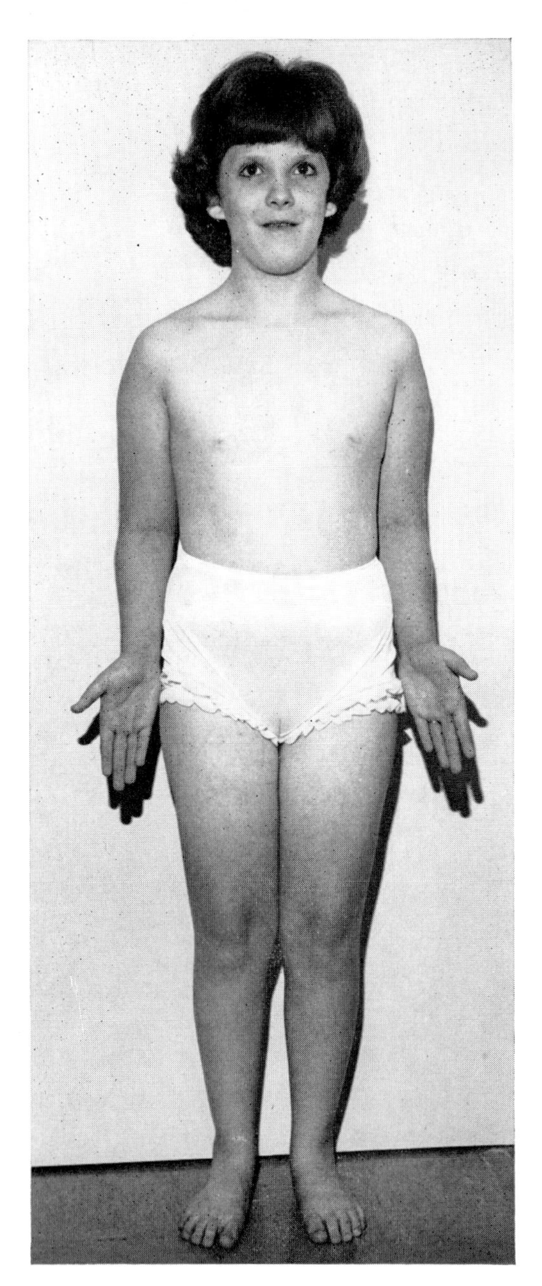

Figure 4-4. Fourteen-year-old girl with XO Turner syndrome. Note that she has the facies and some somatic features of the disorder (shortness, wide-spaced nipples), but like 50% of patients with Turner syndrome she does not have webbing of the neck. Her height at this age was 55 inches.

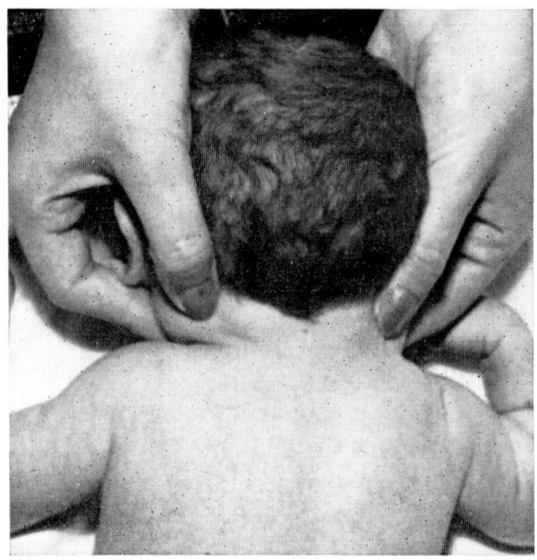

Figure 4-5. Prominent webbing of the neck and low posterior hair-line in newborn with XO Turner syndrome.

modified by the presence of cell lines other than XO. An XO/XX mosaic *may* thus have fewer and less striking Turner stigmata and more of a normal female appearance. Shortness of stature is still found in the vast majority of these patients, but it is not invariable.

STRUCTURAL ANOMALIES OF THE X CHROMOSOME

This group includes isochromosomes of the long-arm [X,i(Xq)] and short-arm deletions (XXp—); isochromosomes of the short arm [X,i(Xp)]; long-arm deletions (XXq—); and ring chromosomes [X,r(X)].

Patients with an isochromosome of the long arm [46,X,i(Xq)] have the normal number of chromosomes, but have a structurally abnormal X chromosome and have sufficient Turner stigmata to suggest a diagnosis of the XO Turner syndrome. Barr bodies are found in the buccal smear, but to the experienced observer they are larger than usual. The karyotype reveals 46 chromosomes, one of which is an unusually large metacentric chromosome that appears to be made up of two long arms of an X chromosome. This chromosome is always inactivated, so on buccal smear this larger amount of chromosomal material is seen as a larger Barr body.

What is absent in this large "inactivated" isochromosome are the loci of the short arm. There is, in effect, monosomy for the short arm of the X chromosome. Ferguson-Smith[5] observed that patients with the isochromosome [46,X,i(Xq)] and the *short-arm* deletion (46,XXp—) were most like the XO Turner syndrome in their overall clinical picture of multiple stigmata and shortness of stature. He hypothesized that Turner syndrome is due to monosomy of the short arm of the X chromosome and further that these missing loci in the short arm of the X chromosome are homologous with loci in the Y chromosome.

Isochromosomes of the short arm [46,X, i(Xp)] have rarely been encountered. Individuals with this anomaly are apparently not short and have few Turner stigmata. However, they have primary amenorrhea. These patients are most like the more commonly found patients with long-arm deletions [46,X,i(Xq)], in whom shortness of stature and multiple Turner stigmata are notably absent but who have streak gonads and infertility. Ferguson-Smith proposes that infertility and gonadal dysgenesis are more a function of monosomy for the long arm and that shortness of stature and the multiple Turner stigmata are more related to monosomy of the short arm of the X chromosome.

Ring-X chromosomes have been found in mosaics [45,X/46,X,r(X)], and in general their features are comparable to those of typical mosaics (45,X/46,XX).

XXY KLINEFELTER SYNDROME

Klinefelter and co-workers[9] recognized a pattern of abnormalities that do not become evident until adolescence. These include small testes, absent spermatogenesis, high urinary excretion of gonadotropins, and, frequently, eunuchoid habitus and gynecomastia (Fig. 4-6). Jacobs and Strong[7] observed an XXY chromosome complement in a patient with Klinefelter syndrome in 1959.

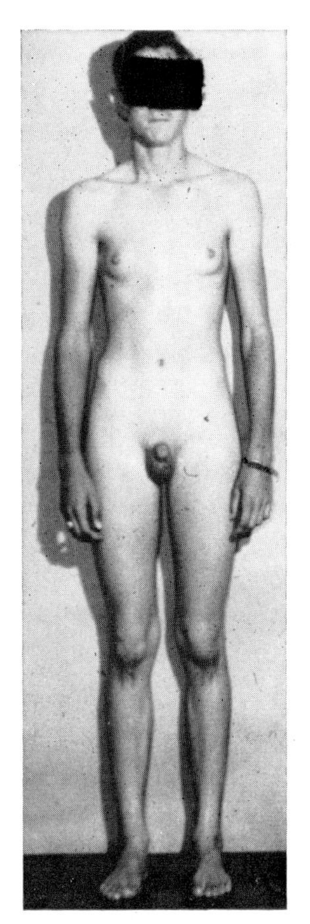

Figure 4-6. Phenotypic features of XXY Klinefelter syndrome (although gynecomastia is found in only 25% of patients with the XXY anomaly).

Although this syndrome is apparently the most common of the X chromosomal anomalies [with a population incidence of perhaps 1:1,000 (1 in 500 males)], it is not diagnosed in infancy or childhood unless the patients are detected through a survey study of buccal smears or karyotypes. The buccal smear is chromatin-positive—a single Barr body is found. Children with Klinefelter syndrome do not look abnormal. It is not until adolescence and young adult life that the syndrome discloses itself, at which time gynecomastia or inadequate sexual development may prompt medical consultation.

The gynecomastia, although occurring in perhaps no more than 25% of XXY patients, may be particularly disturbing. Surgical excision of excess breast tissue is not infrequently required. Body hair is often sparse, and the patient may seldom have to shave. A long-legged, eunuchoid physical habitus is also common. But none of these somatic features is invariable in a patient with an XXY sex chromosome constitution.

What is invariable is the small size of the testes, which usually do not exceed 2 cm in length. Spermatogenesis is rare. Biopsy reveals hyalinized seminiferous tubules or small, immature tubules lined with Sertoli cells and Leydig-cell hyperplasia. Intellectual attainment is generally below that of siblings. Some of these patients may go to college, and some may be found in institutions for the retarded. Patients with Klinefelter syndrome have been ascertained with increased frequency in prison populations. Perhaps the disturbing physical features of their disorder may contribute to sociopathic behavior.

As in most chromosomal anomalies, there are patients who have all of the stigmata of Klinefelter syndrome but who have apparently normal chromosomal constitutions. There are also patients with an XXXY chromosome complement who have the clinical features of Klinefelter syndrome except for a greater degree of retardation and a lesser degree of sexual development. XXXY patients are more severely affected and may have some manifestations of the XXXXY syndrome, such as radioulnar synostosis—which illustrates the finding of increasing disability with increasing numbers of X chromosomes. These patients have a chromatin-positive buccal smear that contains two Barr bodies. In the presence of three X chromosomes, all but one are "inactivated," yielding two Barr bodies.

THE XXXXY SYNDROME

Patients with the XXXXY chromosomal anomaly, atlhough on a continuum of increasing severity of disease with Klinefelter syndrome, present with a distinctive pattern. This relatively uncommon disorder is usually diagnosed in infancy and childhood. The incidence has not been determined.

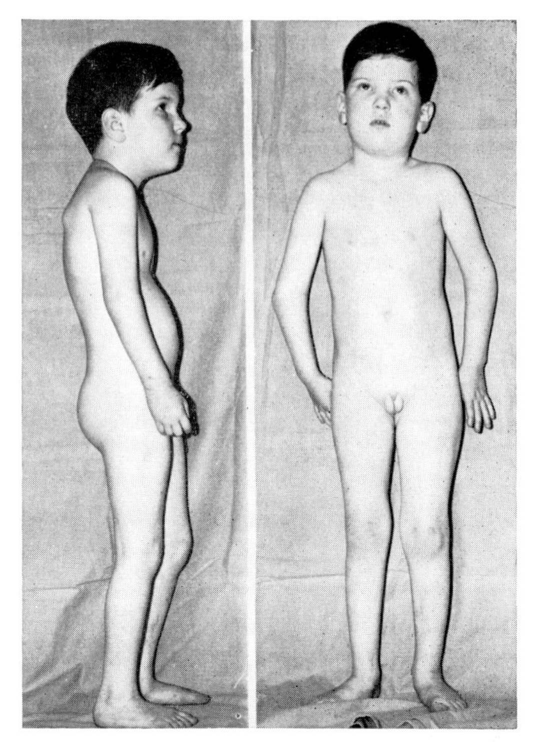

Figure 4-7. Five-year-old boy with the XXXXY syndrome. His IQ was 50. Note the arm deformity of severe radioulnar synostosis.

Retardation is significant and in the same IQ range (25 to 50) as patients who have 21 trisomy. These patients are also sometimes confused with patients having Down syndrome because of certain facial similarities—low nasal bridge, inner epicanthic folds, "mongoloid slant," occasional Brushfield spots (Fig. 4-7)—but can be clearly distinguished by the lack of dermatoglyphic features seen in Down syndrome. They often have cross-eyes, prominent chin, and short neck with occasional webbing. Congenital heart lesions have been observed.

Limited pronation at the elbow is common, and x-rays often reveal radioulnar fusion. Knock-knees and incurved fifth

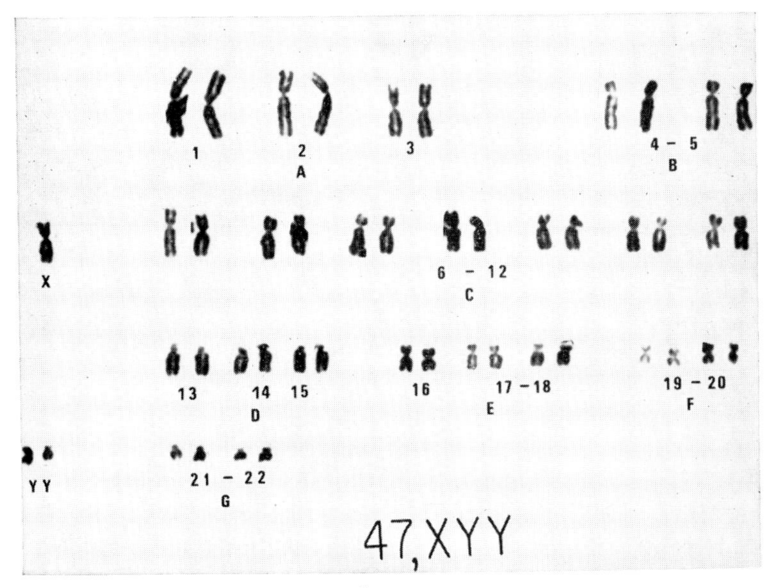

Figure 4-8. Conventional karyotype of XYY male.

fingers are frequently found. The genitalia are underdeveloped. Small penis, small testes, undescended testicles, and occasionally a bifid scrotum may sometimes make the genitalia appear superficially ambiguous. Fertility has not been described.

THE XYY SYNDROME

The first report of a male with the 47,XYY chromosomal constitution was published by Sandberg and co-workers in 1961,[17] but it was not until 1965, when it was discovered that men in maximum-security hospitals had the XYY complement (Fig. 4-8) in numbers that could not be attributed to chance, that attention was attracted to these patients. These XYY males appear to be taller and more aggressive than normal XY males, though it is not clear how consistent this feature is.[7] The aggressiveness and antisocial behavior of XYY males leading to imprisonment may be one of the more important discoveries in human behavioral genetics. Certainly, this aspect of the syndrome has been popularized in the lay press and is even explored in the contemporary novel.

Recent population cytogenetics surveys place the incidence between 1 : 500 and 1 : 3,000. The only reliable ways to diagnose the syndrome in infancy are through a cytogenetics survey or by looking at Wharton jelly or buccal smear specimens for two fluorescent Y bodies (Fig. 4-9).

A young man, 6′8″ tall, referred himself to our clinic having made a self-diagnosis after reading a report of the syndrome in

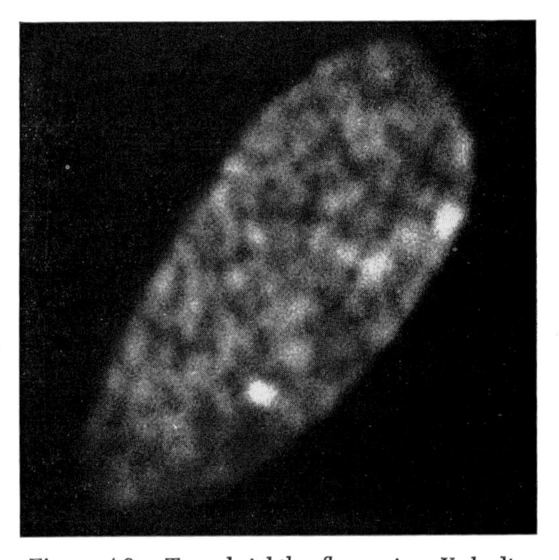

Figure 4-9. Two brightly fluorescing Y bodies from buccal smear of XYY male.

Time magazine. Cytogenetic examination confirmed that he did indeed have an XYY karyotype. He said that he was frankly worried that he might kill someone in a fit of uncontrolled rage. He frequently engaged in barroom brawls, some of which he claimed he did not instigate, but because of his great size, he was a natural target for "small drunks who wanted to feel big." This individual fulfilled the popular conception of the XYY male, and we frankly lacked the courage to request a picture for publication.

The intellectual achievement appears to be fair to good. Although some XYY males have been discovered in institutions for the retarded, others have gone to college. EEG abnormalities have been reported, as have seizures and cardiovascular and skeletal disorders. The adjustment to society is not invariably unfavorable. If the syndrome is as common as recent surveys suggest, there are probably a large number of undiagnosed XYY males who are neither in prisons nor in frequent barroom brawls. A particularly acceptable social adjustment that occurs to the authors would be as a pro-football linebacker. But it is also quite possible that the majority of individuals with XYY chromosomes are not sufficiently large or aggressive for this vocation.

OTHER ABERRATIONS OF SEX CHROMOSOMAL NUMBER

Many combinations of X and Y chromosomal numerical aberrations have been reported: XO, XXX, XXXX, XXXXX, XXY, XXXY, XXXXY, XYY, XXXYY, and so on. In each of these anomalies, the more X chromosomes, the greater the degree of disability. A somewhat complementary proposal with regard to the Y chromosome might also be advanced. At least in terms of adjustment to our contemporary society, an extra Y chromosome appears to predispose to behavioral disability.

Two further syndromes deserve a brief acknowledgment. The first is the XXX syndrome, or "super female." The term "super female" is derived from Drosophila genetics. However, in the human species, the female with three X chromosomes is hardly "super." Retardation and infertility are common but far from invariable. Many of these patients may appear to be perfectly normal intellectually, have no somatic malformations, and may be fertile. Two Barr bodies are observed in the buccal mucosal cells. No XXX daughter of an XXX mother has been reported. The disorder is rare.

The patient with four X chromosomes is more retarded than the XXX individual, and the patient with five X chromosomes, the penta-X syndrome, is severely retarded. Down syndrome was the initial diagnosis attached to patients with penta-X syndrome described in the literature because of the somewhat mongoloid appearance of the face (upward-slanting palpebral fissures). The presence of four Barr bodies and the dermatoglyphic analysis rule against the slightly suggestive somatic features of Down syndrome in these patients; the diagnosis is further confirmed by a chromosomal analysis showing five X chromosomes.

INTERSEX

Hermaphrodites

A true hermaphrodite (Hermaphroditos, the son of Hermes and Aphrodite) is an individual who has both male and female gonadal tissue. The diagnosis is made when testicular and ovarian tissue are recovered either from separate organs or from a single ovotestis. The sex chromatin may be positive or negative, and the chromosomal analysis may reveal mosaicism. The appearance of the external genitalia is variable.

Pseudohermaphrodites

These are individuals who have normal chromosomes and buccal smear for one sex, but have ambiguity of sex characteristics that makes them appear to be of the opposite sex.

Male Pseudohermaphrodites. A 46,XY male with a chromatin-negative buccal smear who is female in external appearance is a male pseudohermaphrodite. An example of this category is the *testicular feminization syndrome*. These patients usually have genitalia and secondary sex characteristics of the female and are an exception to the earlier statement that a Y chromosome produces the male phenotype. Medical attention is sought by these individuals because of infertility, amenorrhea, or sometimes inguinal hernia. The inguinal hernia may contain a testis, or the testis may remain in the abdomen.

Female Pseudohermaphrodites. A 46,XX female with a chromatin-positive buccal smear who has an external appearance suggestive of a male is a female pseudohermaphrodite. Excess circulating sex hormone of maternal origin and the administration of progestational agents to pregnant women to prevent miscarriage may alter the appearance of the external genitalia of the female, producing clitoral hypertrophy and even fusion of the labia majora.

The adrenogenital syndrome (virilizing adrenal hyperplasia) also causes female pseudohermaphroditism. This disease and its different enzymatic forms are discussed in Chapter 6.

SUMMARY

The phenotypic sex of an individual is determined with few exceptions by the presence (male) or absence (female) of a Y chromosome. A buccal smear showing one brightly fluorescing Y body (by a special stain) signifies the presence of one Y chromosome (two Y bodies = two Y chromosomes). A Barr body (sex chromatin) on buccal smear signifies two X chromosomes (two Barr bodies = three X chromosomes, and so on). The single-active-X (Lyon) hypothesis proposes that the Barr body is a heteropyknotic X chromosome that is inactivated, but active genetic loci on the "inactive" X as well as homologous loci on the Y chromosome must be postulated to account for the clinical abnormalities encountered in the XO Turner syndrome, the XXY Klinefelter syndrome, and other aberrations of X chromosomal number. However, much less developmental abnormality results from sex chromosomal anomalies of number and structure than from autosomal anomalies.

REFERENCES

1. Barr, M. L., and Bertram, E. G.: A morphological distinction between neurones of the male and female, and the behavior of the nucleolar satellite during accelerated nucleoprotein synthesis. Nature 163:676, 1949.
2. Brown, S. W., and Chandra, H. S.: Inactivation system of the mammalian X chromosome. Proc. Nat. Acad. Sci. U. S. 70:195, 1973.
3. Caspersson, T., Zech, L., and Johannson, C.: Differential binding of alkylating fluorochromes in human chromosomes. Exp. Cell Res. 60:315, 1970.
4. Davidson, R. G., Nitowsky, H. M., and Childs, B.: Demonstration of two populations of cells in the human female heterozygous for glucose-6-phosphate dehydrogenase variants. Proc. Nat. Acad. Sci. U.S. 50:481, 1963.
5. Ferguson-Smith, M. A.: Karyotype-phenotype correlations in gonadal dysgenesis and their bearing on the pathogenesis of malformations. J. Med. Genet. 2:142, 1965.
6. Ford, C. E., Jones, K. W., Polani, P. E., et al.: A sex-chromosome anomaly in a case of gonadel dysgenesis (Turner's syndrome). Lancet i:711, 1959.
7. Hook, E. B.: Behavioral implications of the human XYY genotype. Science 179:139, 1973.
8. Jacobs, P. A., and Strong, J. A.: A case of human intersexuality having a possible XXY sex-determining mechanism. Nature 183:302, 1959.
9. Klinefelter, H. F., Reifenstein, E. C., and Albright, F.: Syndrome characterized by gynecomastia, aspermatogenesis without A-Leydigism and increased excretion of follicle stimulation hormone. J. Clin. Endocrinol. 2:615, 1942.
10. Lyon, M. F.: Sex chromatin and gene action in the mammalian X-chromosome. Am. J. Hum. Genet. 14:135, 1962.
11. Mukherjee, B. B., and Sinha, A. K.: Single-active-X hypothesis: Cytological evidence for random inactivation of X-chromosomes in a female mule complement. Proc. Nat. Acad. Sci. U.S. 51:252, 1964.
12. Ohno, S., and Hauschka, T. S.: Allocycly of the X-chromosome in tumors and normal tissues. Cancer Res. 20:541, 1960.
13. Painter, T. S.: The Y chromosome in mammals. Science 53:503, 1921.
14. Pearson, P. L., Bobrow, M., and Vosa, C. G.: Technique for identifying Y chromosome in human interphase nuclei. Nature 226:78, 1970.

15. Polani, P. E.: Chromosome anomalies and abortions. Develop. Med. Child. Neurol. 8:67, 1966.
16. Russell, L. B.: Another look at the single-active-X hypothesis. Trans. N. Y. Acad. Sci. 26:726, 1964.
17. Sandberg, A. A., Koepf, G. F., Ishihara, T., et al.: An XYY human male. Lancet ii:488, 1961.
18. Turner, H. H.: A syndrome of infantilism, congenital webbed neck and cubitus valgus. Endocrinology 23:566, 1938.
19. Ullrich, O.: Uber typische Kombinationsbilder multipler Abartungen. Z. Kinderheilk. 49:271, 1930.
20. Wilkins, L., and Felischmann, W.: Ovarian agenesis: Pathology associated clinical symptoms and the bearing on theories of sex differentation. J. Clin. Endocrinol. 4:357, 1944.

Chapter 5

GENETIC BASIS OF HEREDITY

THE STRUCTURE AND FUNCTION OF THE GENE

One of the most exciting discoveries in biology and, indeed, in all science of the last two decades has been the biochemical nature of the gene and how it works, in precise biochemical terms. The details can be found in textbooks on molecular genetics[7]; only a summary of the current view of the gene's biochemical structure and functions will be presented here.

In bacteria, the genetic material is a strand of deoxyribonucleic acid, or DNA. The brilliant work of Watson and Crick showed this material to consist of a double-stranded helix, like a rope ladder in which the ropes are made up of alternating deoxyribose and phosphate molecules and the rungs consist of pairs of nucleotide bases, the ropes being twisted into a double helix. The nucleotides are guanine (G), cytosine (C), adenine (A), and thymidine (T), and the physicochemical restrictions are such that G on one strand can pair with only C on the other, and A with T. Thus the base sequence on one strand bears a complementary relationship to that on the other. When the DNA replicates, the two strands separate, and each lays down a new complementary strand, so that the two new double helices are formed, identical in base sequence with the original (Fig. 5-1).

In higher organisms, the DNA is associated with proteins, particularly histones, to form the microscopically visible chromosome depicted in Chapter 2. The ultrastructural organization of mammalian chromosomes is still not clear—though they look multistranded by the usual methods of inspection, they may contain a single strand of DNA, as in bacteria but intensively folded.

It is now well established that genes act by controlling the amino acid sequences of polypeptides and, thereby, the structures and properties of proteins. For each polypeptide being synthesized there is a corresponding region of a chromosome in which the sequence of base pairs in the DNA determines the amino acid sequence of the polypeptide, and that particular area of the DNA is said to be the gene for the polypeptide. A mutant gene results in an altered amino acid sequence, which may alter the structure of the polypeptide, and hence its properties, thus leading to a genetically determined defect in the corresponding protein, be it an enzyme as in the inborn errors of metabolism, or other protein as in the abnormal hemoglobins.

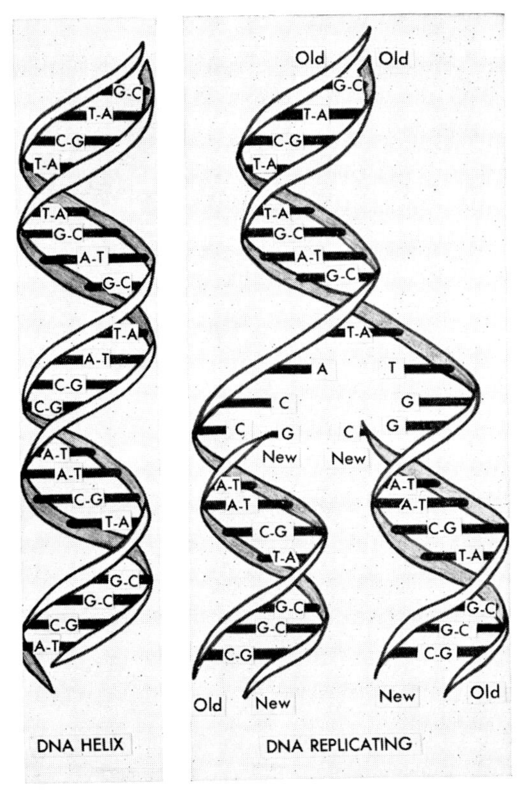

Figure 5-1. Diagram of the DNA double helix, and DNA replicating.

This concept was first suggested by the observation that in sickle-cell anemia the mutant gene causing the disease resulted in an abnormal hemoglobin. Sickle-cell hemoglobin differed from normal hemoglobin only in that the sixth amino acid from the N-terminal end was a valine instead of a glutamic acid. Thus a single gene difference was associated with a single amino acid substitution in a particular polypeptide. Evidence from microbial genetics confirmed that *a gene is that portion of the DNA responsible for the primary structure of a particular polypeptide.*

The means by which the gene determines the amino acid sequence of its polypeptide is, briefly, as follows (Fig. 5-2). The sequence of bases in the DNA constitutes a code for the amino acid sequence of the polypeptide, a triplet of three bases (or codon) corresponding to one amino acid. For instance, the triplet CTT at a particular place on the DNA codes for a glutamic acid at the corresponding place on the polypeptide.

The *translation* of the DNA code into protein is done by means of a special type of ribonucleic acid, or RNA, called messenger RNA, or mRNA. RNA differs from DNA in being single-stranded, with ribose instead of deoxyribose and the nucleic acid uridine (U) instead of thymidine. The mRNA is synthesized on the DNA strand (by the action of the enzyme RNA synthetase) with the same kind of complementary pairing as the two DNA strands; for instance, a CTT triplet in the DNA would correspond to a GAA triplet in the RNA. Thus the mRNA has a sequence of bases determined by that of the corresponding DNA strand. The mRNA migrates from the nucleus to the cytoplasm and becomes associated with a ribosome (which contains another kind of RNA, the ribosomal RNA); there it acts as a mold, or template, on which the amino acids are assembled into polypeptides in the following way.

A third type of RNA, the transfer RNA, or tRNA, exists in the cytoplasm in 20 varieties, one for each amino acid. These serve to bring the amino acids to the messenger, for incorporation into the polypeptide. To do this, the transfer RNA must be able to recognize a specific amino acid, on the one hand, and a specific place on the mRNA, on the other. Thus each species of tRNA has a site—the recognition site—that combines specifically with a particular amino acid, and another site with a particular triplet—an "anticodon"—that can attach to the appropriate codon in the messenger.

As the messenger RNA strand moves along the ribosome, each codon in turn is brought into a position where it can (with the aid of appropriate enzymes) combine with the anticodon on a molecule of the corresponding tRNA so that the amino acid

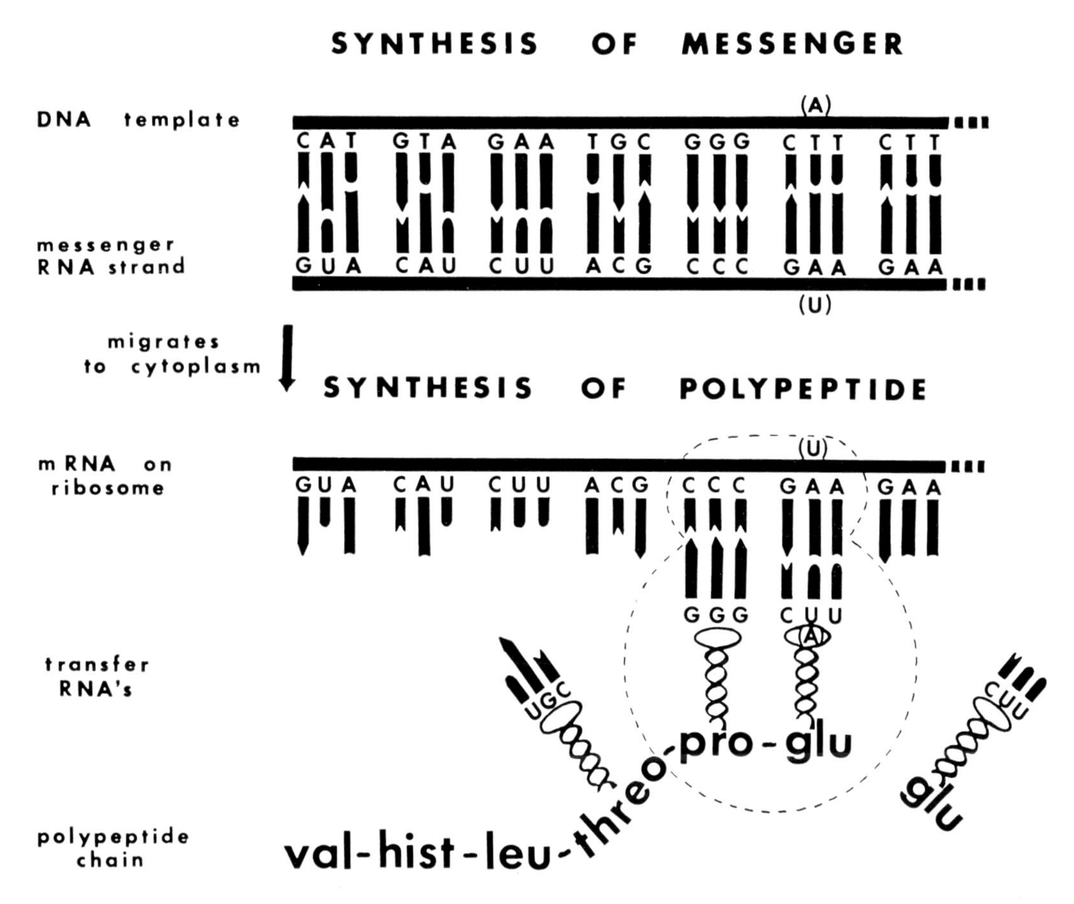

Figure 5-2. Diagram illustrating the synthesis of a polypeptide with a sequence of amino acids corresponding to a sequence of nucleotide triplets in the DNA.

is brought into position to be attached to the growing polypeptide chain. Thus if the codon is a GAA triplet on the messenger RNA, it will combine with a tRNA having the anticodon CUU, which brings a glutamic acid into position to be attached to the growing polypeptide chain. If the next codon triplet is a GUA, it will combine with a tRNA that has a CAU anticodon, and a valine will be brought into position and attached to the chain. In this way the amino acids are lined up on the template in an order specified by the sequence of triplets in the mRNA, which in turn was specified by the sequence of triplets in the DNA. The mutation from the gene for normal hemoglobin beta chains to sickle beta chains presumably involves a change in the sixth triplet of the gene from CTT to CAT, so that the mRNA would carry GUA instead of GAA, and would therefore place valine in the sixth amino acid position instead of glutamic acid.

To recapitulate, the information coded in the DNA sequence of the gene is *transcribed* to the messenger RNA, which carries the information to the ribosome site, where it is *translated* into a specified amino acid sequence in the corresponding polypeptide (Fig. 5-2).

The foregoing description is somewhat oversimplified, but it conveys the basic concept. In fact, there may be more than one tRNA triplet specific for a particular amino acid—that is, the code is redundant. There are specific triplets that initiate transcription at the beginning of the gene, and presumably some that terminate it. Certain regions of the DNA transcribe ribosomal RNA, and others transcribe the transfer RNAs. In higher organisms, there is more DNA than appears to be necessary for the known functions; the reason for the "redundant" DNA is not clear. Perhaps it has something to do with the regulation of gene activity.

Regulation of Gene Activity

Since not all genes are active in all cells, there must be a way of suppressing the activity of certain genes and initiating that of others; the changes may be permanent, as in embryonic differentiation, or intermittent, as in the case of cyclical production of a specific protein by a certain cell type. The first understanding of how this may occur came from bacterial genetics, with the formulation of the "*operon*" concept by Jacob and Monod. The operon is a group of genes, arranged in linear order, that produce a series of enzymes all concerned with the same biosynthetic pathway. The first gene in the series contains the *operator,* which initiates the activity of the whole group. The operator can be activated or suppressed by another gene, the *regulator,* elsewhere on the genome. The product of the regulator gene can be modified by specific molecules in the cytoplasm, so that it will activate or suppress, as the case may be, its own operon. This provides a control mechanism whereby the group of genes responsible for a group of enzymes that metabolize a sugar, for instance, will produce the enzymes only when the sugar is present in the environment, thus making the cell more efficient. Further details are found in Chapter 10.

This is a matter of great importance, because some genetic diseases and defects may result from faulty gene regulation rather than from the production of abnormal proteins, and this suggests an approach to treatment. A disease resulting from inactivity, rather than structural abnormality, of a gene, could be cured by treatment that led to reactivation of the gene. There may also be important implications of gene regulation in the case of neoplastic change. Furthermore, the induction of enzymes by hormones, drugs, and other agents may involve differential gene activation, and it is possible in tissue culture to stimulate the activity of specific genes, for example in the case of orotic aciduria (see Chapter 6, p. 82).

SINGLE MUTANT GENES

Dominant and Recessive Genes

We have said that a gene may be altered by mutation, which changes one of its nucleotide bases, resulting in a corresponding change in its mRNA and the polypeptide for which it codes. Thus a given gene locus can exist in one of several different states. Alternative forms of the same gene are called *alleles.* Each individual carries two sets of genes, one from the mother and one from the father. If the two members of a pair of genes are alike, the individual is said to be *homozygous* for this locus; if they are different, the individual is *heterozygous.* A heterozygous individual will make two kinds of mRNA for that gene, and therefore two kinds of the corresponding polypeptide.

Consider what happens in the case of a gene that codes for an enzymatic protein and a mutation that renders the protein enzymatically inactive. If an individual inherits the inactive allele from both parents, he will not make any active enzyme, the corresponding reaction will not occur, and thus the homozygous mutant individual will be abnormal. On the other hand, an individual who is heterozygous will make about half as much enzyme as one who is homozygous for the normal allele. Reducing the amount of enzyme by half is usually not enough to reduce the rate of the corresponding reaction, and so the heterozygote will function normally. In this case, the mutant allele is said to be *re-*

cessive to the normal allele, since it does not produce any outward effect in the presence of the normal allele. Conversely, the normal allele is said to be *dominant* to the mutant allele. A recessively inherited disease, then, is one that is caused only by a homozygous mutant gene.

If the mutant gene can produce a disease or defect in the heterozygote, the corresponding disease or defect is said to show *dominant* inheritance. This may be because the mutant gene results in the production of an abnormal protein, such as keratin, that, even in the presence of the normal protein produced by the normal allele, results in an abnormal structure. In such cases, an individual homozygous for the mutant gene would probably be much more severely affected than the heterozygote, since there would be none of the normal protein. When the heterozygote is intermediate between the two homozygotes, with respect to the trait in question, dominance is said to be *intermediate*, and if the heterozygote resembles the mutant homozygote, the mutant is said to show *complete* dominance.

Most deleterious dominant genes in man are so rare that homozygous mutants are never observed, since matings between heterozygotes almost never occur, so there is no opportunity to decide whether the given gene shows intermediate or complete dominance. In medical genetics, therefore, the term *dominant* is used for *any gene that is outwardly expressed in the heterozygote*, regardless of whether dominance is intermediate or complete.

Finally, the heterozygote may express the phenotype of both genes. For instance, a person of the AB blood group is heterozygous for an allele that produces antigen A and an allele that produces antigen B. When each gene is expressed, irrespective of the other, they are said to be *co-dominant*.

Models other than a mixture of normal and abnormal structural proteins will also account for the dominance of some mutant genes. For instance, a mutation may render an enzyme insensitive to feedback inhibition, or an operator gene which is normally suppressed by some cytoplasmic regulator may be rendered insensitive to the regulator. In either case there will be excessive activity of the corresponding enzyme(s). Acute intermittent porphyria and certain forms of gout may be examples of this kind. Another possibility would be that the mutant gene alters the specificity of the enzyme, allowing it to attack a different substrate, or to assemble macromolecular material in the wrong way. Much remains to be found out about the biochemical basis of dominance.

It must be made clear that the concept of dominance is an operational one and does not reflect any intrinsic property of the gene. Take, for example, the mutant gene for sickle-cell hemoglobin. At the *clinical level*, the homozygote has a severe anemia but the heterozygous individual is not anemic under normal circumstances, so the mutant gene would be considered recessive. However, when the red blood cells from a heterozygote are put under reduced oxygen tension they become sickle-shaped. Thus at the *cellular level* the mutant gene can express itself when heterozygous, though not as strongly as when homozygous. This would be considered intermediate dominance. Finally, at the *molecular level*, the red cell from a heterozygote contains both normal and sickle hemoglobin, and the alleles are co-dominant. Whether a gene is dominant or recessive may, therefore, depend on the level at which one looks for its effect.

The fact that a mutant gene can be recessive and *not* produce any outward effect in the heterozygote means that two outwardly similar persons may be genetically different. If we consider a gene a^D and mutant form a^R, which is recessive, both homozygous a^Da^D and heterozygous a^Da^R individuals will be outwardly normal, but they will be genetically different. The outward appearance is referred to as the *phenotype,* and the underlying genetic constitution as the *genotype.* Because of recessive genes, and other irregularities to be mentioned later, one cannot always deduce the genotype from the phenotype.

MENDELIAN PEDIGREE PATTERNS

Autosomal Dominant Inheritance

As we have seen, a dominant gene is considered to be one that produces an effect in every individual who inherits it, irrespective of the state of the other allele. Thus the transmission of a dominantly inherited disease in a family is a direct reflection of the transmission of the gene. Each individual who inherits the gene will have the disease. Since each affected individual inherits the gene from an affected parent, the first characteristic of autosomal dominant inheritance is that *every affected individual has an affected parent;* sporadic cases are presumed to have arisen by fresh mutation.

As deleterious mutant genes are rare, the affected individual will almost always inherit the mutant gene from one parent only, and a normal allele from the other parent—that is, he or she (let us assume it is he) will be heterozygous. He will probably marry an unaffected mate. His children will therefore inherit a normal allele from his spouse and either the normal or the mutant allele from him. This then, is the second rule of Mendelian dominant inheritance: If the spouse is normal, *the affected individual's children will each have a 1:1 chance of inheriting the mutant gene and having the disease.*

Figure 5-3 is a pedigree of hereditary "cold urticaria," illustrating the autosomal dominant pedigree pattern (see Appendix B for a description of pedigree symbols). Note that from any affected individual the disease can be traced to an affected parent, grandparent, and so on as far back as information is reliable up to the point of first appearance in the family. Second, the ratio of affected to unaffected offspring of affected individuals is 19:20, which is compatible with a 1:1 expectation for each individual. The proband, for instance, inherited the mutant gene, which we will call a^D, from her mother and a normal allele a^R from her father. Figure 5-4 shows the expectation for her children—each son and each daughter has a 1:1 chance of being affected.

If two heterozygotes mate, the offspring can draw either the normal or the mutant allele from each parent and will be either homozygous normal (1 chance in 4) or

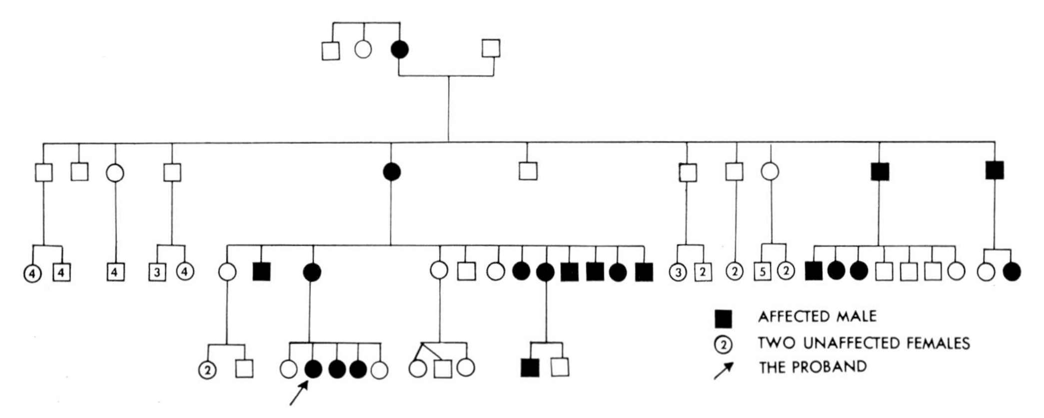

Figure 5-3. A pedigree of hereditary "cold urticaria" illustrating autosomal dominant inheritance. There are approximately equal numbers of males and females (8:12) and among the offspring of affected individuals there are 18 affected and 20 unaffected (excluding the proband), which is close to the expected 1:1 ratio. Carriers of the gene may have episodes of skin blotches, chills and weakness on exposure to cold.

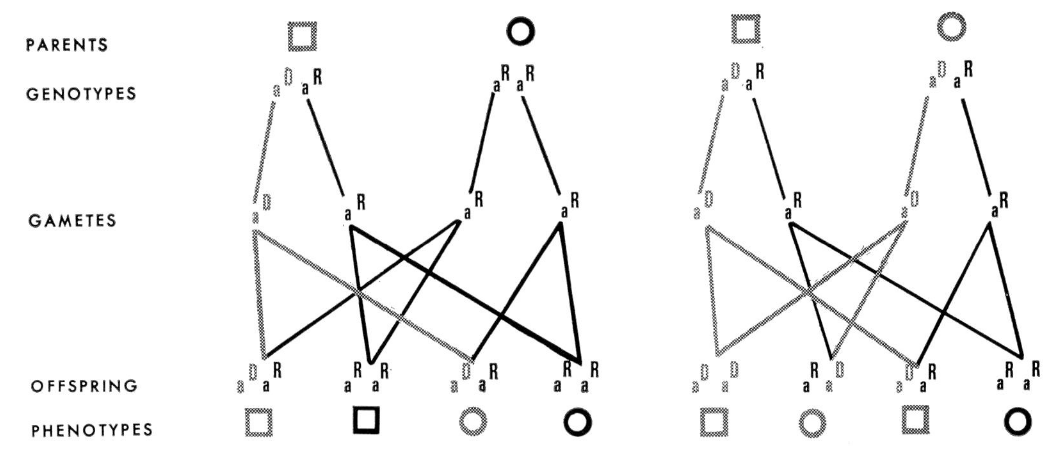

Figure 5-4. Segregation of an autosomal dominant gene in a mating of heterozygous and homozygous normal individuals (left), and between two heterozygotes (right). The shaded symbols represent the mutant gene and phenotype.

heterozygous (2 in 4) or homozygous mutant (1 in 4), since the heterozygotes and homozygous mutants will both be affected (Fig. 5-4).

Finally, in the rare cases where an affected person was homozygous for the mutant gene and had a normal mate, all the offspring would inherit the mutant gene and would be affected.

In summary, the pedigree pattern of autosomal dominant inheritance is characterized by the following features.

1. Each affected individual has an affected parent, to the point in the ancestry where the mutant gene arose by fresh mutation.
2. Each offspring of an affected person (with one affected and one unaffected parent) and a normal mate will have a 50:50 chance of being affected.
3. Unaffected relatives of affected persons will not have affected offspring.

Autosomal Recessive Inheritance

A recessive deleterious gene produces its disease only in the homozygote, so affected individuals must receive one mutant gene from each parent. Since deleterious genes are relatively rare in the population, almost all homozygous affected individuals arise from a mating of two heterozygous unaffected parents. Figure 5-5 illustrates the types of offspring to be expected from a mating of two heterozygotes. The offspring may get the normal allele from both parents and be unaffected, or a normal allele from the father and a mutant allele from the mother and be unaffected but heterozygous, or a mutant allele from the father and a normal allele from the mother, also an unaffected heterozygote, or the mutant allele from both parents and be affected with the disease. Thus *each child of parents who are both heterozygous for a mutant gene has 1 chance in 4 of being homozygous and having the mutant phenotype.*

Since the average family size in most populations is less than 4, and the recurrence risk for siblings is 1 in 4, most cases of recessively inherited disease will be sporadic. That is, the majority of cases will not have affected siblings even though the disease is inherited.

Occasionally, an affected individual may marry a heterozygote, in which case the offspring will have an equal chance of being heterozygous unaffected, or homozygous affected, thus simulating dominant inheritance (Fig. 5-5).

Since affected individuals arise only from

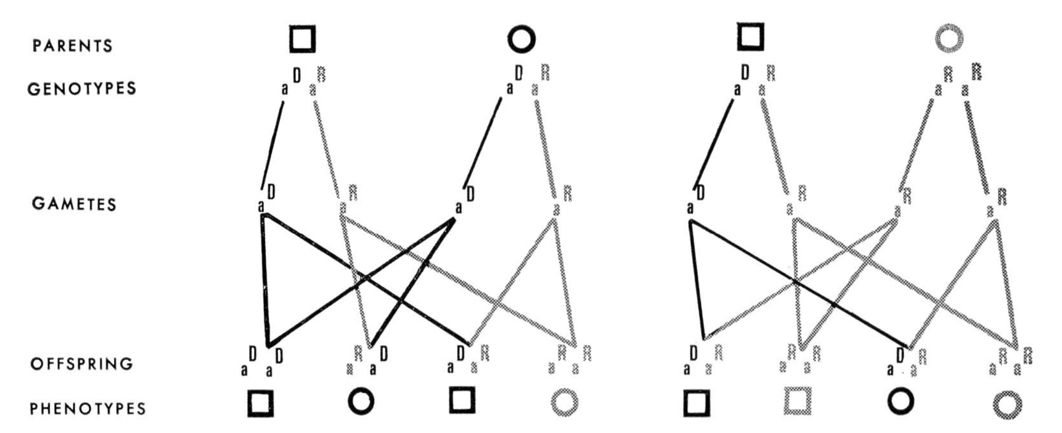

Figure 5-5. Segregation of autosomal recessive genes in a mating between two heterozygotes (left) and between a heterozygote and a homozygous affected individual (right). The shaded symbols represent the mutant genes and phenotypes.

matings between heterozygotes, a recessive mutant may be transmitted through many generations without ever becoming homozygous. Thus it is characteristic of recessive inheritance of rare conditions that *the disease usually does not appear in the ancestors or collateral relatives of affected individuals.*

Exceptions may occur in families where there is inbreeding. The chances of two parents being heterozygous for a mutant allele are increased if they are related and have a common ancestry from which they may inherit the same recessive mutant gene. If such a gene was carried by, say, one of every 50 individuals in the population, the chance that a heterozygote would marry an unrelated heterozygote would be 1 in 50, but if a heterozygote married his first cousin, the chance that she would also carry it would be 1 in 8, a considerably higher risk (Fig. 5-6). It follows that *children with a recessively inherited disease are more likely than average to have related parents.* The rarer the disease, the more likely it is that the diseased individual will have consanguineous parents.

Figure 5-7 illustrates the effect of ancestral consanguinity in the case of cystic fibrosis of the pancreas in a French-

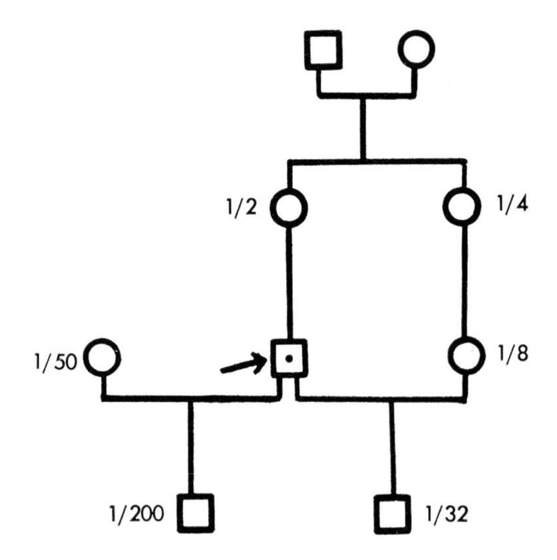

Figure 5-6. The significance of parental consanguinity. If the proband is heterozygous for a recessive gene carried by 1 in every 50 people, the chance that his first child by an unrelated spouse will be affected is $1/4 \times 1/50 = 1/200$. The chance of being heterozygous for the mutant gene is 1/2 for his mother (since he got the gene either from her or his father), 1/4 for his aunt, and 1/8 for his first cousin, so the risk for the first child by his first cousin would be 1 in 32 (ignoring the small possibility that the gene is carried by one of the spouses).

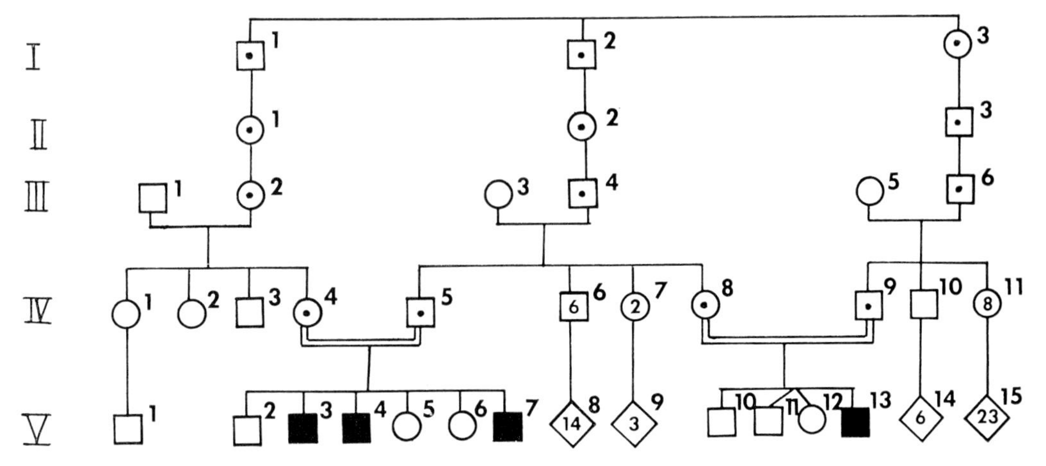

Figure 5-7. A pedigree of cystic fibrosis of the pancreas illustrating the effect of parental consanguinity in bringing recessive mutant genes together.

Canadian kindred. The gene must have been carried by one of the parents of the three sibs in generation I. Its descendants were transmitted to the four individuals in generation IV (IV-4 × IV-5 and IV-8 × IV-9) who married their third cousins, and the disease appeared in the two first-cousin sibships in generation V.

In summary, the autosomal recessive pedigree pattern is characterized by the following features:

1. Almost never is the disease present in the parents, ancestry, or collateral relatives.
2. The sibs of an affected child have 1 chance in 4 of being affected, irrespective of sex.
3. The parents of affected children are more likely to be related to each other (consanguineous) than are parents of normal children; the rarer the disease, the greater the frequency of parental consanguinity.
4. In small families there will be more sporadic cases than familial cases.

Sex Linkage

X-Linked Recessive Inheritance. Genes on the X chromosomes can be dominant or recessive just as those on the autosomes, but the fact that females have two X chromosomes and males only one X and a Y leads to characteristic differences in the pedigree patterns of diseases caused by X-linked genes. In females the dominance relations of mutant and normal alleles are just as they are on the autosomes (with certain exceptions related to Lyonization, as discussed in Chapter 4, p. 45). But the Y chromosome, for the most part, is not homologous to the X; that is, most genes on the X chromosome do not have a corresponding locus on the Y. (Such genes on the X chromosome are said to be *hemizygous*, rather than heterozygous or homozygous.) A mutant gene on the X chromosome will therefore always be expressed in the male, even though it may behave as a recessive in the female. This accounts for the characteristic "criss-cross" pedigree pattern of diseases showing X-linked inheritance, the gene usually being transmitted by unaffected females and producing the disease in males.

The most characteristic mating is that of a female heterozygous for a recessive mutant gene on the X chromosome (X^R) and its normal allele (X^D), mated to a normal male (X^DY). She will give either X^D or X^R to each of her daughters, who

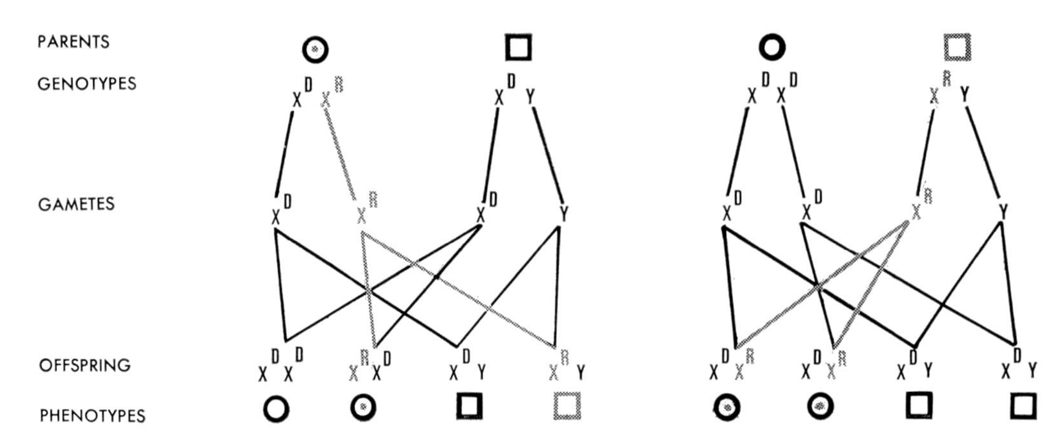

Figure 5-8. Segregation of an X-linked recessive gene in a mating of a heterozygous female by a normal male (left) and in a mating of affected male and normal female (right).

will also receive a normal X from the father and will each therefore have a 50:50 chance of being an outwardly normal carrier (X^DX^R) or a normal homozygote (X^DX^D). The sons will get a Y chromosome from the father and will have a 50:50 chance of being X^DY (normal) or X^RY (affected) (Fig. 5-8).

An affected male mated to a normal female will transmit a Y chromosome to his sons, who will be unaffected, and the X chromosome carrying the mutant gene to all his daughters, who will be unaffected, but carriers (Fig. 5-8).

In the unlikely event that an affected male (X^RY) marries a carrier female (X^DX^R), the daughters will all inherit the mutant gene from their father and will inherit from the mother either the normal allele and be carriers (X^DX^R), or the mutant allele and be affected (X^RX^R).

In summary, diseases showing X-linked recessive inheritance show the following pedigree characteristics, provided the gene concerned is rare:

1. The disease appears almost always in males, whose mothers are unaffected but heterozygous carriers of the mutant gene.
2. Each son of a carrier female has a 1:1 chance of being affected.
3. Affected males never transmit the gene to their sons, but they transmit it to all their daughters, who will be carriers.
4. Unaffected males never transmit the gene.

Thus the gene is usually transmitted through unaffected female ancestors and appears in their male relatives. Therefore one may expect to see it in the patient's brothers, the mother's brothers, the sons of the mother's sisters, or the mother's father. Figure 5-9 is a representative pedigree. In many families, however, the disease may be sporadic—i.e., it may occur in only one person—either because the eligible male relatives are few and by chance have not inherited the gene or because the patient's disease has arisen by fresh mutation (see section on recurrence risks, p. 72ff, for a more detailed discussion).

Independent Segregation

Mendel was fortunate that the traits he chose to study were controlled by genes on different chromosomes, and he was therefore able to formulate the Law of Independent Segregation. This states that genes (on different chromosomes) segregate independently of one another. Thus if there are two Mendelian mutant genes segregating in a family, the risks for a given indi-

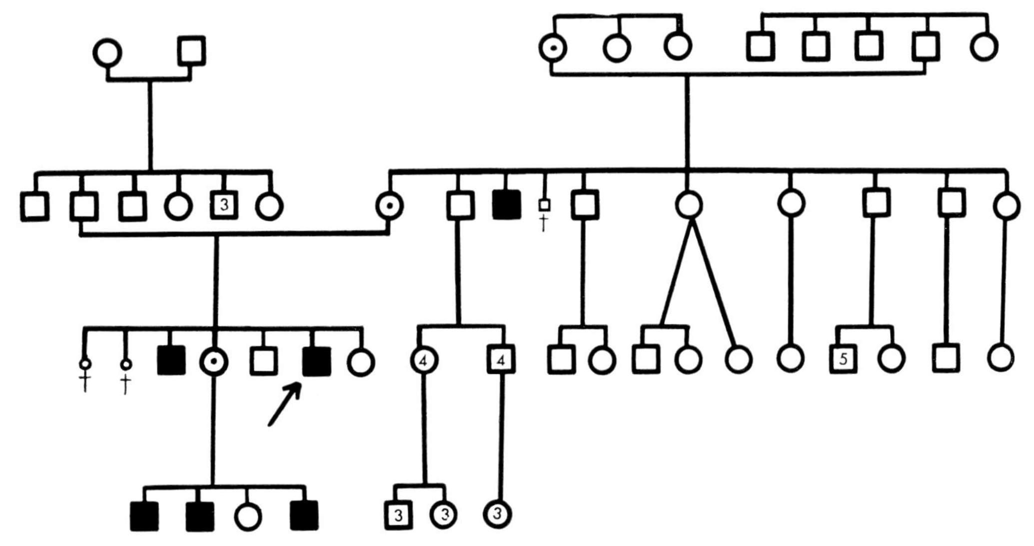

Figure 5-9. A pedigree of hemophilia illustrating the sex-linked recessive pattern of inheritance.

vidual inheriting either or both diseases can be calculated from the law of independent probability. Suppose a couple has had a child with cystic fibrosis of the pancreas, so that each child has a 1 in 4 chance of inheriting this autosomal recessive disease. Suppose one parent also has neurofibromatosis, so that each child has a 1 in 2 chance of inheriting this autosomal dominant disease. Provided the mutant genes are on separate chromosomes, we can say that the chance of the child inheriting both diseases is $\frac{1}{2} \times \frac{1}{4} = \frac{1}{8}$; there is a $\frac{1}{2} \times \frac{3}{4} = \frac{3}{8}$ chance of inheriting neither disease, a $\frac{1}{2} \times \frac{1}{4} = \frac{1}{8}$ chance of inheriting only cystic fibrosis, and a $\frac{1}{2} \times \frac{3}{4} = \frac{3}{8}$ chance of inheriting only neurofibromatosis.

The phenomenon of linkage, where two genes are located on the same chromosome, was discovered much later than the Mendelian laws and has only recently become relevant to genetic counseling.

Linkage and Crossing-over

If two genes occupy the same chromosome, one would expect them not to segregate independently, but to be transmitted together—that is, to be *linked.*

Consider, for instance, a mother who carries on one of her X chromosomes the mutant gene for color blindness and for hemophilia, with the normal alleles for these genes on the other X. Her genotype would then be written $\frac{\text{cb h}}{\text{Cb H}}$. (When the two mutant genes are on the same chromosome they are said to be "in coupling" and when they are on homologous chromosomes they are "in repulsion".) This mother will give one X chromosome or the other to each son, who should be either color-blind and hemophiliac ($\underline{\text{cb h}}$) or neither ($\underline{\text{Cb H}}$). But there is one complication. At first meiotic division the chromosomes may exchange strands—that is, there may be crossing-over. The farther apart the two genes are, the more often will there be crossing-over between them. We know that the genes for color blindness and hemophilia are about 10 cross-over units* apart—that is, there will have been an exchange between the two genes in 10% of the gametes formed. Thus our doubly heterozygous female will form four kinds of gametes—non-cross-over gametes (45% $\underline{\text{cb h}}$; 45% $\underline{\text{Cb H}}$) and cross-over gametes (5% $\underline{\text{cb H}}$; 5% $\underline{\text{Cb h}}$).

*Called centimorgans.

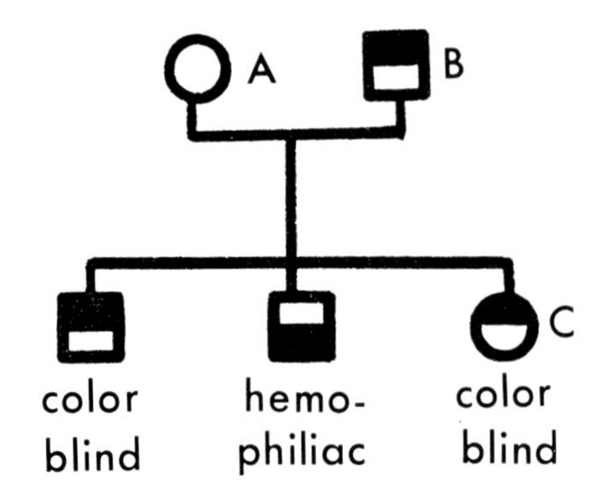

Figure 5-10. A pedigree illustrating linkage of color blindness and hemophilia.

Thus we may measure how far two linked genes are from one another by counting how often they stay together and how often they cross over during transmission from parent to child. Modern methods of searching for and calculating linkages are described elsewhere.[3]

Linkage may occasionally be useful in genetic counseling. Consider the pedigree shown in Figure 5-10. What is the probability that female C is a carrier of the gene for hemophilia? The father (B) is color-blind. The mother (A) is not color-blind

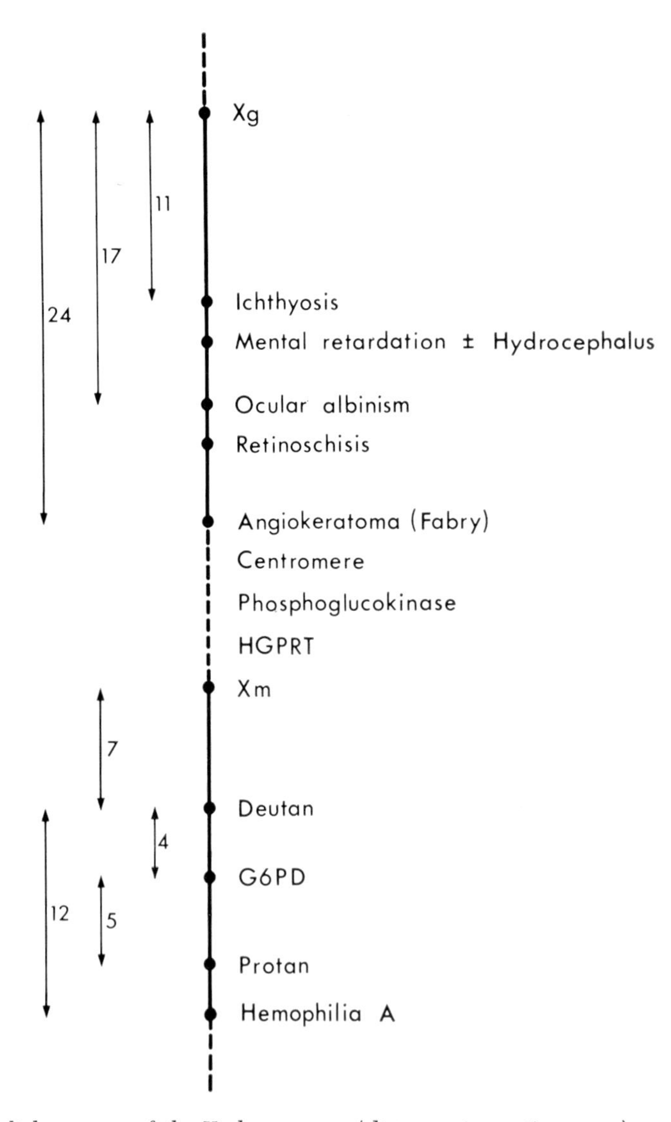

Figure 5-11. A linkage map of the X chromosome (distances in centimorgans).

but must be heterozygous for it, since she has a color-blind daughter (C). The mother has a color-blind son and a hemophiliac son, so in her the two mutant genes are in repulsion—her genotype must be $\frac{\text{cb H}}{\text{Cb h}}$. She will therefore produce four kinds of gametes: 45% cb H, 5% cb h, 45% Cb h, and 5% Cb H. We know that the daughter (C) got either one or other of the first two, since she is color-blind and therefore got the cb gene from both parents. Thus the daughter has a 45/50 or 90% chance of inheriting the cb H chromosome from her mother and only a 5/50 or 10% chance of inheriting the cb h chromosome and being a carrier. Figure 5-11 illustrates the linkage map of the X chromosome.

Irregularities in Mendelian Pedigree Patterns

Unfortunately, not all mutant genes in man display the regularity of transmission and expression shown by the characters of the garden pea that Mendel chose to demonstrate his laws. Neither, as a matter of fact, did some of the other characters that Mendel studied in the pea.

Expressivity. It is well recognized that infection by the same strain of virus or bacteria can produce wide variations in severity of disease in different patients. The same thing is true of genes. *Variable expressivity* is the term used to refer to the variation in severity of effects produced by the same gene in different individuals. For instance, the gene for multiple exostoses, which causes large numbers of disfiguring bone tumors in one person, may produce only a few small exostoses, detectable only by x-ray, in a near relative.

Penetrance. To carry the argument one step further, a gene that expresses itself clinically in one person may produce no detectable effect in another. This failure to reach the clinical surface is referred to as *reduced penetrance.* In statistical terms, penetrance is the per cent frequency with which a dominant gene in the heterozygote or a recessive gene in the homozygote produces a detectable effect. In medical genetics, reduced penetrance is most easily detected in the case of dominant genes, where an individual who must, on genetic grounds, have carried the mutant gene does

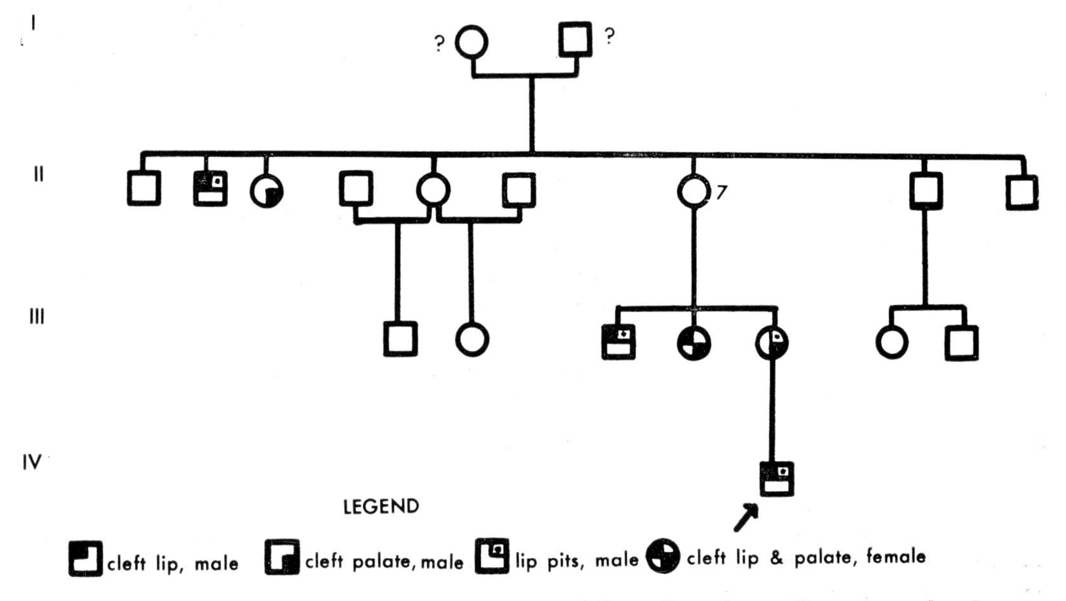

Figure 5-12A. A pedigree of the dominantly inherited "lip-pit" syndrome illustrating reduced penetrance and variable expressivity (see text).

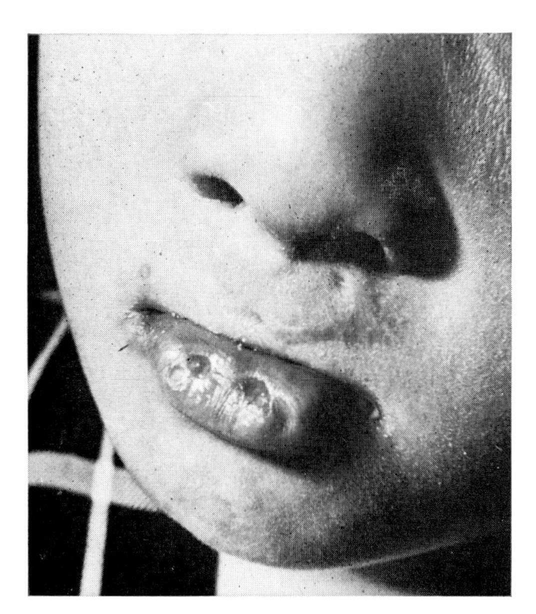

Figure 5-12B. A girl with a (repaired) cleft lip and lip-pits.

not show the mutant phenotype. Figure 5-12*A* illustrates a pedigree of the "lip-pit" syndrome in which an autosomal dominant gene results in two small pits on the lower lip in most, but not all, heterozygotes (Fig. 5-12*B*). These are the openings of accessory salivary glands in the lower lip. The gene may also cause, less frequently, cleft lip or cleft palate, or both. Thus the gene shows both reduced penetrance (II-7 carried the gene but did not show any signs of it) and variable expressivity (in heterozygotes that do show signs of the mutant gene, the signs are variable).

When penetrance is close to 100%, the autosomal dominant pedigree pattern with occasional skips is easy to identify, but the lower the penetrance the more difficult it is to distinguish between dominant inheritance with reduced penetrance and more complicated modes of inheritance. The concept of reduced penetrance has sometimes been used as an "excuse" for the fact that many familial diseases do not fit the expectation for regular Mendelian behavior. Although this explanation has been misused, there is no doubt that reduced penetrance is a fact of life that often poses problems for the counselor.

Failure of a gene to express itself phenotypically may occur for a variety of reasons. If the disease caused by the gene has a variable age of onset, for instance, a person who carried the gene might die before the disease became manifest, and would appear as a "skip" in the pedigree. In other cases the gene may involve some process with a developmental or biochemical threshold, and whether the mutant phenotype is produced will depend on whether the mutant gene has an effect severe enough to prevent the individual reaching the threshold. In this sense variable expressivity and penetrance are closely related phenomena.

As with dominance, the degree of penetrance of a mutant gene may also depend on how hard the observer looks for signs of its presence. The X-linked dominant gene for hypophosphatemic rickets, for instance, may produce full-blown rickets in some individuals, but only a low blood phosphorus level in others. Almost nothing is known about what makes the difference. In some cases, the effects of the mutant dominant gene in an unaffected carrier cannot be detected by any known means. Presumably, the number of such cases will decrease as our biochemical skills increase but will probably never disappear entirely.

CALCULATION OF SEGREGATION RATIO

In estimating the genetic component of any disease, the first question to be asked is likely to be: "What is the frequency of the disease in the near relatives of patients?" Family histories are then collected, and the number of affected and unaffected relatives counted. However, analysis of the data must take into account certain biases inherent in the collection of such data in man. To begin with, there are the problems of determining accurately which relatives are affected and which are not. We will not discuss these here. Second, the more striking the family history, the more likely it is to come to the attention of the investigator if care is not taken to avoid this bias. This is particularly true if one is using cases from the literature, but may also result from the efforts of well-meaning colleagues to refer "interesting" families. Furthermore, and quite apart from the tendency of striking families to be preferentially published or referred, if one is ascertaining families by identifying an affected individual, families with more than one affected member may be more likely to be ascertained than families with only one. We will return to the question of ascertainment bias shortly.

The basic question is this: In a group of individuals of a given relationship to an affected person what proportion are themselves affected? If the relation is anything other than sib, the situation is reasonably straightforward. In a group of children ascertained through one affected parent, for instance, one simply counts the number of affected and unaffected children. If, on the other hand, one is interested in the *sibs* of affected persons, the situation is more complicated.

To begin with, if we are measuring the frequency of the condition in the sibs of the proband, we must *omit the proband* from the calculation since, by definition, the probability of the proband being affected is 100%. If one were measuring the risk of contracting tuberculosis in the sibs of tuberculous patients, for instance, one might ascertain 20 tuberculous patients, the probands, and find that they had a total of 60 sibs, none of whom was affected. The frequency of tuberculosis in the whole group is 20/80 = 1 in 4, but this does not support Mendelian inheritance—the recurrence risk in the *sibs* of the probands is 0/60, and the probands must be omitted since they were selected *because* they were affected. This seems almost too obvious to mention, but nevertheless the mistake does appear in the literature from time to time, leading to gross overestimates of recurrence risk for poliomyelitis, asthma, and congenital heart disease, to cite three examples.

Complete Ascertainment

Second, there is the question of ascertainment bias. In the case of *complete ascertainment*, every affected case in the given population is ascertained. If so, every case is a proband. (By definition, the proband is an affected individual through whom the family is ascertained.) We wish to estimate the frequency of the disease in the sibs of probands, so for each family we omit one proband and count the family as

TABLE 5-1. Methods of Counting the Proportion of Affected to Unaffected Sibs with Different Assumptions about the Mode of Ascertainment of Probands[a]

	Ascertainment					
	Complete		*Single*		*Incomplete*	
Family	Aff.	T	Aff.	T	Aff.	T
1. 0 0 0 ↗●	0	3	0	3	0	3
2. 0 ↗● ● 0	2	6	1	3	1	3
3. ↗● 0 0 0	0	3	0	3	0	3
4. ● ↗● ↗● 0	6	9	2	3	4	6
5. 0 0 ↗● 0	0	3	0	3	0	3
Ratio	8	24	3	15	5	18
% affected	33.3		20.0		27.8	

[a] The probands, in incomplete ascertainment, are indicated by arrows.

many times as there are probands. Thus a sibship in which there were three affected and five normal children would be scored as 2 affected out of 7, three times, or 6 out of 21. This method also applies in other situations where each affected person has an equal probability of being ascertained.

Table 5-1 presents a hypothetical example in which there are five families of four siblings each. In the column headed "complete ascertainment" each family is counted as many times as there are affected individuals, omitting one affected each time. This estimates the probability of an affected sib as 8/24, or 33%. If ascertainment is, in fact, *not* complete, this method will overestimate the recurrence risk.

Single Ascertainment

At the other extreme, *single ascertainment,* the probands are chosen in such a way that each family is ascertained only once. In this case, the more affected individuals there are in the family, the more likely it is to be ascertained, so families with more than one affected would be over-represented in the sample as compared to their frequency in the population. In single ascertainment, this bias is exactly compensated for by the above procedure—that is, omitting the proband from the calculation and counting only the sibs (Table 5-1). This estimates the probability of an affected sib as 3/15 or 20%. If, in fact, ascertainment is not single, this method will underestimate the recurrence risk.

Incomplete Multiple Ascertainment

In practice, the situation is usually somewhere in between complete and single ascertainment. Some families with several affected sibs may be ascertained only once, the other affected sibs being identified only secondarily, through the family history. Other families may be ascertained separately and independently by each affected sib; in still others some affected sibs may be ascertained independently and others discovered only secondarily. In this case, the same rule is followed: count the family once for each proband, omitting the proband each time. Table 5-1 demonstrates the procedure in the column headed "incomplete ascertainment," more properly called "incomplete multiple ascertainment." The probability of a sib being affected is estimated by this method as 5/18 or 28%.

Sometimes, particularly with data from the literature, it is not clear which affected individuals are probands and which are secondary cases. In this case, one can at least get a rough estimate by making the limiting assumptions. Calculate the frequency in sibs assuming single ascertainment, which will underestimate the real value if ascertainment is not single. Then calculate the value assuming complete ascertainment, which will overestimate the real value if ascertainment is incomplete. The true value should lie somewhere in between.

A method that avoids the ascertainment bias is to calculate the recurrence risk only on children born after the proband, but this has the disadvantage of losing about half the data. Do *not* use children born after the first affected individual; unless ascertainment is complete, this will grossly overestimate the real value (33.3% for the data in Table 5-1).

The *a priori* Method

If the data are being tested for goodness of fit to a Mendelian ratio, an *a priori* method can be used. In families of parents who are both heterozygous for an autosomal recessive gene, for instance, some will have several affected, some one affected, and some none affected, just on the basis of chance. If the families are ascertained by an affected child, the families in which there were no affected children will not be included in the data. (This is known as "truncate" selection—the normal families are "cut off.") It is possible to calculate, from the binomial distribution, for any family size, the expected number of families omitted because they contained no affected children, and the data can be tested to see if, when due allowance is made for these families, a satisfactory fit to a 1 in 4 ratio is obtained. Details can be found in many textbooks of human genetics. However, the method is relatively insensitive and can be misleading unless a fairly large sample size is available. Furthermore, it cannot be used to calculate empirical recurrence risks in cases that do not fit the Mendelian ratios.

Other Methods of Segregation Analysis

A number of more sophisticated methods of analyzing segregation ratios have been developed[3] but are beyond the scope of this book. Again, they are most useful in cases where a large amount of data is available.

CALCULATING RECURRENCE RISKS

Mendelian Inheritance

Calculating recurrence risks is simple when the disease in question shows regular Mendelian inheritance and the genotypes of the parents are known. Predictions are then made on the basis of the segregation ratios described above. For instance, if the disease shows autosomal dominant inheritance, the risk for any child of an affected (heterozygous) parent with a normal spouse is ½. For autosomal recessive diseases, the normal parents of an affected child must both be heterozygous, so the risk for each subsequent child is ¼. And so on.

It may be, however, that the genotypes of the parents are not known but must be estimated from the family data at hand.

Dominant Inheritance

Variable Age of Onset. Consider, for instance, a man whose father has Huntington's chorea. This disease shows autosomal dominant inheritance and has a highly variable age of onset. Our consultand (i.e., the person whose genotype we are evaluating) is still unaffected at age 30. On the basis of the Mendelian law, the probability that he inherited the gene from his father is ½, and the probability that he did not is ½. This is the *a priori* probability based only on his antecedents. But we have another source of information. He has reached the age of 30 without developing the disease. We can see intuitively that the longer he lives, unaffected, the greater the probability that he did not inherit the gene. Thus his age, still unaffected, contributes additional information about the probability that he carries the gene. Previous studies have shown that about one third of those who inherit the gene have developed signs of the disease by this age, so there is a ⅔ probability that, if he inherited the gene, he will still be unaffected at age 30. This is the *conditional* probability. What, then, is his overall (posterior) probability? There are three possible outcomes:

1. He did not inherit the gene–probability ½
2. He did inherit the gene but is unaffected at age 30–probability $\frac{1}{2} \times \frac{2}{3} = \frac{1}{3}$
3. He inherited the gene and is affected at age 30–probability $\frac{1}{2} \times \frac{1}{3} = \frac{1}{6}$

We exclude the third outcome, since he is unaffected. So the chance that he is a

TABLE 5-2. Risk for Offspring of a Parent with Huntington's Chorea Developing It after Surviving Without Signs of the Disease to Various Ages

Age of Offspring, Still Healthy	*Age of Onset, Plus Ogive*	*Risk of Becoming Affected*
0	0	50
20–24	20	44
25–29	31	40
30–34	50	33
35–39	66	25
40–44	80	14
45–49	90	9

carrier ($\frac{1}{3}$) compared to the chance that he is not is as $\frac{1}{3}$ to $\frac{1}{2}$, or $\frac{2}{6}$ to $\frac{3}{6}$, which is 2 to 3, or 2 out of 5 = 40%.

That is, he has a 40% chance of being a carrier, rather than the 50% chance he had at birth. Risks for other ages are presented in Table 5-2.

Reduced Penetrance. A similar line of reasoning can be used in the case of a dominant gene with reduced penetrance. Suppose an affected father has an unaffected daughter who wants to know whether she may pass the gene on to her children. Assume that the gene shows 90% penetrance. There are the following possible outcomes:

1. The daughter did not inherit the gene—probability $\frac{1}{2}$
2. The daughter inherited the gene but does not show the phenotype—probability $\frac{1}{2} \times \frac{1}{10} = \frac{1}{20}$
3. The daughter inherited the gene and shows the phenotype—probability $\frac{1}{2} \times \frac{9}{10} = \frac{9}{20}$

We can exclude the third outcome; the probability of outcome 2 out of the remaining two outcomes is $\dfrac{\frac{1}{20}}{\frac{1}{20} + \frac{1}{2}} = \frac{1}{11}$

Therefore the daughter has 1 chance in 11 of being a carrier, so the chance of her first child being affected is $\frac{1}{11} \times \frac{1}{2} \times \frac{9}{10} = \frac{9}{220}$, or about 4%.

Sex-Linked Recessive Inheritance

Female with Carrier Mother and Normal Sons. This kind of reasoning is useful in cases where a female relative of a male with a sex-linked recessive disease wants to know the chances that she is a carrier. Consider the situation in Fig. 5-13. Female A is almost certainly heterozygous for the gene since she transmitted it to two children.

The *a priori* probability of female B inheriting the gene is $\frac{1}{2}$, but each normal son she has makes it less likely that she is a carrier, and she has three normal sons. The probability that she is a carrier and has three normal sons is

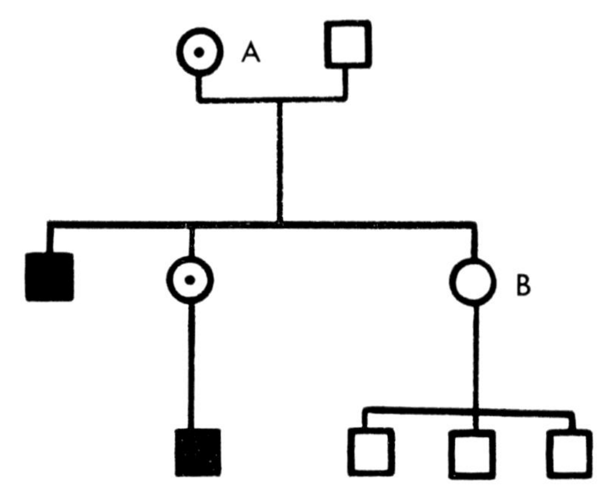

Figure 5-13. A hypothetic pedigree of a sex-linked recessive condition.

$$\frac{1}{2} \times (\frac{1}{2})^3 = \frac{1}{16}$$

So of the following possible outcomes

1. She did not inherit the gene ($\frac{1}{2}$)
2. She did inherit the gene and has three normal sons ($\frac{1}{16}$)

the probability of outcome two is

$$\frac{\frac{1}{16}}{\frac{1}{16} + \frac{1}{2}} = \frac{1}{9}$$

The Sporadic Case. An appreciable number of cases of X-linked recessive diseases have a negative family history. Consider, for instance, the family of a boy with Duchenne-type muscular dystrophy with no brothers, no maternal uncles and a negative family history. If he represents a fresh mutation, the risk for subsequent brothers is negligible, but if his mother is heterozygous, his brothers have a 50 : 50 chance. What is the probability that his mother is heterozygous? We can calculate the proportion of mutant to nonmutant cases from the equation (Chapter 8):

$$m = \frac{1\text{-}f}{3}x \text{ or } x = \frac{3m}{1\text{-}f},$$

where m is the mutation rate and f the fitness of the mutant phenotype (in this case 0). Therefore, in the case of a single affected male, with no affected male maternal relatives, it may be that either:

1. The mother is heterozygous, with a probability of $\frac{6m}{1\text{-}f} - 2m$, and passed the mutant gene to her son, probability $\frac{1}{2}$
2. The mother is not heterozygous; the son received a fresh mutation—probability m

So the posterior probability that the mother is heterozygous is

$$\frac{\frac{1}{2}\left(\frac{6m}{1\text{-}f} - 2m\right)}{\frac{1}{2}\left(\frac{6m}{1\text{-}f} - 2m\right) + m} = \frac{f+2}{3}.$$

For a lethal gene, where $f = 0$, this works out to $\frac{2}{3}$. That is, the chances are 2 to 1 that the mother is heterozygous for the gene, with a 1 in 2 risk for each subsequent male child. Thus the total risk for the brother of a sporadic case will be $\frac{2}{3} \times \frac{1}{2} = \frac{1}{3}$.

This risk will be modified downwards the more unaffected sons, brothers, mother's brothers, and so on, there are, and upwards if the gene is not lethal and therefore more common, with proportionately fewer cases due to mutation. Since the algebra may get complex, the risks have been presented in table form (Table 5-3, modified from Murphy and Mutalik[4]).

For more complicated situations, the reader is advised to consult one of several recent expositions on the subject.[4,6]

Autosomal Recessive Inheritance

In the great majority of cases where the disease involved shows autosomal recessive inheritance, the parents have identified themselves as being heterozygous by the fact that they have had an affected child. The risk for each subsequent child is therefore 1 in 4. Other questions may arise, however, that usually involve the question of whether near relatives of the affected person might have affected children.

Children of Near Relatives of Affected Persons. The sib (B) of an affected person (A), married to an unrelated spouse (C), wants to know the risk that his children will get the disease. The risk depends on both B and C being heterozygous. The risk of B being heterozygous is $\frac{2}{3}$. C's risk can be calculated from the Hardy-Weinberg Law (Chapter 8). For cystic fibrosis of the pancreas, for instance, if the disease frequency is 1 in 2000, then

1. The frequency q of the cf gene will be $\sqrt{\frac{1}{2,000}}$ or $\frac{1}{45}$
2. The frequency of the heterozygote, $2pq$ will be about $\frac{1}{22}$
3. The frequency that B and C will both be heterozygous is

$$\frac{2}{3} \times \frac{1}{22} = \frac{1}{33}$$

4. The probability of the first child being affected is

$$\frac{1}{4} \times \frac{1}{33} = \frac{1}{132}$$

TABLE 5-3. Probability That the Mother of an Isolated Case of an X-linked Trait is Heterozygous (Assume $m = 1/10^5$)

Disease frequency	*Number of unaffected sons* 0	1	2	3	4	5
1/100,000	0.67	0.50	0.33	0.20	0.11	0.06
1/10,000	0.92	0.85	0.73	0.58	0.40	0.26
1/5,000	0.98	0.96	0.92	0.86	0.76	0.61
1/1,000	0.99	0.98	0.96	0.93	0.86	0.76

As before, this probability would decrease with each unaffected child born to these parents, and an affected child would, of course, raise the risk to 1 in 4.

Matings between Affected Persons. In matings between two affected individuals who are homozygous for mutants at the same locus, all the offspring will be affected. However, the situation may be complicated by genetic heterogeneity. In a number of cases, e.g., congenital deafness and albinism, the same phenotype can be caused by mutations at different loci. If a couple who are both congenitally deaf carry mutations at different loci, their children will be unaffected. Counseling in this situation depends on the relative frequencies of mutations at the various loci, which is usually not known for a given population. In Northern Ireland, it has been shown[6] that in 2 out of 3 matings between congenitally deaf partners (excluding those that have a known syndrome), none of the children is deaf; in 1 out of 6 all the children are deaf; and in 1 out of 6 some are deaf and some are not. The data suggest that there must be several fairly common recessive genes leading to congenital deafness and that some of the sporadic cases (perhaps half) are phenocopies.

Matings between Consanguineous Partners. The question of marriage between cousins is one that the counselor meets from time to time, often because of the violent opposition that the prospect arouses in the family concerned. This may stem partly from the opposition of certain religions to consanguineous marriages and partly from the popular opinion that such unions lead to deformity, insanity, idiocy, and degeneracy of all sorts. These opinions are based more on fancy than fact.

The idea that cousin marriages are genetically disastrous may come from the observation that children with rare recessive diseases often have consanguineous parents, without taking into account the consanguineous parents that have children without such diseases. What are the facts?

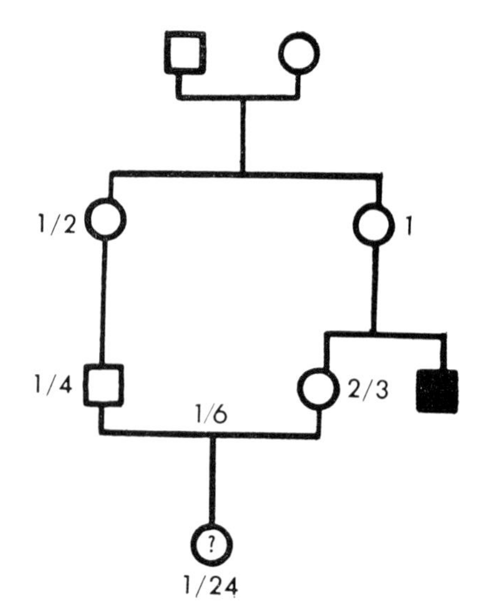

Figure 5-14. Hypothetic pedigree of an autosomal recessive condition.

If there is a recessively inherited disease in the family, the risk of that disease occurring in the children of the cousin mating can be calculated from Mendelian principles. For example, in Figure 5-14, if III-2's brother, III-3, had Tay-Sachs disease (see Chapter 6), III-2 would have a $\frac{2}{3}$ chance of being heterozygous, and the other probabilities can be calculated as shown. The chance of the first child being affected would be $\frac{1}{24}$.

If there are no recessive diseases in the family, the estimate becomes much less precise. If in Fig. 5-14, one of the common grandparents carried a recessive mutant gene then II-1 and II-2 each have a one in two chance of inheriting it, III-1 and III-2 have a one in four chance, and the chance that they are both heterozygous for the gene is $\frac{1}{4} \times \frac{1}{4} = \frac{1}{16}$. The chance that their first child would be affected is $\frac{1}{4} \times \frac{1}{16} = \frac{1}{64}$. Since there are two common grandparents involved, the risk for the child is twice this, or $\frac{1}{32}$. But we do not know the probability that the grandparent carries a recessive mutant gene. The available data on offspring of consanguineous matings record a rather low frequency of diseases known to show

Mendelian recessive inheritance, probably less than 1%.[5] This suggests that the number of genes causing recessively inherited diseases carried by the "average" person is somewhat less than 1.

There are also significant, but small, effects of inbreeding on infant mortality, IQ, stature, and other multifactorial traits,[5] but these are not large enough to be much of a deterrent to the individual couple.

For incestuous matings, between first-degree relatives, the theoretical risk would be considerably greater than for first-cousin matings, and the limited amount of data available suggests that this is indeed so.[1]

In summary, the facts suggest that genetic counseling for cousins contemplating marriage should be more optimistic than the advice they often get from their church, friends, families and, alas, some genetic counselors. Their risk of having a child with a recessively inherited disease is increased perhaps a hundredfold, but in absolute terms is still small, being of the order of 1%.

SUMMARY

The genetic material is DNA, and a gene is that portion of the DNA responsible for the primary structure of a particular polypeptide. DNA consists of a double helix, which may be visualized as a rope ladder made up of paired ropes of alternating deoxyribose and phosphate molecules and rungs of nucleotide bases. Protein is produced as the result of DNA-synthesizing messenger RNA, which migrates to the cytoplasm where it becomes associated with ribosomes and transfer DNA and is translated into a polypeptide chain having the sequence of amino acids corresponding to the sequence of nucleotide triplets (codons) in the DNA.

Alternative forms of the same gene (alleles) may exist at the same locus. Each individual carries two sets of genes (one from the mother and one from the father). If both members of the pair of genes are alike, the individual is homozygous for the locus; if the genes are different, he is heterozygous. A gene (of a gene pair) that is outwardly expressed in the heterozygote is dominant; if it is not expressed in the heterozygote, it is recessive; and if both genes at a locus are expressed, they are codominant.

In autosomal dominant inheritance, the mutant gene produces the trait or disease in the heterozygote; each offspring has an equal chance of inheriting the mutant gene and being affected, or of inheriting the normal allele and being unaffected. In autosomal recessive inheritance, both parents are usually heterozygotes, and unaffected; one-fourth of their offspring will be homozygous and have the trait or disease. X-linked recessive disorders are manifest mainly in males who receive the gene from carrier mothers (who are unaffected or minimally affected and who transmit the gene to half of their sons, who will have the disease, and to half of their daughters, who will be carriers). An affected son will not transmit the gene to his sons, but will transmit it to all of his daughters, who will be carriers. Mutant genes close together on the same chromosome tend to segregate together and are said to be linked. A mutant gene may express itself differently in different individuals (variable expressivity) or may not produce any effect where it would be expected to (reduced penetrance). Methods are described for calculating segregation ratios and recurrence risks that take into account the special problems of human pedigree data, including ascertainment bias, variable age of onset, reduced penetrance, mutations, and consanguinity.

REFERENCES

1. Adams, M. S., and Neel, J. V.: Children of incest. Pediatrics 40:55, 1967.
2. Levitan, M., and Montague, A.: Textbook of Human Genetics. London, Oxford University Press, 1971. Chapter XI.
3. Morton, N. E.: Segregation and linkage. *In* W. J. Burdette (ed.), Methodology in Human Genetics. San Francisco, Holden-Day, 1962.
4. Murphy, E. A., and Mutalik, G. S.: The application of Bayesian methods in genetic counselling. Hum. Hered. 19:126, 1969.

5. Schull, W. J., and Neel, J. V.: The Effects of Inbreeding in Japanese Children. New York, Harper and Row, 1965.
6. Stevenson, A. C., and Davison, B. C. C.: Genetic Counselling. London, Heinemann, 1970.
7. Watson, J. D. The Molecular Biology of the Gene. 2nd ed. New York, Benjamin, 1970.

Chapter 6

BIOCHEMICAL GENETICS

Genes control the structure of polypeptides and their corresponding proteins. A gene mutation, in which a single nucleotide base is changed to another, leads to a change in amino acid in the corresponding protein. Depending on the nature of this amino acid substitution, and where it is in the molecule, the function of the corresponding protein may be altered. Biochemical genetics deals with the biochemical changes resulting from substituting mutant for normal proteins and—by inference—with the functions of the normal proteins. Genetic defects in enzymes (which may also cause "inborn errors" of metabolism or transport) and in other proteins will be considered first.

INBORN ERRORS OF METABOLISM

Biochemical genetics began with the concept of the inborn error of metabolism, thanks to the insight of the English physician Sir Archibald Garrod. In 1909, through his studies of alkaptonuria, he defined the characteristics of this group of diseases resulting from lack of a functional enzyme to carry out a particular step in a chain of metabolic reactions. The original inborn errors of metabolism were alkaptonuria, albinism, pentosuria, and cystinuria (which subsequently turned out to be a transport defect instead). Thus Garrod was far ahead of his time in describing the essence of the "one gene—one enzyme" hypothesis so elegantly demonstrated experimentally by Beadle and Tatum in the bread-mold, Neurospora, some thirty years later.

The characteristics Garrod specified were that the diseases resulting from inborn errors of metabolism had an increased frequency of parental consanguinity, tended to recur in sibs (in fact, alkaptonuria was the first human trait shown to fit the expectation for a Mendelian autosomal recessive gene), appeared early in life, showed marked deviations from normal, and were not subject to marked fluctuations in severity. The rapid growth of biochemical knowledge, and particularly enzymology, since then has led to the identification of over sixty inborn errors and to rational means of treatment for a dozen or more[3,5,12] (Table 6-1).

We have said that inborn errors of metabolism result from lack of a functional enzyme. Several mechanisms can account for this reduction in enzymatic activity. When the gene coding for a particular enzyme is changed by a mutation, this can lead to a functional deficiency of the en-

TABLE 6-1. Some Genetic Metabolic Diseases Susceptible to Treatment

Disease	*Treatment*	*Efficacy of Treatment*
Amino Acid Metabolism		
Phenylketonuria	Phenylalanine-restricted diet	Good if started in first two months of life
Maple syrup urine disease	Diet restricted in leucine, isoleucine, and valine	Fair if started in neonatal period
Homocystinuria	Vitamin B_6 and cystine supplement. Diet restricted in methionine	Not yet known
Histidinemia	Histidine-restricted diet	Not yet known
Tyrosinemia	Diet restricted in phenylalanine and tyrosine	Not yet known
Cystinosis	Diet restricted in methionine and cystine; kidney transplantation (symptomatic)	Not yet known
Cystinuria	Alkali, high fluid intake, D-penicillamine	Good for prevention of kidney stones
Diseases of the urea cycle (some forms)	Protein-restricted diet	Fair, but limited experience
Glycinemias (some forms)	Protein-restricted diet	Fair, but limited experience
Carbohydrate Metabolism		
Galactosemia	Galactose-free diet	Good if started in neonatal period
Fructosemia	Fructose-free diet	Good if started in early infancy
Malabsorption of disaccharides and monosaccharides	Monosaccharide-free or disaccharide-free diet	Good
Other Metabolic Pathways		
Wilson's disease	D-penicillamine, potassium sulfide, copper-restricted diet	Fair or better
Primary hemochromatosis	Removal of Fe by phlebotomy, desferrioxamine	Fair
Pyridoxine dependency	High doses of pyridoxine	Can be good if started in neonatal period
Familial hyperlipoproteinemias	Fat restriction, use of medium-chain fatty acids, cholestyramine, clofibrate	Fair
Familial defective synthesis and delivery of thyroid hormone (familial goiter)	Levothyroxine or desiccated thyroid	Good
Adrenogenital syndrome	Cortisone; mineralocorticoids in patients subject to salt loss	Good
Cystic fibrosis	Pancreatic extracts, diet, bronchial mucolytics, etc.	Short-term prognosis much improved; long-term prognosis unknown
Crigler-Najjar syndrome	Blood exchange transfusion, glucuronyl transferase stimulation by phenobarbital	Unsatisfactory long-term results
Nephrogenic diabetes insipidus	High fluid intake of low osmolarity, saluretics	Good if started in early infancy
Rickets refractory to vitamin D	Vitamin D and phosphate salts	Fair or better
Renal tubular acidosis (Butler-Albright syndrome)	Alkali therapy	Good

Reprinted with permission from World Health Organization.[13]

zyme in several ways. In homozygotes for the mutant gene, the enzyme coded for by that gene may (a) not be produced at all or (b) be produced in an abnormal form with reduced activity. Third, (c) the mutation may involve a gene that regulates the rate of production of the enzyme, leading to an inadequate amount of normal enzyme. So far, there are no known examples of this type in man. Fourth, (d) the enzyme may be degraded at an excessive rate leading to a deficiency of active enzyme, as in the case of certain types of G6PD deficiency. Finally, optimal activity may depend on association with a cofactor, and mutations that (e) interfere with absorption or biosynthesis of the cofactor or (f) alter the binding site on the enzyme to impair binding with the cofactor may reduce the activity of the enzyme. The vitamin-dependencies are the outstanding example. Thus one might expect that the activity of a particular enzyme could be reduced by mutations at several different loci and that each locus might have several different allelic mutations. This is the basis for the *genetic heterogeneity* so well recognized in many inborn errors of metabolism and elsewhere.

One useful way of classifying the diseases resulting from inborn errors of metabolism is according to the pathological effects of the block in the metabolic pathway. Consider a prototype metabolic pathway converting substrate S1 through a series of enzyme-catalyzed reactions to an end-product P (Fig. 6-1). Disease may result from absence of end-product, pile-up of substrates in the pathway proximal to the block, presence of excessive amounts of metabolites, and secondary effects of the above metabolic distortions on regulatory mechanisms in the same or other pathways. Although many inborn errors show several of these results, one of them usually accounts for the major features of the child's disease. It must be admitted, however, that the precise mechanisms by which enzyme defects produce clinical defects are still a major area of ignorance.

Defects in membrane transport, although they may represent inborn errors involving enzymes of the transport process, present such special features that they will be considered separately from inborn errors of intermediary metabolism.

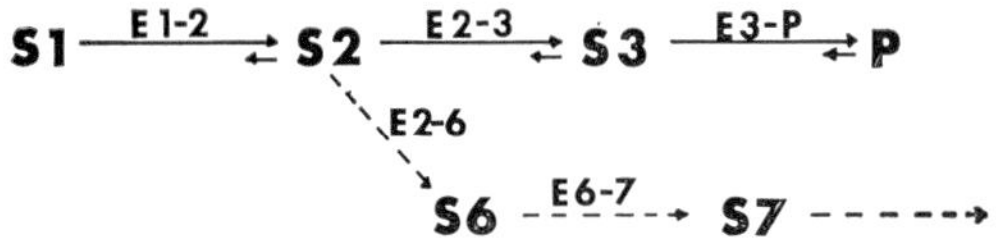

S1 —E1-2⇄ S2 —E2-3⇄ S3 —E3-P⇄ P
E2-6
S6 --E6-7--→ S7 -----→

Figure 6-1. A hypothetic metabolic pathway converting substrate S1 to end-product P through the successive actions of enzymes E1-2, E2-3, and E3-P. An alternative minor pathway is indicated.

Diseases Resulting from Absence of End-Product

One of the original inborn errors of metabolism, albinism, is a good example of a disease in which the major clinical problems result from absence of the end-product of a metabolic pathway. In our archetypical diagram (Fig. 6-2), enzyme E2-3 is indicated as missing, or inactive. All substrates beyond the block are therefore absent, including P.

In the classical type of albinism, lack of tyrosinase in the melanocyte blocks the pathway leading from tyrosine through DOPA (3,4-dihydroxyphenylalanine) to melanin (Fig. 6-3, block C). Note that the mutant gene affects only melanocyte tyrosinase, and not that in the liver and elsewhere, showing that there must be at least two separate loci for this enzyme. Furthermore, genetic heterogeneity exists, since there are several other ways of blocking the pathway. In fact, there are at least seven genetically different forms of albinism (Fig. 6-4).[14]

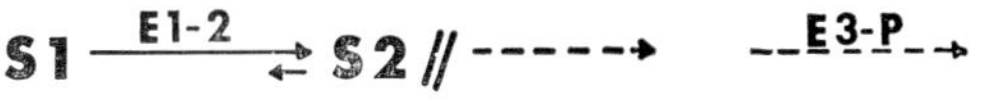

S1 —E1-2⇄ S2 // -----→ --E3-P--→

Figure 6-2. The pathway shown in Figure 6-1 blocked by the absence of enzyme E2-3.

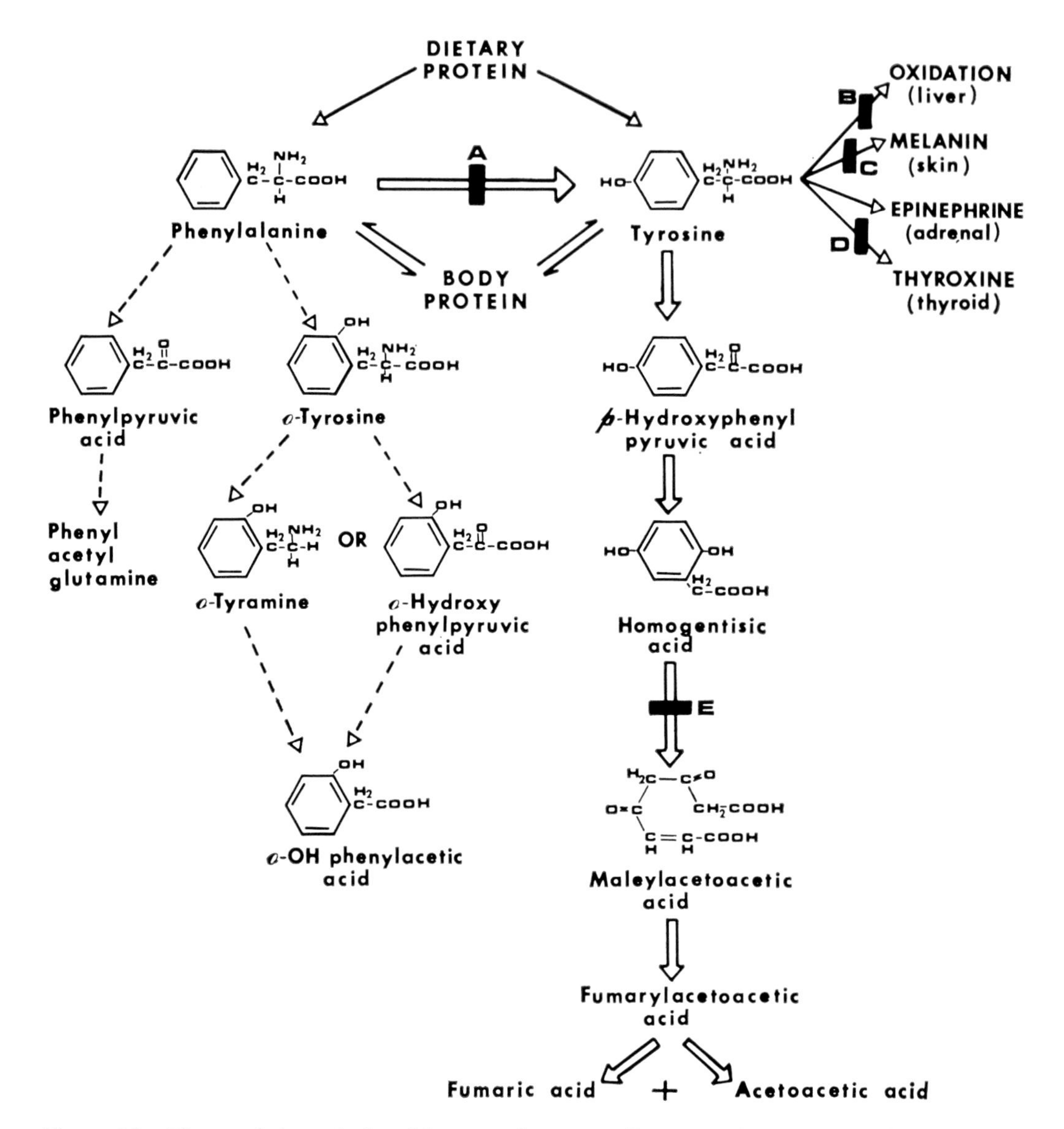

Figure 6-3. The metabolism of phenylalanine and tyrosine, illustrating the diseases produced by various enzyme deficiencies (see text).

Other examples of this class of disease include the various types of recessively inherited goitrous cretinism (dwarfism, mental retardation, coarse features; see Fig. 6-5) in which the pathological effects result from a lack of thyroid hormone (Fig. 6-3, block D), pitressin-sensitive diabetes insipidus (drinking and excreting excessive amounts of water) in which the pituitary does not produce antidiuretic hormone, and the adrenogenital syndromes, in which part of the trouble results from a deficiency of cortisol. However, the latter syndrome will be considered below (p. 82) as an example of interference with regulatory mechanisms.

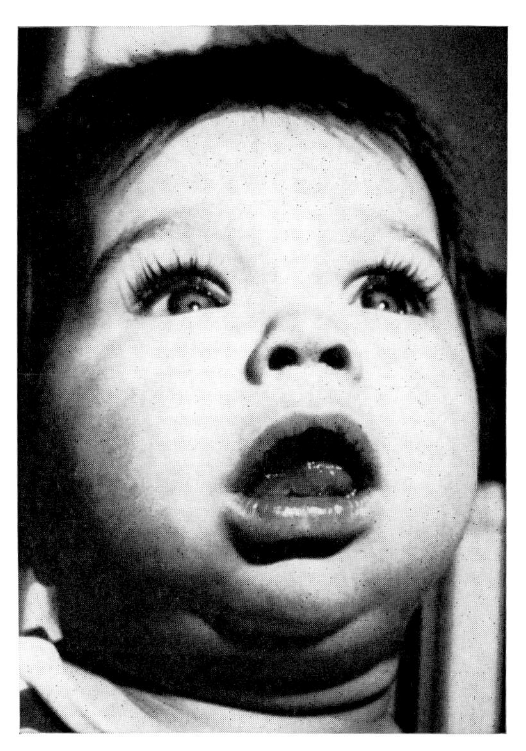

Figure 6-5. Cretinism, a result of deficiency in thyroxin, the thyroid hormone.

Diseases Resulting from Pile-up of Substrate(s)

In some cases the substrate just before the block (in this case substrate S2), not being converted to substrate S3, will increase in concentration and may appear in abnormal quantities in blood and urine. Since most enzymatic reactions are reversible, substrate S1 may also pile up and be excreted (Fig. 6-6).

An example is galactosemia, in which the defective enzyme is galactose-1-phosphate uridyl transferase, which normally converts galactose-1-phosphate to glucose-1-phosphate (Fig. 6-7). In the mutant homozygote this step cannot occur, and galactose-1-phosphate accumulates in the blood cells, liver, and other tissues, damaging the liver, brain, and kidney. Other effects will be considered in the following section.

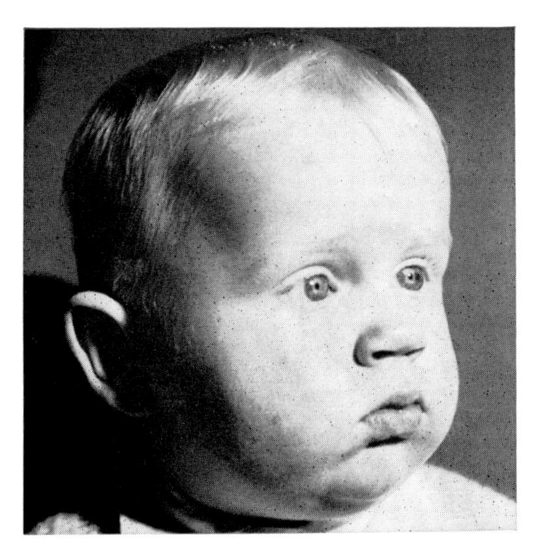

Figure 6-4. Albinism—lack of melanin pigment in hair, skin, and eyes.

Alkaptonuria is another example: The homogentisic acid accumulates in the blood (Fig. 6-3, block E) and, in polymerized form, is deposited in cartilages (ochronosis), leading to degeneration and arthritis. It also forms a polymer in the urine, which turns black on exposure to air. Some stor-

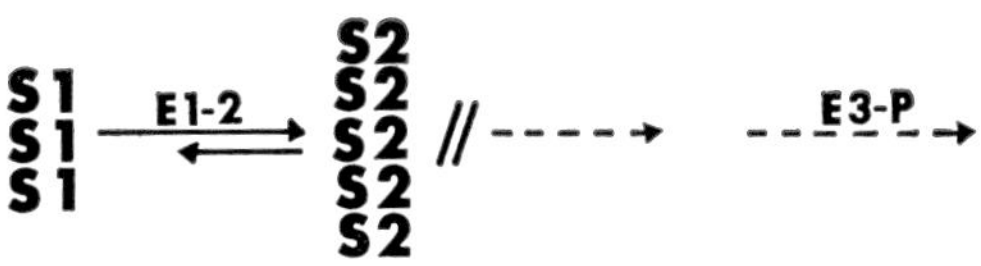

S1 S1 S1 —E1-2⇄ S2 S2 S2 S2 S2 //----→ --E3-P→

Figure 6-6. The hypothetic pathway of Figure 6-1 showing pile-up of precursor S2 when enzyme E2-3 is lacking.

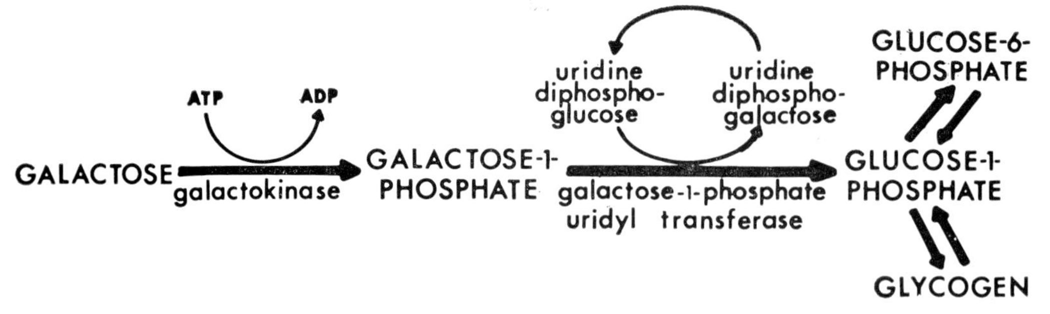

Figure 6-7. The metabolic pathway of galactose.

age diseases also fit into this class, but they will be considered separately in a later section (p. 83).

Diseases Resulting from Excessive Amounts of Metabolites

In this category, the damage is done not so much by the excessive amounts of precursors behind the block as by excessive amounts of metabolites produced by the breakdown of these precursors through alternate pathways that are normally used only slightly, but are called upon to deal with the abnormal situation (Fig. 6-8). An example is classical phenylketonuria, which is characterized by mental retardation, seizures, and decreased pigmentation. Here the enzyme phenylalanine hydroxylase is inactive or missing, and phenylalanine is not converted to tyrosine. Phenylalanine therefore accumulates in the blood and is broken down to phenylpyruvic, phenylacetic, and phenylactic acids, which may be toxic substances (Fig. 6-3, block A). For instance, they may inhibit the enzyme 5-hydroxytryptophan decarboxylase, resulting in decreased synthesis of serotonin (5-hydroxytryptamine)—important to brain function—and inhibit the melanocyte tyrosinase, accounting for the decreased pigmentation.[6] This effect could be considered an example of diseases covered in the next section, illustrating that the same disease may fall into several categories.

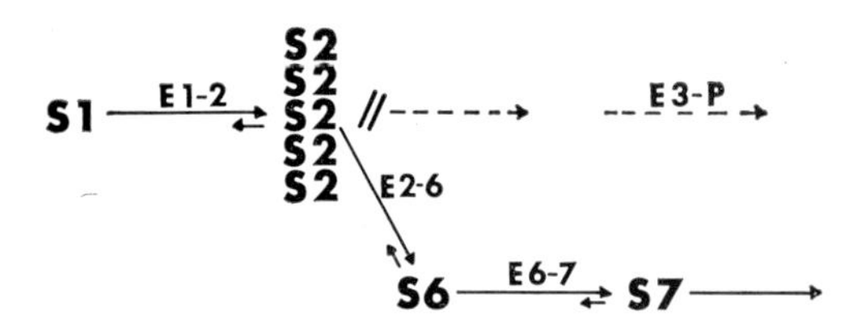

Figure 6-8. The hypothetic pathway of Figure 6-1 showing increased use of an alternate pathway.

Diseases Resulting from Interference with Regulatory Mechanisms

A fourth category of pathological effects resulting from genetic blocks in metabolic pathways are those in which lack of the end-product, or excessive amount of a substrate, interferes with feedback or other regulatory mechanisms.

In the *adrenogenital* syndromes, for instance, there is a block in the biosynthesis of cortisol by the adrenal cortex. This deficiency stimulates excessive production of ACTH by the pituitary, since the level of cortisol normally regulates the output of ACTH by a negative-feedback mechanism. The increased ACTH levels, in turn, stimulate the adrenal cortex to increase synthesis of the cortisol precursors but, of course, only as far as the block. Breakdown of the accumulated precursors by alternative pathways leads to the androgenic effects.[1] In a female fetus this may result in masculinization of the external genitalia (Fig. 6-9).

Orotic aciduria is another example of an inborn error of metabolism resulting in a regulatory mechanism defect. The pathway leads from aspartic acid and carbamyl

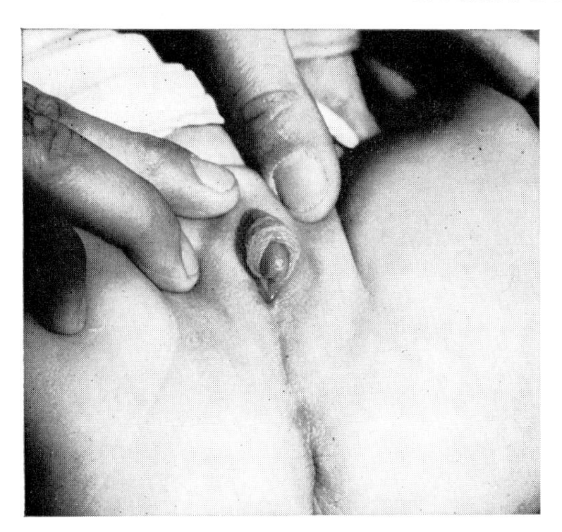

Figure 6-9. Masculinization in a patient with the adrenogenital syndrome (female pseudohermaphrodite).

phosphate through a series of steps to uridine monophosphate (UMP). In homozygotes for orotic aciduria, two enzymes—orotidylic acid pyrophosphorylase and decarboxylase—are absent, and the proximal precursor piles up and appears in the urine. (It is unlikely that this represents an operon; more probably the two enzymatic functions are carried by a multifunctional protein complex.) The end-product, UMP, is a feedback inhibitor of the first enzyme in the pathway. In its absence, the reaction appears to proceed more rapidly, leading to a great excess of orotic acid. Adding uridine to the diet inhibits the first enzyme and leads to a sharp decrease in the production of orotic acid.[11]

The Storage Diseases

In a number of inborn errors of metabolism, one of the substrates that accumulates is deposited in abnormal quantities in the cells and may cause damage merely by its presence.

The glycogen storage diseases are classic examples.[4] Glycogen is a polymer of alpha-D-glucose, assembled into a multibranched treelike structure. The glycogen molecules are constantly being degraded and resynthesized to meet the varying metabolic demands of the individual. There can be errors at various steps of mobilization or synthesis. For instance, cleavage of the outer branches of the glycogen molecule, when glucose is to be mobilized, is done by a phosphorylase. Absence of this enzyme in muscle cells results in accumulation of glycogen in muscle, causing painful muscle cramps on exercise (glycogen storage disease type V).

Many storage diseases fit into a special category, only recently recognized—the *lysosomal diseases*. The lysosomes are intracellular organelles consisting of a lipid membrane enclosing a variety of acid hydrolytic enzymes. If a special lysosomal enzyme is missing, the corresponding substrate may accumulate in the lysosome, and the cell becomes laden with the resulting storage vacuoles. The first example to be discovered was Pompe's disease, one of the glycogen storage diseases (type II) Homozygotes are deficient in α-1,4-glycosidase, a lysosomal enzyme that can hydrolyze the outer chains of glycogen to give glucose. Apparently, fragments of glycogen are constantly being taken up by the lysosomes and degraded. In Pompe's disease, uptake goes on but degradation does not; the lysosomes become swollen with glycogen in the cells of the heart, muscle, liver and other tissues. The heart and liver enlarge, and the child usually dies within the first six months of life.[4]

Another well-known example of a lysosomal disease is Tay-Sachs disease, or G_{M2} gangliosidosis. An autosomal recessive gene, particularly frequent in Ashkenazi Jews, causes a deficiency of an enzyme, hexoseaminidase A, that is involved in the metabolism of a class of nervous system lipids called gangliosides. In the absence of the enzyme, one of the gangliosides, G_{M2}, accumulates in the ganglion cells of the brain (and other organs and tissues) leading to progressive retardation in development, paralysis, dementia, blindness, and death by the age of 3 to 4 years. The enzyme is normally present in leukocytes and cul-

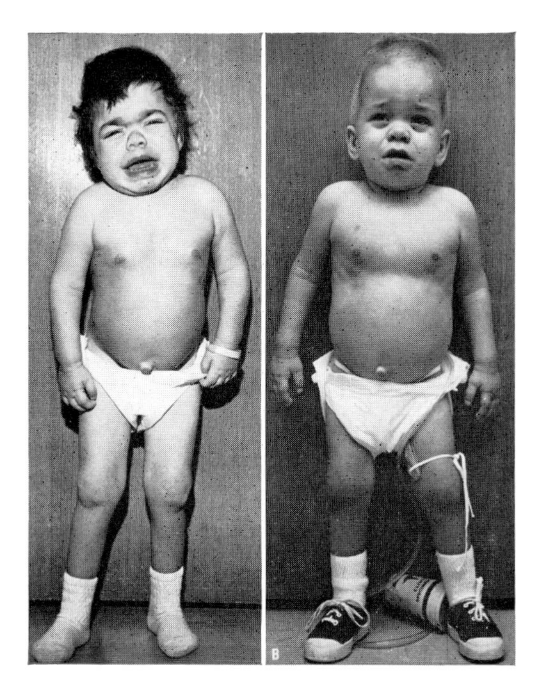

Figure 6-10. Two- and three-year-old sibs with Hurler syndrome, showing the "gargoyle" facies and claw hands.

tured fibroblasts, diminished in heterozygotes and almost absent in homozygotes, so that the disease is a suitable candidate for population screening and prenatal diagnosis.[8]

Hurler syndrome (Fig. 6-10), or gargoylism, is one of several disorders involving storage of mucopolysaccharide. The defective enzyme is α-L-iduronidase, a lysosomal enzyme that cleaves the side chains from the acid mucopolysaccharides (or glycosaminoglycuronoglycans in the modern terminology), an important component of connective tissue ground substance. Storage of mucopolysaccharide in various tissues leads to the progressive mental retardation, coarsening of features, stiff joints, coronary artery insufficiency and heart valve thickening, and death, usually by the age of 10 years. Interesting new developments in the study of this group of storage diseases are mentioned on p. 87 and in Chapter 21.

INBORN ERRORS OF TRANSPORT

Our increasing understanding of diseases resulting from genetically determined errors of membrane transport is a good example of how specific mutations can be used as tools to probe the biology of the normal organism.[10]

Inborn Errors of Amino Acid Transport

Cystinuria was first classified (erroneously) as an inborn error of metabolism of cystine, but when it was found that blood levels of cystine are not elevated in patients with the disease, Dent proposed that the condition was an inborn error of membrane transport rather than of intermediary metabolism. That is, the cystine was not being reabsorbed by the renal tubule from the glomerular filtrate and therefore appeared in abnormal amounts in the urine. This implied a specific transport mechanism across the tubule membrane, which was defective in the cystinuria patients. The cystine, being relatively insoluble, may form "stones" in the kidney. The discovery that not only cystine, but the structurally similar dibasic amino acids, lysine, arginine, and ornithine, were being excreted in abnormally large quantities led to the idea that there was a membrane transport system that would accept all four of these amino acids, but not others.

Harris[3] demonstrated genetic heterogeneity for the disease when he showed that in some families the heterozygotes showed mild degrees of the relevant aminoacidurias, while in other families they did not. It was then found that some cystinuria homozygotes had the same defect in transport across the intestinal membrane, and this led to further definition of genetic heterogeneity. All homozygotes have similar urinary findings, but some (type I) have greatly impaired transport of cystine and the dibasic amino acids from the intestine into plasma; others (type II) have only moderate impairment of intestinal membrane transport, and a third group (type III) have mild impairment. To make matters still more complicated, kinetic studies suggest that the renal system common to the four amino acids is an efflux system and that the renal influx system has separate sites for cystine and the dibasic amino acids, respectively.

Heterozygotes for the type I mutant have normal urinary amino acids—that is, the gene is "completely recessive," whereas those of types II and III have an excess of the relevant amino acids in the urine, somewhat more marked in type II. Matings between heterozygotes of different types produced offspring with the full-blown homozygous urinary phenotype, showing that the three mutants are allelic.

The hereditary *iminoglycinurias* provide another example of genetic determination of a specific transport mechanism and of genetic heterogeneity. Homozygotes have decreased tubular resorption of proline, hydroxyproline, and glycine, with normal plasma levels. This suggests a renal tubular transport mechanism specific to these substances. As with cystinuria, there is heterogeneity when intestinal transport is examined; in some homozygotes it is impaired and in others it is not. Again, family studies suggest allelism. The condition is probably harmless.

Hartnup disease is characterized by defective transport of the neutral amino acids (other than the iminoacids and glycine), suggesting a membrane site specific to the transport of these molecules. The relation to the clinical manifestations (intermittent attacks of ataxia and a pellagralike skin rash) remains unknown.

Finally, there are mutant genes that interfere with membrane transport of a wide variety of amino acids and other substances. The *Fanconi syndromes* (rickets, glucosuria, and aminoaciduria) are well-documented examples. In some cases, this may be secondary to impairment by some toxic metabolite of the energy supply necessary for transport; in others, there may be a defective component in the transfer mechanism beyond the binding site. Perhaps further study of mutant phenotypes will throw more light on the nature of the transfer process, as it has on the nature of the binding sites.

Why is it that these mutant genes produce only partial defects in tubular transport of the affected amino acids? In the iminoglycinuria homozygote, for instance, about 80% of the ability to reabsorb proline, hydroxyproline, and glycine is retained. Kinetic studies both *in vitro* and *in vivo* suggest that there are at least two kinds of system involved in amino acid tubular transport. One type is represented by the mutant phenotypes we have been describing. They have "group" specificity, high capacity, and low affinity; they appear to operate at concentrations that exceed the usual physiological range. Another type of transport site is characterized by low capacity, high affinity, and specificity to a particular amino acid. If so, one would expect to find mutant phenotypes involving failure to transport specific amino acids, and this is so. Siblings have been reported with excessive amounts of cystine, but not of the dibasic amino acids in the urine. There are also gene-determined hyperdibasic aminoacidurias in which the transport of dibasic amino acids, but not of cystine, is impaired. Similarly, there is a "blue diaper" syndrome that involves defective intestinal transport of tryptophan, and there is a methionine malabsorption syndrome.

Inborn Errors of Transport of Other Than Amino Acids

Site specificity of transport is not limited to the amino acids. Genetically determined defects of membrane transport have been found for many other substances, and, again, study of the disorder has often been the first evidence that there is a specific site for the transport of that particular substance. These include

Renal glucosuria—failure to reabsorb glucose; harmless (except when misdiagnosed as diabetes); pattern of inheritance varies from family to family

Glucose-galactose malabsorption—diarrhea after ingestion of these sugars or disaccharides and polysaccharides that give rise to them; autosomal recessive

Hypophosphatemic rickets—failure to reabsorb phosphate leads to loss of calcium and to rickets; X-linked dominant.

Renal tubular acidosis—increased permeability of distal tubule cells to hydrogen; autosomal dominant in some families

Chloridorrhea, congenital—chloride lost in intestine; autosomal recessive

Hereditary spherocytosis—impaired transport of sodium across the blood cell membrane; autosomal dominant

Diabetes insipidus, nephrogenic—impaired tubular resorption of water; X-linked recessive

PRINCIPLES OF TREATMENT OF INBORN ERRORS OF METABOLISM

Rational methods of treating the disorders resulting from the inborn errors of metabolism depend on understanding the biochemical nature of the error and of the resulting disease processes. Correction of the end-results of a genetic defect has been termed "environmental engineering," as opposed to "genetic engineering," which attempts to modify the genetic material itself.[9] An increasing number of genetically determined metabolic diseases are now susceptible to treatment.

Part of the problem is the organization of medical resources in order to get the patient with a (usually rare) genetic disease to a source of expert diagnosis and treatment, or better still, in some cases, to get the management to the patient.[2]

The following approaches to therapy may be considered:

Substrate Restriction

Diseases where the metabolites proximal to the enzyme block interfere with development or function could logically be treated by restricting the supply of substrate. This has been quite successful when the substrate comes primarily from the diet. Thus, reduction of dietary phenylalanine in phenylketonuria, of lactose and galactose in galactosemia, and of fructose in fructosemia are relatively effective in preventing the pathological consequences of the genetic defect.

When the substrate is synthesized endogenously, it may be much more difficult to control its accumulation by simple dietary restriction. Sometimes it is possible to impose a block elsewhere in the pathway, where the results may be less harmful. For instance, the accumulation of oxalate in oxalosis can be reduced by treatment with calcium carbamide, which inhibits aldehyde oxidase and thereby reduces the synthesis of glycolate and its conversion to glyoxalate and oxalate.

Removal of Toxic Products

An alternative to restricting substrate would be to remove the accumulating toxic product. This approach is taken in the treatment of Wilson's disease by removing excess copper with penicillamine and of hemochromatosis by bloodletting and desferrioxamine, for example. One might also place in this category the prevention of hemolytic disease of the newborn due to Rh− isoimmunization by treating Rh-negative mothers of Rh-positive babies with gamma globulin containing anti-Rh antibody, which neutralizes any Rh antigen the mother may have received from the baby.

Product Replacement

When the pathology results from lack of a product in the metabolic pathway distal to the block, it would seem logical to replace the product. This is the rationale for treating, for instance, inherited defects of thyroid hormone synthesis with thyroxin, the adrenogenital syndromes with cortisol, orotic aciduria with uridine, hemophilia with antihemophiliac globulin, and cystic fibrosis of the pancreas with pancreatic enzymes. However, replacement of product may present technical difficulties, particularly if the product is intracellular, as in albinism, for instance.

In some diseases, product replacement is used together with substrate restriction. In homocystinuria, for instance, methionine is restricted, but cystine must be added, since the homozygote cannot synthesize it from methionine.

Cofactor Supplementation

Diseases in which the reduced enzyme activity results from a defective or lacking coenzyme, or from a change in the enzyme's capacity to bind to the coenzyme, may be treated by supplying large amounts of the coenzyme. For instance, more than half of the tested cases of homocystinuria respond well to large doses of pyridoxine (vitamin B_6), a cofactor for the enzyme cystathionine synthetase, which is deficient in this disease. Vitamin B_6 is also a cofactor for cystathioninase, kynureninase, and glutamic acid decarboxylase and is effective in the treatment of the corresponding inborn errors, cystathioninuria (one form), xanthurenic aciduria, and convulsions due to vitamin B_6 dependency. Other examples are a form of methylmalonicaciduria that responds to vitamin B_{12}, and a form of propionicacidemia that responds to biotin.

TREATMENT OF INBORN ERRORS OF TRANSPORT

Most hereditary transport defects of man are rather benign, and treatment is often limited to what might be called second-order clinical manifestations. For instance, cystinuria is a serious disorder only when urinary stones are formed. Keeping the urine diluted prevents stone formation (all that is needed is a glass of water and an alarm clock to awaken the patient at the appropriate hour), and solubilization of cystine with penicillamine reduces cystine excretion and causes stones to dissolve, though, unfortunately, there are problems with toxicity.

In nephrogenic diabetes insipidus, the logical approach is to replace the water lost by inadequate tubular resorption, and in renal tubular acidosis (Butler-Albright's disease), where hydrogen ion clearance is inadequate, leading to excessive excretion of bicarbonate, sodium, potassium, and calcium, treatment with alkali adjusts the imbalance quite well.

NEW TREATMENT APPROACHES

In theory, the best way to treat a disease resulting from an enzyme deficiency would seem to be replacement of the enzyme, and this is already true for such things as congenital trypsinogen deficiency and some of the clotting disorders. This approach seems promising when the deficiency involves an extracellular enzyme, but intracellular enzymes present problems. Rapid inactivation, failure to reach the site of reaction with the substrate, and the development of antibodies to the "naked" enzyme complicate this approach. For instance, attempts to treat metachromatic leukodystrophy (a degenerative brain disease) with arylsulfatase A infusion failed, since there was no increase in enzyme in the brain, and infusion of alpha-glucosidase in patients with type II glycogenosis led to severe immunologic intolerance.

Perhaps inclusion of the enzyme in a semipermeable, inert microcapsule may avoid some of these problems, and this approach has already achieved temporary correction of the biochemical phenotype of acatalasemic mice.

One intriguing new development is the treatment of mucopolysaccharidoses. Studies of somatic cell cultures showed that when fibroblasts from a patient with Hurler syndrome were cultured with normal fibroblasts, the mutant cells stopped storing acid mucopolysaccharides. Furthermore, when cells from a patient with Hurler syndrome were cultured with cells from a patient with another form of acid mucopolysaccharidosis, each type of cell corrected the metabolic defect of the other. The active factors are probably enzymatic. Application of this work to therapy led to the discovery that injection of normal serum led to striking improvement in patients with Hunter and Hurler syndromes. Injection of white cells from nonmutant donors may be an even better approach. However, there may be trouble with antibody formation against the "foreign" protein factor, and long-term results are not as promising as anticipated.

In some cases, it may be possible to induce synthesis of the missing enzymes; for example, treatment with phenobarbital induces synthesis of glucuronyl transferase and lowers the bilirubin level in the Crigler-Najjar syndrome (congenital nonhemolytic unconjugated hyperbilirubinemia). Stabilization of a defective enzyme by the addition of an appropriate compound is another, so far theoretical, possibility. Since the necessary amount of enzyme activity is often far below the normal amount, a relatively small increase in activity may be therapeutic.

Another way of correcting an enzyme deficiency would be by organ transplantation. This is complicated by the problem of graft rejection, but progress is being made. Transplantation of bone marrow and thymus is being attempted in some of the immune deficiency syndromes. Liver transplantation has been performed in Wilson's disease, with short-term success, and will no doubt be attempted in many other diseases involving hepatic enzymes. Renal transplantation has had some success in correcting the phenotype of Fabry disease and in correcting the uremic nephropathy in cystinosis (though not the cystine storage in other tissues).

Looking farther into the future, directed gene change appears as a way of providing the missing enzyme in a mutant individual. Already it has been possible to prepare a bacteriophage that will incorporate into itself the genetic material that codes for a particular enzyme (galactose-1-phosphate uridyl transferase) and transfer it to the cells of a mutant (galactosemic) individual lacking the enzyme, where it restored enzymatic activity—an example of genetic transduction. If such cells could safely be grafted back into the mutant individual, the metabolic defect would be corrected.

In this discussion of principles of treatment of the inborn errors of metabolism and transport, we have shown how an understanding of the nature of the basic error, and of the mechanisms by which the error leads to the specific features of the disease, leads to rational methods of treatment. We have also shown how imaginative applications of new advances in modern biology hold promise of further exciting advances in therapy. The future—in this respect—looks bright.[9]

THE GENETICS OF PROTEIN STRUCTURE

We have said that mutation of a gene results in the substitution of one amino acid for another in the corresponding polypeptide chain. This is not always true. The genetic code is redundant, and a mutation that changed one triplet to another coding for the same amino acid would not produce any change in the structure of the protein—though it might change its rate of synthesis, as we shall see in Chapter 10. Nevertheless, the majority of mutations in genes coding for polypeptides would be expected to result in an amino acid substitution.

The Hemoglobinopathies

Hemoglobin was the first molecule in which an association was shown between a mutant gene—for sickle-cell disease—and a specific amino acid substitution. Sickle-cell disease is a form of chronic hemolytic anemia characterized by the presence of elongated filiform or crescent-shaped red blood cells. Family studies in the 1940's showed that (with certain exceptions that later proved the rule) the disease fitted the segregation ratio expected for autosomal recessive inheritance. Heterozygotes showed the sickle-cell trait—sickle cells

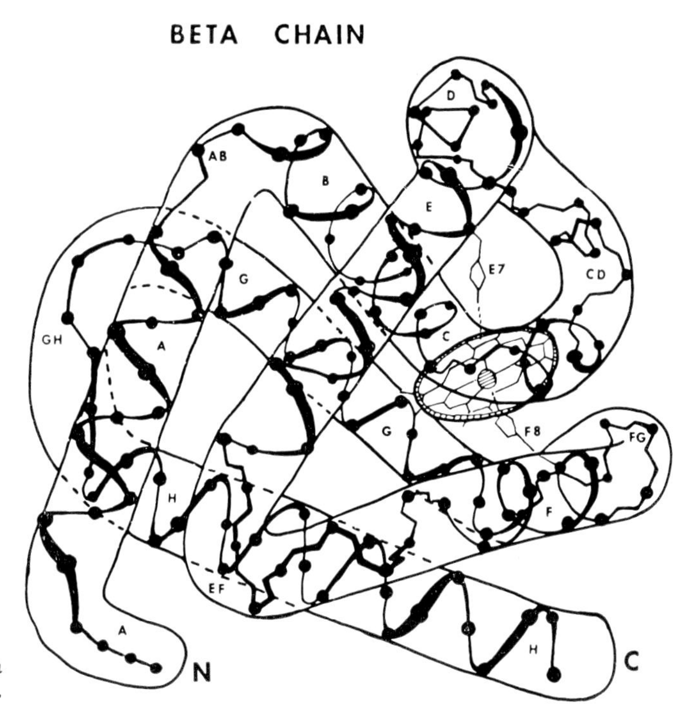

Figure 6-11. Diagram of the beta chain of hemoglobin.

Figure 6-12. The hemoglobin molecule.

were not normally present in blood smears, but when the red cells were made hypoxic—by incubation or treatment with sodium metabisulfite—sickling would occur.

The disease has an extraordinarily high frequency in populations of West African origin, occurring in about 1 in 400 U.S. Negroes. In addition to the effects of chronic anemia, the patients may suffer from intravascular sickling resulting in thrombi and local infarcts in the intestine (sometimes mistaken for appendicitis), lungs, kidneys, or brain.

After Pauling's discovery that the disease resulted from a physicochemical difference in the hemoglobin molecule (the first "molecular disease") and Ingram's demonstration that sickle-cell hemoglobin differed from normal hemoglobin by a single amino acid—a valine substituted for a glutamic acid—progress was rapid. The molecule, already known to be a tetramere, was shown to consist of two alpha and two beta chains. The sickle-cell substitution involved the sixth amino acid from the N-terminal end of the beta chain. The high frequency in West Africans appears to result from an increased resistance of heterozygotes to falciparum malaria.

By electrophoretic and chromatographic procedures, a large number of other "mutant" hemoglobins have been identified, almost all associated with a single amino acid substitution. By convention, the normal molecule is assigned the formula alpha_2^A beta_2^A and the mutant forms are designated according to the amino acid substitution. For instance, sickle-cell hemoglobin is alpha_2^A $\text{beta}_2^{6\text{glu-val}}$, or alpha_2^A beta_2^S for short.

One of the triumphs of molecular biology has been the use of these amino acid substitutions in specific regions of the molecule, and the resulting alteration of charge and bonding, to elucidate the functional properties of the molecule in physicochemical terms.[7]

The molecule is a tetramere consisting of four polypeptide chains, two alpha and two beta chains. Each chain is coiled and folded in a complex but characteristic manner and has a pocket that contains a heme group—a porphyrin ring with an iron atom at its center that combines with oxygen (Fig. 6-11). The molecule is allosteric with respect to its affinity for oxygen; as a molecule of deoxyhemoglobin moves into a region of increasing oxygen tension, an oxygen atom will bind to an iron atom in the heme group of one chain, causing the iron atom to move slightly, which results in a slight twist in the chain where the heme group is attached to it. This in turn causes a change in the conformation of the other chains in the tetramere that increases the affinity of their heme groups for oxygen and decreases the affinity for carbon dioxide. As the next heme group combines with oxygen, the affinity of the other two changes still more, and so on. Thus when the hemoglobin arrives in the lung, where the oxygen tension is high, its affinity for oxygen increases, and it readily picks up oxygen, but as it moves to the periphery where the oxygen tension is low, it begins to lose oxygen and its affinity decreases so it more easily releases the oxygen where it is needed.

The alpha chain has 141 amino acids, and the beta chain 146. The amino acid sequence is similar, though not identical, in the two chains. The sequence of amino acids is the *primary structure* of the polypeptide chain. Much of the chain is in the form of a helix, named by the protein chemists the alpha helix (not the same alpha as the chain), but some segments are not coiled. The helical coiling is the *secondary structure*. The (mostly) coiled chain is folded in a complex way (Fig. 6-12), forming a pocket for the heme group and surfaces for relating to the other three chains of the tetramere. This is referred to as its tertiary structure. The tertiary structure is very similar in the two chains. Figure 6-12 indicates that there are eight helical segments, designated alphabetically, with nonhelical portions at the bends, designated by the letters of the segments they join. Specific amino acids can be numbered consecutively from the N-terminal end, or by their position in the segment. For instance, the sickle-cell substitution involves amino acid 6 from the N-terminal end, or A3, since the first three amino acids are nonhelical. (The advantage of the helical nomenclature is that it allows more meaningful comparisons between corresponding amino acids in different chains. Thus the histidine F8 is linked to the heme group in the alpha chain, the beta chain, and myoglobin.) Finally, the four chains are associated to form a more or less globular molecule, and this association is the *quaternary structure* of the molecule.

It is now clear that the primary structure of the chain determines its secondary, tertiary, and quaternary substructures. It does so by means of the side chains on the amino acids, which form bonds with other side chains; the sequence of amino acids determines the positions of the bonds, and thus the folding. The nature of the folding determines the external shape of the chain and thus the way in which it associates with other chains. The short side chains such as oxygen and nitrogen tend to be polar, or hydrophilic, and the longer radicals, such as phenyl rings, tend to be nonpolar, or hydrophobic, or "greasy." It appears that the internally situated side chains, those lining the heme pockets or binding one helical segment to another, are hydrophobic, whereas external side chains can be either hydrophilic or hydrophobic.

Amino acid substitutions can cause abnormalities of the hemoglobin molecule in several ways.[7] First, they may affect contact between the chain and the heme group. For instance, proline cannot form part of the alpha helix, so if it replaces an amino acid somewhere in a helix, the helix will bend (disturbing the tertiary structure) or break, and may thus lose contact with the heme group. This causes instability of the molecule and precipitation resulting in hemolysis and anemia. [Examples are hemoglobins Bibba (alpha H19 leu-pro), Genova (beta B10 leu-pro), and Santa Ana (beta F4 leu-pro).]

If the substitution involves an amino acid of different size in a part of the chain lining the heme pocket, this may allow water to enter the pocket, the heme to fall out, and the molecule to precipitate. Again, a chronic hemolytic anemia may result. [Examples are hemoglobins Torino (alpha CD1 phe-val) and Hammersmith (beta CD1 phe-ser), Sydney (beta E11 val-ala), Kahn (beta FG5 val-met), and Zurich (beta BE7 his-arg).] The latter is unstable only in the presence of sulfonamide. There are no known substitutions involving amino acids with hydrophilic bonds on the walls of the heme pocket—presumably, they would be too disruptive to allow survival of the molecule.

Another possible effect of an amino acid substitution is an alteration in oxygen-carrying function, and this can happen in two ways. In hemoglobins M Boston and M Saskatoon, the histidine in the E11 position (on the alpha and beta chain, respectively) is changed to a tyrosine. This histidine is adjacent to the heme group, and the tyrosine forms an ionic bond with the iron, changing it from the ferrous to the ferric state, forming methemoglobin, in which oxygen-binding capacity is impaired. Heterozygotes have cyanosis, but no harmful effects other than the risk of being investigated for congenital heart disease.

Alternatively, an amino acid substitution can affect relations between the four subunits. For instance, arginine 92 (FG4) in the alpha chain forms part of the bridge from chain alpha 1 to beta 2. In hemoglobin Chesapeake, this arginine is changed to a leucine. The resulting change in spatial relations between the chains leads to an increase in affinity for oxygen, an oxygen deficit in the peripheral tissues, and a compensatory polycythemia.

Most substitutions on the external surface of the hemoglobin molecule do not seem to affect function. However, sickle-cell hemoglobin may be an exception. The substitution involves the sixth amino acid in the A helix, which is internal when the molecule is oxygenated but becomes exposed with the allosteric shift as the molecule loses oxygen. The sickle-cell mutation causes a substitution of a nonpolar for a polar residue; it has been suggested that in the deoxygenated state, the exposed nonpolar bond can attach to a binding site on another sickle-cell hemoglobin molecule, and this to a third, thus leading to the formation of the long chains that are the basis for the sickling phenomenon.

The above studies are an elegant example of the fruitful interaction between genetics and biochemistry. Family studies identify gene mutations affecting the molecule, and these can be used by the protein biochemist to elucidate the relation between the structure of the molecule and its function.

Further examples will be found in Chapter 10 on developmental genetics, where the other hemoglobin chains and mutations affecting their synthesis will be discussed.

SUMMARY

Biochemical genetics deals with the biochemical changes resulting from substituting mutant for normal proteins. If the abnormal protein is an enzyme, an inborn error of metabolism may result. Diseases caused by abnormal enzymes may result from (1) absence of end-product, (2) pile-up of substrate, (3) excessive amounts of metabolites, (4) interference with regulatory mechanisms, and (5) abnormal storage. Other genetic defects of enzymes may result in inborn errors of membrane transport of amino acids and a varicty of other substances such as glucose and phosphorus.

Treatment of inborn errors of metabolism and transport include (1) substrate restriction, (2) removal of toxic products, (3) product replacement, (4) cofactor supplementation, and (5) general supportive measures. New approaches include enzyme replacement, induction of synthesis, transplantation, and incorporation of the missing genetic information into mutant cells using devices such as bacteriophage transduction.

Genetic defects of protein structure have been extensively studied in the hemoglobinopathies, with sickle-cell disease as the prototype. A single amino-acid substitution results in a mutant polypeptide. The changes in the primary structure may lead to alteration of secondary, tertiary, and quaternary structure, with consequent changes in function. Much has been learned about the structure and function of proteins by study of the effects of specific amino acid substitutions.

REFERENCES

1. Bongiovanni, A. M.: Disorders of adrenocortical steroid biogenesis (the adrenogenital syndrome associated with congenital adrenal hyperplasia). *In* J. B. Stanbury, J. B. Wyngaarden, and D. S. Fredrickson (ed.), The Metabolic Basis of Inherited Disease. 3rd ed. New York, McGraw-Hill Book Co., 1972.
2. Clow, C. L., Fraser, F. C., Laberge, C., and Scriver, C. R.: On the application of knowledge to the patient with genetic diseases. *In* A. G. Steinberg and A. Bearn (ed.), Progress in Medical Genetics. Vol. 9. New York, Grune and Stratton, 1973.
3. Harris, H.: The Principles of Human Biochemical Genetics. Frontiers of Biology. Vol. 19. New York, Elsevier, 1970.
4. Howell, R. R.: The glycogen storage diseases. *In* J. B. Stanbury, J. B. Wyngaarden, and D. S. Fredrickson (ed.), The Metabolic Basis of Inherited Disease. 3rd ed. New York, McGraw-Hill Book Co., 1972.
5. Kelley, V. C. (ed.): Metabolic, Endocrine, and Genetic Disorders of Children. (3 vols.) Harper and Row, New York, 1974.
6. Knox, W. E.: Phenylketonuria. *In* J. B. Stanbury, J. B. Wyngaarden, and D. S. Fredrickson (ed.), The Metabolic Basis of Inherited Disease. 3rd ed. New York, McGraw-Hill Book Co., 1972.
7. Lehmann, H., and Huntsman, R. G.: The hemoglobinopathies. *In* J. B. Stanbury, J. B. Wyngaarden, and D. S. Fredrickson (ed.), The Metabolic Basis of Inherited Disease. 3rd ed. New York, McGraw-Hill Book Co., 1972.
8. O'Brien, J. S.: Ganglioside storage diseases. *In* H. Harris and K. Hirschhorn (ed.), Advances in Human Genetics. Vol. 3. New York, Plenum Press, 1972.
9. Scriver, C. R.: Enzyme therapy and induction in genetic disease: pox or pax. *In* A. G.

Notulsky and W. Lenz (ed.), Birth Defects. Amsterdam, Excerpta Medica, 1974.

10. Scriver, C. R., and Rosenberg, L. E.: Amino Acid Metabolism and Its Disorders. Philadelphia, W. B. Saunders, 1973.
11. Smith, L. H., Huguley, C. M., and Bain, J. A.: Hereditary orotic aciduria. *In* J. B. Stanbury, J. B. Wyngaarden, and D. S. Fredrickson (ed.), The Metabolic Basis of Inherited Disease. 3rd ed. New York, McGraw-Hill Book Co., 1972.
12. Stanbury, J. B., Wyngaarden, J. B., and Fredrickson, D. S. (ed.): The Metabolic Basis of Inherited Disease. New York, McGraw-Hill Book Co., 1972.
13. World Health Organization: Genetic Disorders: Prevention, Treatment and Rehabilitation. WHO Tech. Rep. Ser. No. 497, 1972.
14. Witkop, C. J., White, J. G., Nance, W. E., Jackson, C. E., and Desnick, S.: Classification of albinism in man. Birth Defects VII(8):13, 1971.

Chapter 7

NORMAL TRAITS

Nearly everyone is interested in the inheritance of physical features, and it is rather disappointing that so few of them show clear-cut Mendelian pedigree patterns. One difficulty is that normal physical differences often do not fall into sharply different classes, so that it is difficult to know how to classify individuals in the overlap range. Nevertheless, there is a great deal of data about the inheritance of normal features, as well as a good many misconceptions. The reader is referred to Amram Scheinfeld's book, *Your Heredity and Environment,* for an interesting and detailed coverage of inherited normal variations.[2] We will touch only lightly on the subject.

EYE COLOR

Probably the example of "simple Mendelian inheritance" in man most frequently cited in elementary texts and popular articles is eye color. This has the advantage of being a trait with which almost everyone is familiar, but it also has a disadvantage. It is *not* an example of *simple* Mendelian inheritance, as a modicum of observation and a little thought will tell you. Eye color is clearly a graded character, with many possible shades of color as well as innumerable patterns. That it is genetically determined is clear from the striking resemblances in color and pattern between monozygous twins. The color is determined by the amount and distribution of melanin in the iris. Albinos have none at all, so the iris appears red because it transmits light reflected from the fundus. "True blue" eyes have virtually no pigment in the anterior part of the iris, but some in the posterior layers, and darker colors have progressively more melanin (yellow or brown) present. The structure of the iris will also modify the shade.

In general the genes for the darker colors tend to be dominant to those for the lighter, but the situation is complex: A child with eyes darker than those of both parents is not necessarily cause for divorce! Remember that the iris may darken considerably for some months, or even years, after birth.

HAIR COLOR

The innumerable shades of hair color also attest to complex inheritance, as well as a considerable amount of environmental modification, at least in some populations. Again, the various shades of blonde through black are determined by the concentration and type of melanin, and the genes causing the darker colors tend to dominate those for the lighter ones.

Red hair results from another pigment, which appears to be under the control of a separate gene locus. The gene for presence of the red pigment is recessive to its "not-red" allele, but of course the difference between red and not-red is visible only if the hair is fair. That is, the dark-hair genes are epistatic to the red hair locus.

HAIR FORM

The form or texture of the hair depends on its cross-sectional shape, which is round in straight hair and elliptical in curly hair. A case has been made for a single locus, with curly hair resulting from homozygosity for one allele, straight hair from homozygosity for the other allele, and the heterozygote having wavy hair, but the situation is hardly as simple as that, as again there are many degrees of waviness. Kinky hair in Caucasians shows dominant inheritance, and the straight hair of Orientals is also said to be dominant, but there is a lack of critical data.

BALDNESS

Hair loss in older age presumably is determined multifactorially. Pattern baldness, with onset before about thirty years of age, is one of the few common traits that seem to fit Mendelian expectations. It is caused by an autosomal gene that expresses itself in the heterozygote in males but not in females. Presumably, androgen makes the difference, and its lack may also prevent expression of the gene in homozygous females; otherwise, there should be more pattern-bald women than there appear to be.

SKIN COLOR

Once again, it should be evident that skin color is multifactorially determined, since there is a continuous range of shades from "black" to "white." Davenport's original proposal that the Negro-Caucasian skin color difference is due to two independent loci, each showing intermediate dominance, is an oversimplification. Gate's scheme, involving three loci contributing different amounts of melanin (dark, beige, and dark brown), allows for 18 possible shades and fits observed family patterns reasonably well. Either scheme implies that a "dark" person and a "white" mate cannot have a baby much darker than the dark parent, thus disposing of the myth that Negro ancestry on only one side of the family can result in a "black" baby even though both parents are light-skinned.

Little information is available on the genetic control of skin color in the "red-skinned" and "yellow-skinned" peoples.

ATTACHED EAR LOBES

Most ear lobes extend below the point of attachment of the ear, but some merge with the facial skin along the anterior border, making it difficult to wear earrings. The attached lobe is said to be recessively inherited, but in some people it is difficult to decide whether the lobe is attached or not.

EAR PITS

Small pits in the skin of the ear lobe, as if it had been not quite pierced for earrings, or in the skin just anterior to the attachment of the ear are said to show recessive inheritance.

TONGUE ROLLING

The ability to roll the tongue into a trough, or even tube, is said to be dominant to the inability, but there are exceptions—e.g., occasional discordant monozygotic twins.

HANDEDNESS

Left-handedness is certainly familial, but how much of the tendency to resemble parents is cultural is not at all clear. In one study, the frequency of left-handedness was about 6% when both parents were right-handed, 17% when one parent was left-handed, and 50% when both parents were left-handed. This can be made to fit a single-locus scheme if the right- or left-handedness of heterozygotes is postulated as depending on subtle environmental variations.

HAND CLASPING

When you fold your hands, which thumb is on top? This is a sharply determined characteristic, with about 50% of Caucasians preferring one hand to be uppermost, and 50% the other. It does not seem to be related to handedness and has only a slight tendency to be familial, though there are racial variations in frequency. It is curious that this very discrete difference does not seem to have a simple genetic basis.

"HITCH-HIKERS' THUMB"

The ability to extend the terminal phalanx of the thumb more than 30° from the axis of the second phalanx is said to be recessive, but a number of people fall close to this value and are therefore hard to classify.

DENTAL ANOMALIES

Inherited variations in tooth morphology are numerous and cannot be reviewed adequately here. One of the most striking is the dominant gene that causes peg-shaped or missing lateral incisors.

WEBBED TOES

Partial webbing of varying degrees is an anomaly so frequent that it may be included among the normal variations. Autosomal dominant inheritance is the usual pattern, though in some families it appears only in females.

NORMAL PHYSIOLOGICAL VARIATIONS

In addition to "normal" morphological variants, a number of interesting physiological variants have been identified. We will exclude the biochemical polymorphisms here.

PTC Taste Threshold

The ability to taste phenylthiocarbamide, or related goitrogenic chemicals with the N-C-S group, shows marked variation between individuals. The majority of people can taste very weak concentrations of the compound (tasters), but (in Caucasians) about one out of three people can taste only much higher concentrations (nontasters). This striking physiological difference is determined by a single locus, the nontaster allele being recessive. It is not related to taste acuity in general. If taste thresholds are carefully measured, a few individuals fall into the intermediate range, but if due allowance is made for general taste sensitivity, quite good discrimination can be achieved, and the heterozygotes can be shown to have somewhat lower taste thresholds for PTC than the homozygous tasters.[1] As with other polymorphisms, there are wide variations in gene frequency in different populations, and there is some evidence to suggest that the nontaster genotype predisposes to the development of toxic goiter.

Ear Wax

Almost all Caucasians and Negroes have brown, wet, sticky ear wax, but in the Japanese the common type is gray, dry, and flaky. The dry type is also frequent in American Indians and Eskimos and appears to be recessively inherited.

Color-blindness

Lack of the chlorolabe pigment in the retinal cone cells results in inability to discriminate green colors, or *deuteranopia.* The responsible mutant gene is on the X chromosome and is carried by about 5% of Caucasian males. A second allele, with a frequency of about 1.5%, causes a partial defect in green discrimination or *deuteranomaly.*

Similarly, a lack of the erythrolabe pigment, necessary for discrimination in the red end of the spectrum, results in *protanopia* (1% of males), and a partial defect results in *protanomaly* (1%). This gene is also X-linked and appears to be quite close to the deuteranopia locus.

Beetroot Urine

An autosomal recessive gene results in the appearance of red pigment in the urine after eating beets.

CONCLUSION

It must be admitted that much of the data on normal variations cited above is uncritical and should not be taken too seriously. Part of the difficulty is the quantitative nature of many of these traits, and the situation may improve as specific components of the total variation are identified.

SUMMARY

The inheritance of physical features and normal variation is a subject of general interest attended by a considerable amount of misconception. A number of physical features are briefly discussed.

REFERENCES

1. Kalmus, H.: Improvements in the classification of taster genotypes. Ann. Hum. Genet. 22:200, 1958.
2. Scheinfeld, A.: Your Heredity and Environment. Philadelphia, Lippincott, 1964.

Chapter 8

THE FREQUENCIES OF GENES IN POPULATIONS

Man is blessed with numerous advantageous genes and plagued by deleterious ones. What determines their proportions? The question is important for several reasons. Gene frequencies are being changed by the effects of radiation and other environmental mutagens. Are these changes large enough to worry about? Gene frequencies are also changing because of marked changes in population structure resulting from the widespread use of contraceptives, for instance, and many other factors influencing the birth and death rates of various social and ethnic groups. Are our tax structures dysgenic insofar as exemptions increase with family size? Medical advances have improved the fertility of patients with many hereditary diseases. Will we thereby become a species of malformed morons? Finally, the ability to estimate the frequencies of heterozygotes for genes causing recessively inherited diseases may be useful to the genetic counselor.

In this context, we must stop thinking of genes segregating in families and consider the population as a pool of genes, from which any individual draws two for each locus.

THE HARDY-WEINBERG EQUILIBRIUM

Consider a particular locus "D," with two alleles D and d, in a population in which, for simplicity's sake, we will assume there is no mutation and no selection. Suppose that the dd genotype causes a recessively inherited disease. Assume also that the frequency of the d allele is 1%, and that of the D allele is 99%. If mating is at random (except with respect to sex, of course), each individual can be considered as drawing two of the "D" locus genes (either D or d), one from the father and one from the mother, and will have one of three possible genotypes—DD, Dd, or dd.

What is the probability that the individual will draw two d alleles and have the disease? There is a 1% chance that he will draw a d allele from his mother, and also a 1% chance that he will draw one from his father, so the probability that he will do both and be homozygous dd is

$$1/100 \times 1/100 = 1/10{,}000.$$

Note that this is the frequency of the disease, which can be measured. Thus we have developed an important rule:

> In a population in equilibrium the frequency of a disease caused by an autosomal recessive gene is the square of the frequency of the recessive gene.

In practice, we usually proceed in the other direction—that is, we measure the frequency of the disease and take the square root of this to estimate the gene frequency. Thus we may state the rule conversely:

> The frequency of a gene is the square root of the frequency of the homozygote for that gene.

Similarly, we may deduce the frequency of heterozygotes—a question that sometimes comes up in genetic counseling. An individual may draw a d allele from the mother (1/100) and a D allele from the father (99/100), so the probability that he will do both is 1/100 × 99/100, or 99/10,000. But it can also happen that he draws a d allele from his father and a D allele from his mother, and the probability that he does both is again 1/100 × 99/100. Since these are alternative possibilities, the total probability of the individual being heterozygous (Dd) is found by adding the two alternative probabilities and is thus 2 × 1/100 × 99/100, or 198/10,000. (For convenience, this may be rounded off to 200/10,000, or 1/50). Our second rule will therefore be

> The frequency of the heterozygote for two alleles is the frequency of one allele multiplied by the frequency of the other allele, times two.

These principles were first developed, independently, by an English mathematician, Hardy, and a German ophthalmologist, Weinberg, shortly after the rediscovery of Mendelism, and they are known as the *Hardy-Weinberg law.* In algebraic terms, this law states that, in a population in equilibrium, if a genetic locus has alleles D and d, with frequencies p and q, *the frequencies of the genotypes* DD, Dd, *and* dd *will be* p^2, $2pq$, *and* q^2, *respectively.*

The significance of this relationship in genetic counseling was illustrated in Chapter 5, page 73.

FACTORS ALTERING THE FREQUENCIES OF GENES

Mutation

A mutation may be defined as a change in the genetic constitution from one stable state to another and, strictly speaking, can be either a chromosomal alteration or a point mutation—that is, a change in the DNA from one nucleotide to another, resulting in a change in the messenger RNA from the locus involved, and a corresponding amino acid substitution in the polypeptide chain for which the gene codes. It is the latter sense that is commonly used in population genetics.

Mutations are beneficial in the sense that they provide genetic variation, without which evolution could not take place. On the other hand, most mutations with overt effects are harmful, since they are random changes in a system that has already incorporated most of the possible improvements. Throwing a monkeywrench into a running motor would hardly ever improve its function.

Selection

A mutation that is harmful, through causing a deformity or disease or otherwise impairing function and thus reducing fertility, will have less chance of being passed on to the next generation than its normal allele. In other words, it will be selected against and will have a lower frequency than the normal allele.

Selection can be expressed mathematically as the probability of the mutant gene being passed on to the next generation, relative to that of the normal allele. If this probability is low, there is strong negative selection. The converse of this is "fitness."

The stronger the selection against a genotype, the less "fit" it is.

The Balance between Mutation and Selection

The more harmful a mutation is, the stronger the selection against it, and the less frequent the gene will be. On the other hand, the higher the mutation rate, the more frequent it will be. Thus *the frequency of any given allele reflects a balance between the rate at which alleles of this kind are being removed from the population by selection and the rate at which new ones are being created by mutation.*

Consider a locus A at which a mutation to an allele A^L occurs in one of every 100,000 gametes that contribute to the next generation. Suppose that A^L is dominant and causes a disease that causes death before puberty, or produces sterility. Thus the gene would not be passed on to the next generation—a fitness of 0. What will be the frequency of the disease? Since 100,000 gametes give rise to 50,000 people, if 1/100,000 gametes carries the mutant gene, 1/50,000 people will have the disease. Thus,

For a dominant lethal gene the disease frequency will be twice the mutation rate—or in algebraic terms

$$x = 2m,$$

where x is the frequency of the disease and m is the mutation rate.

Suppose circumstances now changed so that the allele had a *fitness* of 0.5—that is, the mutant allele had half as much chance of contributing to the next generation as the normal allele. Since selection would remove fewer genes than before, but mutation would still be providing new ones, the frequency of the allele would increase, and the disease frequency would increase (Fig. 8-1). When the frequency of A^L reached 2 per 100,000 genes, there would be 4 per 100,000 diseased individuals. Since only half the mutant genes would be transmitted to the next generation, the other half, or 1 per 100,000, would be lost. Thus selection would remove 1 per 100,000 A^L mutant genes per generation, and mutation would create 1 per 100,000 new mutant alleles. A new equilibrium would have been reached at a higher frequency of the mutant gene, where the loss of A^L alleles by selection was balanced by the input of new A^L al-

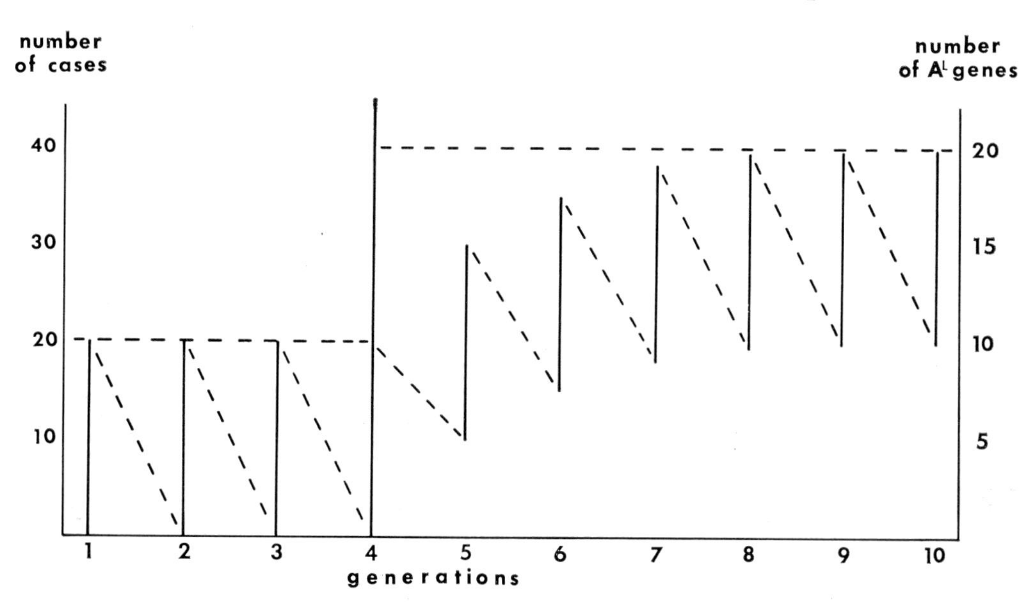

Figure 8-1. Diagram illustrating the relation of gene frequency to mutation and selection, assuming a population of 1×10^6 and a mutation rate of 1 per 10^5 gametes per generation.

leles through mutation. Algebraically the relationship is

$$x = 2m/(1-f),$$

where f is the fitness of the mutant allele, measured as the proportion of mutant to normal alleles that are transmitted to the next generation.

For recessive genes, in which selection acts only on the homozygote, each death due to the mutant gene removes two mutant alleles, rather than one, and the equation is

$$x = m/(1-f).$$

Heterozygote Advantage

The above discussion has assumed that in the case of a dominant mutation the mutant allele is so rare that the homozygotes can be ignored, and that in the case of recessive mutations the mutant gene does not affect the fitness of the heterozygote. It sometimes happens, however, that the heterozygote is fitter than either homozygote. The best known example is sickle-cell anemia.

The gene for sickle-cell disease, beta^s, produces a serious and often lethal disease in the homozygote and only mild symptoms, if any, in the heterozygote (Chapter 6, p. 90). However, the heterozygous individual is more resistant to falciparum malaria than the normal homozygote, so that in countries where falciparum malaria is endemic, the beta^s gene in heterozygous individuals has a better chance of being transmitted to the next generation than the beta^A gene. This provides a mechanism for increasing the frequency of the mutant alleles and explains why the sickle-cell gene is so frequent in certain races of African origin.

Heterozygote advantage is therefore a means of increasing the frequency of a given gene other than by mutation. It may explain why certain diseases, such as cystic fibrosis of the pancreas or Tay-Sachs disease (in Ashkenazi Jews), are so frequent. Cystic fibrosis of the pancreas may involve 1 in 3,600 births or more in some populations, and since fitness is close to 0 it would require a mutation rate of about 1 in 3,600 to maintain the gene frequency if selection occurred only in the homozygote. This is higher than any known mutation rate. But a fairly small heterozygote advantage—too small to detect by the usual family studies—would be enough to maintain this high a frequency without involving mutation at all. For instance, it can be calculated that the high frequency of cystic fibrosis of the pancreas could be maintained if the heterozygote for the mutant gene had a fitness about 1.7% greater than the normal homozygote.

There are, however, other explanations for disease frequencies higher than expected on the basis of a balance between mutation and selection. In particular, there are genetic heterogeneity and segregation ratio advantage.

Genetic Heterogeneity

When the same clinical disease can be caused by mutants at each of several different loci, there is genetic heterogeneity. The more refined our methods of analysis become, the more examples are found. Cystic fibrosis, for instance, can be caused by homozygous mutant alleles at each of at least three different loci. Twenty such loci, each with a mutation rate of 1/60,000, would maintain a disease frequency of 1 in 3,000.

Segregation Ratio Advantage

If a mutant allele altered the segregation ratio, by influencing meiosis or by selective gamete survival, for instance, so that the mutant allele was more likely to be passed on to the next generation than its normal partner, this would tend to increase the frequency of the mutant. Examples are known in lower organisms—for instance, the T locus in the mouse—but so far no examples in man have been recognized where the mutant allele has a segregation ratio advantage.

Genetic Drift

Finally, variations in gene frequency can result from genetic drift, which refers to random fluctuation in gene frequency resulting from small sample size. To take a ridiculous example, suppose that Adam was homozygous for the blood group A gene, that Eve was heterozygous AB, and that all the descendants arose from four of their children. Each child would have an equal chance of being AA or AB. There would be $(\frac{1}{2})^4$ or 1 chance in 16 that all children would be homozygous AA. Thus by chance the B gene could have been lost altogether. In populations that arose from small groups of ancestors, quite wide differences in gene frequency may arise simply from this kind of random variation. The high frequency of Ellis-van Creveld syndrome in the Amish isolate is a well-documented example. (See also the section below on "Our changing gene frequencies.")

MEASURING MUTATION RATES

In 1927 the great geneticist H. J. Muller, working with Drosophila, showed for the first time that an environmental factor, radiation, could cause mutation. Since then, a variety of physical and chemical agents have been shown to be mutagenic in a variety of organisms. There is no reason to suppose that man is immune to environmental mutagens, but the difficulty of measuring precisely the spontaneous mutation rate in man has precluded any direct demonstration of mutagenicity in man. How are human mutation rates measured?

The Direct Method

The most obvious method is a direct count. In the case of a disease that shows dominant inheritance, any case that had unaffected parents would presumably represent a fresh mutation. Thus the mutation rate could be estimated directly from the incidence of sporadic cases. This approach assumes that there is no illegitimacy, that there is full penetrance, that there are no phenocopies, and that all cases of the disease represent mutations at the same locus. It is recognized that these assumptions are not entirely valid. For instance, the estimates for conditions such as tuberose sclerosis may be biased upward by reduced penetrance (since more cases will be counted as mutations than there really are), and the estimates for achondroplasia may be biased upward by the inclusion of cases of recessively inherited chondrodystrophy erroneously diagnosed as achondroplasia. At best, the method will permit only a very rough estimate of the mutation rate. Estimates of rates for a number of dominantly inherited diseases more or less appropriate for this method range from 3 to 68×10^{-6} per locus per generation.

The Indirect Approach

Another approach, and the only one possible for recessive mutations in man, makes use of the assumption that, in a population in equilibrium, the input of new mutant genes by mutation is balanced by the loss of genes through selection. As we have seen, the mutation rate for recessive genes is related to disease frequency and fitness according to the formula

$$x = m/(1-f), \text{ or } m = (1-f)x.$$

The disease frequency and fitness can be measured, and the mutation rate calculated. Thus for a mutant that is lethal in the homozygote, the mutation rate should be equal to the disease frequency. This method also assumes full penetrance, no phenocopies, and no genetic heterogeneity, and further assumes no effects of the gene on fitness in the heterozygote. The method can also be used for X-linked recessive diseases, using the formula

$$m = (1-f)x/3.$$

Estimates for X-linked recessive and autosomal recessive mutants range from 11 to 95 $\times$ 10^{-6} per locus per generation. These are reasonably consistent with those for dominant mutations, which is somewhat reassuring. However, it should be pointed out that the diseases selected for study are those of sufficient frequency to permit reasonably accurate measurements of frequency, and so these figures probably represent the higher ranges of the spectrum of mutation rates.

THE GENETIC LOAD

The burden of disease and death that is created by the effects of our deleterious genes is termed the genetic load. A great deal of effort, stimulated largely by concern over the harmful genetic effects of radiation, has gone into estimating the size and nature of this load.

Two concepts relevant to a discussion of the genetic load are "genetic death" and "lethal equivalent."

Genetic Death

This phrase refers to the fact that a gene that impairs a person's ability to reproduce will eventually fail to be transmitted to the next generation, and thus cease to exist, or suffer genetic death. A dominant lethal "dies" in the generation in which it arose. A dominant gene that causes a 20% reduction in fitness may fail to survive after one generation or many, but on the average will survive for five generations. Furthermore, there is evidence from Drosophila and other organisms that a recessive gene causing severe disease in the homozygote may have a small detrimental effect in the heterozygote. If a gene is rare, it is far more likely to exist in the heterozygous than the homozygous state. Thus small deleterious effects in the heterozygote may actually cause more genetic deaths than a major effect in the homozygote. The main aim of this discussion is to point out that the harmful effects of a mutation may be spread over many individuals and many generations and to show how difficult it is to predict the extent of mutational damage.

Lethal Equivalent

To estimate the mutational load by attempting to count the number of mutations at every locus and measure their effects would be an impossible task. The concept of lethal equivalents was developed in an attempt to measure the total impact of recessive mutant genes on mortality, rather than trying to distinguish them individually. A lethal equivalent is a gene or group of genes that, when homozygous, would bring about the death of an individual. Thus one lethal equivalent could be one recessive lethal mutant, or two recessive genes each of which had a 50% chance of causing death, or 10 recessives each of which had a 10% chance of causing death, and so on. Estimates based on the effect of inbreeding on mortality have suggested that the average number of lethal equivalents in man may be about 3.

Note that this is not the same as saying that each individual carries an average of three genes causing lethal recessive diseases. Some genetic advice for cousins contemplating marriage has unfortunately been based on this misinterpretation. If this were so, the frequency of recognizable recessively inherited diseases in the offspring of cousin marriages would be about 10%, but in fact the frequency seems to be much lower than this. The rest of the lethality attributable to increased homozygosity in offspring of cousin marriages is made up by small increases in abortion, stillbirth, and mortality from various common diseases.

The Mutational Load

We have already seen that there are two major factors tending to maintain the frequencies of deleterious genes in the population—mutation and heterozygote advantage. If there were no mutation, there would be less disease and death. The excess of disease and death due to mutation, above the level expected if there were no mutation, is termed the *mutational load.*

The Segregational Load

On the other hand, some deleterious genes are maintained in the population because they are at an advantage in the heterozygote, but they cause disease and death when they segregate into homozygotes. The amount of disease and death due to this kind of gene, in excess of what it would be if there were no heterozygote advantage, is called the *segregational load.*

There has been much discussion, and no little argument, about the relative importance of the two kinds of genetic load. If the segregational load is the major one, the importance of radiation as a genetic hazard is correspondingly less. Unfortunately, the question is still not settled, but it would seem that both components of the genetic load are appreciable.

THE MUTAGENIC EFFECTS OF IONIZING RADIATION

The concern over the genetic effects of ionizing radiation aroused by the advent of the Atomic Age has stimulated a great deal of research on the subject. We have seen that the measurement of mutation rates in man is so subject to error that it would be impossible to detect an increase of the size one might expect from observed exposures to radiation. Almost all our information comes by extrapolation from data on other animals.

The Dose-Respose Relationship

It is generally assumed that the relation of dose to mutation rate is linear in the low-dose range and that there is no threshold dose, since one "hit" is enough to cause a mutation. However, the picture is complicated by the following features:

1. Different loci may have different mutation rates, and the differences may not be the same for induced as for spontaneous rates.
2. Dose-response relationships vary from species to species, so extrapolations from lower organisms to man require caution.
3. Both spontaneous and induced mutation rates vary with sex and the stage of gametogenesis.
4. The frequency of induced mutation varies with the dose rate. Low dose rates produce much lower mutation frequencies than the same dose at high dose rates.

The Doubling Dose, for Point Mutations

Attempts to predict the effects of exposure to ionizing radiation in man involve estimating the "doubling dose"—that is, the dose of radiation that will increase the mutation rate to double the spontaneous rate, assuming a linear dose-response curve. A permanent doubling of the mutation rate would eventually double the frequency of diseases normally maintained in the population by spontaneous mutation. For dominant lethals, the effect would be immediate; for dominant nonlethal mutants and recessives, the final effect would be spread over many generations. For genes maintained by heterozygote advantage, the effect would be negligible.

Original estimates of the doubling dose suggested that the most probable value for man was about 30 rad per generation. In the light of more recent information, a value of 120 rad is more likely, at least for chronic, low-intensity irradiation of male mice. It is somewhat less for females.

The genetically significant (gonadal) exposure of human populations to background and cosmic radiation varies widely, with an average of about 3 rem per generation (a rem is roughly equivalent to a rad). Medical uses of radiation add another 0.5-5 rem. Atomic testing may have contributed about 0.05 rem. Thus the contribution of medical practice to our mutation rate is not negligible. By calculations too extensive to describe here it has been estimated, for instance, that an exposure of 5 rad to a population of 100 million people would cause about 40,000 embryonic and neonatal deaths, 16,000 infant and childhood deaths, 8,000 gross mental or physical defects, and 100 cases of achondroplasia in the first generation.[3] The corresponding figure for cases appearing in subsequent generations would be 760,000, 384,000, 72,000, and 20. Obviously, every effort should be made to reduce exposure to irradiation to the absolute minimum compatible with good medical care.

Almost nothing is known of the other effects of radiation on the genetic material. At least two studies have suggested that a history of diagnostic or therapeutic pelvic irradiation is found more frequently in the parents of children with chromosomal aberrations than in parents of control children. There is no doubt that ionizing radiation causes chromosomal breaks and rearrangements that may persist in the individual for many years. It also increases the frequency of leukemia and other forms of neoplasia. Prenatal diagnostic irradiation increases the risk of the unborn child to develop leukemia or cancer postnatally. It seems likely that the chromosomal breaks are causally related to the neoplasia, but this remains to be clarified. In any case, these findings are further good reasons for avoiding irradiation as much as possible.

OTHER ENVIRONMENTAL MUTAGENS

Many environmental agents, including a number of drugs, viruses, and food additives, cause chromosomal breaks and rearrangements, particularly *in vitro*. Many of them are mutagenic in lower animals. However, the significance of these findings with respect to genetic damage in man is difficult to evaluate. Although it seems reasonable to suppose that agents that break chromosomes may also be mutagenic, and this has been demonstrated in a number of cases in lower organisms, we have little idea of the quantitative relationships involved. Certainly, the fact that an agent will break chromosomes *in vitro* does not necessarily mean that it will do so *in vivo*, or that it will be mutagenic. On the other hand, it would be foolhardy to ignore the possibility. Further data, particularly from the field of somatic cell genetics, may help to resolve some of the uncertainty.

OUR CHANGING GENE FREQUENCIES

A world increasingly concerned about the rapid growth of human populations must also be concerned about the quality of its genes. When widespread family planning becomes a necessary fact of life, parents will want their limited number of children to be free of genetic disease. And if there are likely to be changes in the frequency of genetic disorders, public health administrators and others concerned with forecasting the need for health care facilities will want to know ahead of time. What, then, is likely to happen to our gene frequencies?

The frequency of a mutant gene in a population at equilibrium depends on the mutation rate and the strength of selection for or against the gene in the heterozygote and homozygote, as outlined previously. This relationship, and particularly the effect of changes in selection pressure, varies with the mode of inheritance.

At present, most diseases caused by mutant genes are rare, as there has been strong selection against them. Nevertheless, they constitute a considerable burden of disease and death. In one North American pediatric hospital, for instance, they account for about 7% of admissions, and in Great Britain they account for about 11% of pediatric deaths.

Certain genes have a much higher frequency in some populations than in others. There are two possible explanations. In some geographically or ethnically isolated groups (isolates), the present population may have descended from a relatively small group of ancestors. A mutant gene present in the original group may, by chance, either be lost or be transmitted to a relatively high proportion of descendants (genetic drift) and then spread among the descendants, being maintained at a high frequency by lack of outbreeding. In some cases, the mutant gene can be traced to a single ancestor, in which case the genetic drift has been termed the "founder" effect. Tyrosinemia in a French-Canadian isolate and the Ellis-van Creveld syndrome in the Amish (recessive) and the South African type of porphyria variegata (dominant) are examples where the founder has been identified.

There are other examples of an unusually high frequency of a deleterious gene that cannot be accounted for by founder effect, since the populations involved are large and the founder effect depends on small sample variation. Pancreatic cystic fibrosis in Europeans, sickle-cell anemia in West Africans, and beta-thalassemia in Italians and Greeks are well-documented examples. In the case of sickle-cell disease, the high gene frequency resulted from an increased resistance of heterozygotes to falciparum malaria, so that heterozygotes living in a malarial region had a reproductive advantage over those not carrying the gene. Heterozygote advantage is the most reasonable explanation for the other examples too, but the mechanism is not known. Because most mutant genes exist in the heterozygous state, a very small heterozygote advantage will exert a relatively large effect on the gene frequency. For instance, it has been estimated that a recessive lethal disease could be maintained at a frequency of 1 per 1,000 births if heterozygotes had about 3% more children than normal homozygotes. This would be virtually impossible to detect.

In populations of intermediate size it may not be possible to establish whether the unusually high frequency of a deleterious gene results mainly from genetic drift or heterozygote advantage, or from both—for example, Tay-Sachs disease in Ashkenazi Jews and congenital nephrosis in Finnish populations.

EFFECTS OF RELAXED SELECTION

As we have said, in hereditary disease where the mutant gene is not protected by heterozygote advantage, "natural" selection keeps the gene frequency low by preventing affected individuals from contributing their genes, good as well as bad, to the next generation. Medicine is devoted to the opposite task—that is, ameliorating the effects of our mutant genes and thus increasing the probability that these genes will be passed on to the next generation. Thus medical care is dysgenic and, in the absence of countermeasures, will lead to an increase in frequency of genetic diseases. Will this increase be large enough to cause concern?

In the case of a lethal (in the sense of preventing reproduction) recessively inherited disease for which a treatment was found that fully restored fertility, the frequency of the gene would slowly increase at a rate equal to the mutation rate. For example, if the original gene frequency was 0.01 (resulting in a disease frequency of 1 in 10,000) and selection was completely relaxed, it would take 100 generations to double the gene frequency, assuming a mutation rate of 10^{-4}. This would result in a fourfold increase in the disease frequency—to 1 in 2,500.[4] The slowness of the rise may be reassuring, but the more diseases for which successful treatments are found, the greater the cumulative effect would be.

For dominant lethals, completely relaxed selection would again lead to an increase in gene frequency equal to the mutation rate, and this will result in a linear increase in disease frequency. If the lethal trait had a frequency of 1 in 10,000, and selection were completely relaxed, it has been calculated that the disease frequency would double in the first generation and rise to 102 in 10,000 after 100 generations (assuming a mutation rate of 5×10^{-5}).[4] There would also, of course, be an increase in the proportion of familial to sporadic cases, and this has already been observed in the case of retinoblastoma, even though in this case the relaxation of selection is far from complete. Thus the effects of relaxed selection would be much more worrisome for dominant than for recessive traits.

Diseases showing X-linked recessive inheritance have an intermediate position, since the gene is exposed to selection in hemizygous males as well as homozygous females. Within four generations of completely relaxed selection, the disease frequency would double.

For diseases showing a multifactorial etiology, the results of relaxed selection are harder to predict, since environmental factors are involved and we know virtually nothing about the nature of the underlying genetic factors and the selective factors acting upon them. In the case of myelomeningocele, for example, improved treatment is allowing many more individuals to reach the reproductive age. The frequency of the malformation in the offspring of affected children is likely to be about 3%, and an increasing number of these affected children will reproduce. It seems likely that, following completely relaxed selection, the frequency of the disease would increase by about 3 to 5 per cent per generation over the next few generations, assuming that there is no change in the relevant environmental factors.

PREVENTION OF GENETIC DISEASE

A program aimed toward reducing the frequency of harmful mutant genes would be termed negative eugenics (as contrasted to positive eugenics, which is aimed at increasing the frequency of beneficial genes). Ignoring, for the present purpose, the unpleasant connotations of the term "eugenics," let us consider the possible results of such programs, taking the extreme case in each example, and realizing that in practice the theoretical limits are not likely to be met.

For an autosomal dominant gene causing a disease that could be diagnosed before puberty, if all heterozygotes were dissuaded from mating, the frequency of the disease would fall in one generation to twice the mutation rate. Obviously, such a program would have no effect on the frequency of a gene that was already lethal (in the sense that it killed or prevented reproduction of the affected individual) since selection would already be maximal. At the other extreme are diseases such as Huntington's chorea, which usually appear after puberty. Assuming a disease frequency of 1 per 1,000 births, a program of prenatal diagnosis and selective termination would reduce the frequency to 1 per 100,000, assuming a mutation rate of 5 per million. As there is as yet no way of diagnosing the disease much before its onset, the only means of reducing the disease frequency is through genetic counseling, and the effect would depend on how successful such a program was in persuading the offspring of affected individuals not to have children.

For achondroplasia, assuming a fertility 20% that of normal, a program of intrauterine diagnosis and selective termination would reduce the frequency to 80% of its original figure.

In the case of an autosomal recessive gene, it is more difficult to lower the gene frequency, since usually only homozygotes are exposed to selection, and the great majority of the genes are heterozygous. The effects of a counseling program would depend on the distribution of family size. Assuming the family size distribution of the United States, a counseling program that persuaded all parents of an affected child to have no more children would reduce the disease frequency by about 15%. Prenatal diagnosis of homozygotes and selective termination of pregnancy would have a similar effect on disease frequency, but the gene frequency would increase slightly since heterozygous pregnancies would not be terminated.

A population with a high frequency of a mutant gene provides the opportunity to screen for heterozygotes premaritally, in the hope that heterozygotes would avoid marrying other heterozygotes. Such programs already exist for sickle-cell disease in populations of West African descent, for thalassemia in Mediterranean races, and for Tay-Sachs disease in Ashkenazi Jews (where prenatal diagnosis removes the necessity for selective mating). If they are completely successful, they will result in the disappearance of the disease altogether. However, we know virtually nothing about the psychological effects of discovering that one carries a "bad" gene, or the kinds of social pressures that may be brought to bear on an individual so identified. Any such program should be accompanied by a well-designed public education campaign, and the early stages of such programs should include intensive study of their psychological implications.

PREVENTION OF CHROMOSOMAL DISORDERS

Prenatal screening of all pregnancies, with selective termination, would remove a major portion of our load of chromosome disorders, but this would place an impossible burden on our health resources. Screening high-risk populations, however, could be justified in terms of a favorable cost-benefit ratio. For instance, nondisjunctional events are more likely to occur in older mothers—thus the birth frequency of Down syndrome could be reduced by about half by a program of prenatal screening and selective termination, in the 10 per cent of pregnancies occurring in women over 34 years of age. The cost of the program would be substantially less than that of institutional care for this fraction of the trisomic population.

In summary, it seems that the means are available for protecting future generations against the dysgenic effects of relaxed selection. We would add our hope that any program for reducing the frequencies of deleterious genes would be by education and voluntary co-operation, not by any form of coercion.

SUMMARY

Genes may be considered at the level of the individual, the family and the population. The Hardy-Weinberg law ($p^2 + 2pq + q^2 = 1$) defines the frequency of genotypes (AA, Aa, and aa) in a population in equilibrium for alleles A and a with respective frequencies of p and q. Gene frequencies may be altered by mutation, which adds new alleles, and by selection, which reduces the frequency of deleterious genes. Certain mutations that may be disadvantageous or even fatal to the homozygote are beneficial to the heterozygote (heterozygote advantage) and are maintained in the population through this mechanism. Genetic heterogeneity, genetic drift, and possibly segregation ratio advantage may complicate the Hardy-Weinberg equilibrium.

Genetic load is the burden of disease and death produced by deleterious genes. If a mutant gene sufficiently impairs reproductive fitness that it sooner or later prevents an individual from transmitting it to the next generation, "*genetic death*" of that gene occurs. The total impact of recessive mutant genes on mortality may be estimated in terms of *lethal equivalents;* the average number of lethal equivalents in man is estimated as being about 3. The excess of disease and death due to mutation is the *mutational load,* and the excess morbidity and mortality in homozygotes resulting from genes maintained in the population by heterozygote advantage is the *segregational load.* Radiation, viruses, drugs, and other chemicals represent potential mutagens in man, although the evidence for their mutagenicity is derived from lower animals.

Programs for the prevention of genetic and chromosomal disorders require knowledge of genetic mechanisms, appropriate diagnosis, counseling, and public education.

REFERENCES

1. Cavalli-Sforza, L. L., and Bodmer, W. F.: The Genetics of Human Populations. San Francisco, W. H. Freeman, 1971.
2. Levitan, M., and Montagu, A.: Textbook of Human Genetics. London, Oxford University Press, 1971.
3. Newcombe, H. B.: Effects of Radiation on Human Populations. *In* J. de Grouchy, F. J. G. Ebling and I. W. Henderson (ed.), Human Genetics. Amsterdam, Excerpta Medica, 1972.
4. World Health Organization. Genetic Disorders: Prevention, Treatment and Rehabilitation. WHO Tech. Rep. Ser. No. 497, 1972.

Chapter 9

RACE

To introduce the subject of race, let us begin by reviewing some elementary principles of evolution. Within any species of organism there exists virtually infinite variety, much of which is provided by genetic variation. That is, new mutations are continually arising, and selection and other forces act to increase or decrease their frequencies by determining which ones are more or less likely to be transmitted to the next generation. If groups of individuals are separated by some barrier, such as a mountain range, so that they do not interbreed, their genetic compositions will become progressively more different. If the groups are small, they may have different gene frequencies to begin with, which may be perpetuated or magnified in subsequent generations (genetic drift). Furthermore, environmental differences may exist that lead to different selective effects in the two groups. Thus the two populations will have progressively increasing differences in their arrays of gene. In man, divergencies in language and other cultural characteristics will arise even more rapidly.

Furthermore, chromosomal rearrangements will occur, and some may have a selective advantage and thus increase to become the "normal" condition. Over a period of eons, the karyotypes of the two groups will thus become progressively more different. (Turleau *et al.*[11] have recently shown how evolution of man and the great apes from a common ancestor can be traced by their karyotypes, which appear to have undergone a series of centric fusions and translocations.) Eventually, the groups may become so different that they cannot interbreed, and it is supposed that this is the way species originate.

Before this point, the two groups may be different enough that an observer from another planet might recognize the *groups* as different, though he might not be able to distinguish every individual as belonging to one group or the other. Such groups might be termed "races."

ORIGINS OF "RACES"

To illustrate in more detail the probable origins of racial groups, let us suppose that the world became entirely unpopulated, except for, say, New York City. To repopulate the world, groups of several hundred people were taken at random from New York and deposited in various regions where they were left to multiply for several thousand generations with relatively little interbreeding between some groups and more between others; at the end of this time, an anthropologist and geneticist from Mars arrive upon the scene. What will they observe?

To answer this question, let us predict what we would expect to happen. First, there would be differences in gene frequencies between the original groups arising purely by chance—for instance, one group of 200 might happen to lack any individuals with blood group B and another might have five or even ten. These original differences in gene frequency would tend to be perpetuated in the descendant populations; thus we would have differences arising by *genetic drift.* Second, the different environments in which the groups lived would exercise *selection* on the various gene pools. This would cause differences in the frequencies of genes with adaptive value. In most cases, the basis for the adaptive nature of the genes selected for would be too subtle to be readily identified. In some cases, the adaptive value of a gene might even be negative in the homozygote and positive in the heterozygote in some environments (*heterozygote advantage*). Third, the groups with appreciable degrees of interbreeding or common descent would be more similar to one another than would isolated groups. Fourth, the different environments and cultural isolation between groups would lead to differences in dietary and living habits, language, and social organization.

Our anthropologist would therefore notice that, on the average, individuals of one group tended to resemble one another more closely than they resembled members of other groups, though there would usually not be any one character that distinguished all members of one group from all members of another. Furthermore, groups living in similar environments, particularly if they shared a common ancestry, would tend to resemble each other more closely than they resembled groups from different environments. For instance, groups from tropical regions might have higher frequencies of dark-skinned individuals. The anthropologist might wish to classify similar groups together as constituting *ecotypes,* which would separate them from dissimilar groups of groups. Note that he would be able to classify only groups, not each individual, as belonging to a particular ecotype. Since the amount of dissimilarity that would justify classification into separate groups would be a matter of judgment, differing anthropologists would probably come up with different classifications.

The geneticist would note that the frequencies of various marker genes differed from group to group and that the degree of difference tended to be greater for unrelated than for related groups. He would usually not be able to tell whether any particular difference in gene frequency had arisen by selection or genetic drift, since most "normal" variations with obvious adaptive value do not result from single gene differences.

Both the anthropologist and geneticist would notice variations in disease frequencies between different groups, but again it would be difficult to account for the origins of these differences. There would be some rare deleterious recessive genes with unusually high frequencies in a few groups, but with no demonstrable environmental difference to prove heterozygote advantage. Common, familial (multifactorial) diseases would also have different frequencies in different groups, but with etiologies so complex that it would usually be difficult to tell whether the differences resulted from selection or nongenetic cultural factors.

If intelligence tests were designed by members of one group and given to members of other groups, there would be differences in average performance between groups. No doubt, some members of some groups would proclaim their group superior to others on the basis of such results, but there would be no little discussion, not to say controversy, about whether these differences in performance reflected innate (i.e., genetic) or cultural (environmental) differences between groups, or ecotypes.

So it is with races. We can distinguish groups of people from one another by differences in physical appearance, arrays of genes, and cultural characteristics, although by none of these criteria are the distinctions always complete. Certain differences between groups appear to be adaptations to environmental differences—for instance, the preponderance of dark-skinned peoples in tropical areas and of light-skinned groups in Northern climes. Frequencies of specific genes tend to be more similar in geographically close than distant groups. There are, indeed, certain diseases that occur with inordinately high frequencies in certain groups—for instance, Tay-Sachs disease in Ashkenazi Jews—and there are differences in behavioral characteristics, such as performance on intelligence tests, between certain groups.[8] By now, we hope we have illustrated the basis for the following definition of race, proposed by Stern.[10]

> *Race.* A geographically or culturally more or less isolated division of mankind whose corporate genic content (gene pool) differs from that of all other similar isolates.

How different two groups should be in order to be considered separate races is a matter of individual judgment. Whether certain differences (particularly behavioral) are cultural or genetic is often difficult or impossible to determine, but that does not seem to prevent certain people from dogmatically adopting one side or the other and defending it by bitter argument or even action. The tendency to assume that one's own racial characteristics are superior, as well as inborn, is, we hope, not itself inborn, but it seems to be prevalent. To offset this, we should try to remember that although geneticists and anthropologists are forced to emphasize differences, which are their stock-in-trade, the differences between individual men, or races of man, are far outnumbered by the similarities that make us all human beings, with all the rights, privileges, and respect pertaining thereto.

CLASSIFICATIONS OF RACES

We have said that separation of groups of people into different races on the basis of dissimilarities in arrays of features is a matter of judgment, and this is borne out by the variety of classifications in use by anthropologists and geneticists. One very rough grouping divides man into Caucasoids, Negroids, Mongoloids, and Australoids. Another possible classification, based largely on differences in blood group frequencies (see Chapter 18), has been proposed by Boyd[1]:

A. *European Groups*
   1. Early European. Represented by the Basques and possibly North African Berbers. High incidence of Rh-negative and no B.
   2. *Lapps.* High N, Fy^a, A2; low B.
   3. *Northwest Europeans.* Fairly high A, A2, Rh-negative; low B.
   4. *Eastern and Central Europeans.* B higher than 0.1. Rh-negative somewhat less than in Northwest Europeans.
   5. *Mediterraneans.* Southern Europeans, Middle East, much of North Africa. Higher B, lower Rh-negative than Northwest Europeans; R° (cDe) increased, suggesting some relationship with Africans.

B. *African Group*
   6. *African.* Very high R°; high Fy, P, A2, B. Includes populations of Black Africa, many local subgroups.

C. *Asian Groups*
   7. *Asian Race.* High A and B; low A2 and Rh-negative.
   8. *Indo-Dravidians.* Includes various subgroups of the India-Pakistani subcontinent. Intermediate in many respects between Europeans and Asiatics. (Many Indian tribes do not fit well into the classification and vary widely among themselves.)

D. *American Group*

9. *American Indian Race.* Absent A2, B, Rh-negative. Very high M, R^2. Some tribes have high Diego (nearly absent in Europeans). High secretor and "taster." Includes Eskimos.

E. *Pacific Group*

10. *Indonesian Race.* High R^1, absent A2, fairly high A, B, M.
11. *Melanesian Race.* A and B somewhat higher than Indonesians. A2 absent, low M. High R^1. Micronesians are similar.
12. *Polynesians.* A1 high, B low. M somewhat high. R^1 and R^2 about equal, low R°.
13. *Australian* (aboriginal). High A, no B, S, or Rh-negative; low M; R^1 lower than in Pacific races.

Thus all of these geographically distinguishable races also have distinct blood-group gene profiles. One can also note gradients in the frequencies of various genes across geographic areas.

This is just one suggestion for dividing groups of people into separate "races." Other experts prefer finer subdivisions: One group recognizes nine major geographical races, with over 30 local races within them.[4] These are well illustrated by Goldsby.[5]

Table 9-1 demonstrates some racial variations in a selection of immunologic, biochemical, and physical traits.

DIAGNOSIS OF RACIAL IDENTITY

Although we said that our anthropologist might have difficulty in assigning any one individual to a particular group (race) on the basis of his morphological features, the geneticist might be able to estimate the *probability* of his belonging to one race or another by using a number of genetic markers that differ in frequency between two races.

To take a somewhat unrealistic example, suppose that we have determined from bloodstains at the scene of a crime that the criminal is blood group MN,kk. Two men are suspected, one of race X and the other of race Y. We know that the frequency of MN is .48 in race X and .32

TABLE 9-1. Frequencies of Various Characteristics in Some Ethnic Groups[a]

	Proportion of		*Proportion of high BAIB[b]*	*Haptoglobin types*				*Proportion of PTC*	*Fingerprint patterns*	
Population	*Diego+*	*Diego−*	*secretors*	0	11	21	22	*tasters*	*% arch*	*% whorls*
Australian aborigines	0	1.00		0	0.12	0.68	0.20	0.27		
Chinese	0.05	0.95						0.93	0.03	0.50
English	0	1.00	0.09	0.03	0.10	0.55	0.32	0.69	0.07	0.25
Eskimos	0	1.00	0.23					0.59		
North American Indians	0.02	0.98	0.59					0.97	0.05	0.50
South American Indians	0.86	0.15								
Japanese	0.07	0.93						0.91		
North American Negroes			0.29	0.04	0.26	0.48	0.21	0.91		

a Adapted from Lerner.[7]
b Beta-aminoisobutyric acid in the urine.

in race Y. For kk, the respective frequencies are .81 and .49. Then:

The probability of MN,kk in race X is $.48 \times .81 = .3888$

The probability of MN,kk in race Y is $.32 \times .49 = .1568$

$.3888 + .1568 = .5456$

The relative probability that the criminal is race X is $\frac{.3888}{.5456} = .7126$

The relative probability that the criminal is race Y is $\frac{.1568}{.5456} = .2874$

In other words, on the basis of this evidence alone, there is a 71% chance that the criminal belongs to race X.

Note that if one of the blood groups is rare in one race and more common in the other, the probability of assignment to a particular race can be quite high. For instance, suppose the criminal is group Di^a Di^a, which has a frequency of .0036 in race X and .0001 in race Y. On the basis of this blood group alone, then, there would be 35 chances out of 36, or a probability of .973, of his belonging to race X. Taking all three blood groups into account:

Probability of MN,kk,Di^aDi^a in race X is $.3888 \times .0036 = .00140$

Probability of MN,kk,Di^aDi^a in race Y is $.1568 \times .0001 = .00001$

$.00140 + .00001 = .00141$

Relative probability criminal in race X is $\frac{.00140}{.00141} = .993$

Thus, there is a 99.3% probability that the criminal is of race X.

GENE FLOW BETWEEN "RACES"

In our hypothetical example, we postulated varying degrees of restriction on the amount of interbreeding between racial groups. These restrictions may be geographic or cultural. Geographic barriers are rapidly breaking down as travel increases, but Eskimos still rarely have the opportunity to marry Maoris. There are also strong cultural barriers. In the United States, for instance, the number of matings between whites and nonwhites is only 2.3% of what it would be if mating were random. Nevertheless, there is always some gene flow between races that are not geographically separated.

If one knows that a particular gene differed in frequency in two races before there was any gene flow, then one can estimate the amount of flow from one to the other. For instance, if

P_c is the frequency of a gene, say Fy^a, in a U.S. Caucasian population,

P_a is the frequency in the African ancestors of U.S. blacks,

P_b is the frequency in U.S. blacks, and

M is the percent admixture of genes from whites to blacks, then

$$P_b = M \cdot P_c + (1-M)\ P_a$$

and thus

$$M = \frac{P_b - P_a}{P_c - P_a}.$$

For instance, if Fy^a has a frequency of 0.429 in U.S. whites and 0.020 in Africans, and 0.094 in U.S. blacks, then

$$M = \frac{.094 - .020}{.429 - .020} \text{ or } 18.1\%.$$

Estimates of admixture of U.S. whites and blacks vary, depending on the populations sampled and the genetic markers used, but in general they range from about 5% to 30%.[9]

GENETICS OF "RACIAL" DIFFERENCES

In spite of the large number of studies of offspring resulting from crosses between races, very little is known of the genetic basis of the differences in physical features that occur between races. This is partly because a proper genetic analysis of such differences requires accurate observations not only of the first-generation hybrids but of the offspring of matings between hybrids, and preferably of back-cross offspring as well. In general, the differences appear to involve several gene loci and varying degrees of dominance.

For instance, in crosses between "pure-blooded" Negroes and Caucasians, the kinky hair of the Negroes appears in the offspring, suggesting dominance of one or more genes. In crosses between such hybrids, a range of different hair types appears, so there must be more than one gene difference involved.[10] The straight hair of American Indians appears in the offspring of Indian $\times$ Negroes matings, but a variety of types appears in later generations. On the other hand, the skin of the Negro/Caucasian hybrid (mulatto) is intermediate in pigmentation, but again there is a wide range of skin color in the "F_2" generation.

It has sometimes been claimed that racial mixing (miscegenation) can lead to disharmonious combinations of genes. For example, the offspring of a dachshund and a St. Bernard has a large heavy body that drags on the ground because it is too big for the short legs. Though this argument no doubt appeals to those who oppose miscegenation, it is not supported in man by any convincing evidence known to us. On the other hand, it is perfectly clear that racial intercrossing can give rise to some very fine specimens of humanity indeed, as anyone who has been to Hawaii will testify.

The characteristic that gives rise to most controversy with respect to the genetic basis of racial differences is, of course, intelligence. The fact that so much has been written on the differences between blacks and whites in performance on IQ tests and their possible genetic basis stems partly from the difficulties inherent in the interpretation of the data and partly from the emotional content of the issue. The problem is exceedingly complex and can only be summarized here.

There can be no doubt that

1. Children from lower socioeconomic classes make lower average scores on IQ tests than do children from upper socioeconomic classes.
2. U.S. blacks make lower average scores on IQ tests than do U.S. whites.
3. A much greater proportion of blacks than whites come from the lower socioeconomic groups.
4. It is also clear that intelligence, as measured by IQ tests, has a substantial heritable component but is also influenced by environmental factors. For a recent review of the literature, see Jensen.[6]

It has been argued that the demonstrated racial differences are meaningless since IQ tests are not adequate measures of intellectual potential. Admittedly, IQ tests are not entirely culture-free, but they are still reasonably good predictors of performance. Would you not be concerned if your child had an IQ of 85?

The problem is to determine how much of the difference between the performance of blacks and whites results from the differences in socioeconomic class and how much, if any, represents genotypic differences between the two groups.

How big are the differences? When the results of a large number of studies are averaged, blacks score about 1 standard deviation, or 15 IQ points, lower than whites. In a population composed of equal numbers of blacks and whites, this difference between racial groups would account for only about 23% of the total variance, whereas differences within the groups would account for 77%. Within grossly comparable socioeconomic groups, there is still a difference of about 11 points between blacks and whites, which might seem to suggest that the differences are inborn, rather than cultural, but the counterargument is that blacks and whites of grossly comparable socioeconomic class may still have appreciable differences in their cultural environment.

The differences in mean IQ between the highest and lowest socioeconomic class (at least in England) is about 50 points, much greater than the spread between blacks and whites.[2] Offspring of each parental class show "regression to the mean," as expected if the differences in IQ between classes are genetically determined.

What then, can be said about the nature of the black-white differences? We have pointed out that races, by definition, differ in their arrays of genes, and we have given numerous examples of this for genetic markers that can be individually identified. There is no *a priori* reason to doubt that races may differ also in the arrays of genes influencing intelligence. The questions remaining are how much, in which direction, and in what ways?

There are also reasons to think that environmental factors may account for a major part, if not all, of the difference. For instance, blacks scored higher on IQ tests administered by other blacks than on those administered by whites—a reminder that IQ tests are not culture-free. Maternal malnutrition during pregnancy can probably impair mental development of the offspring. In one study of interracial crosses, the offspring of white mothers had higher IQ scores than the offspring of black mothers, again suggesting the importance of environmental factors in the lower performance of blacks.[13]

Figure 9-1 emphasizes that racial (ethnic) differences in performance on intelligence tests are not limited to those between blacks and whites.[8] It also shows that even when overall IQ's are similar, there are differences in scores on tests that measure more specific mental abilities. Furthermore, in a low socioeconomic environment,

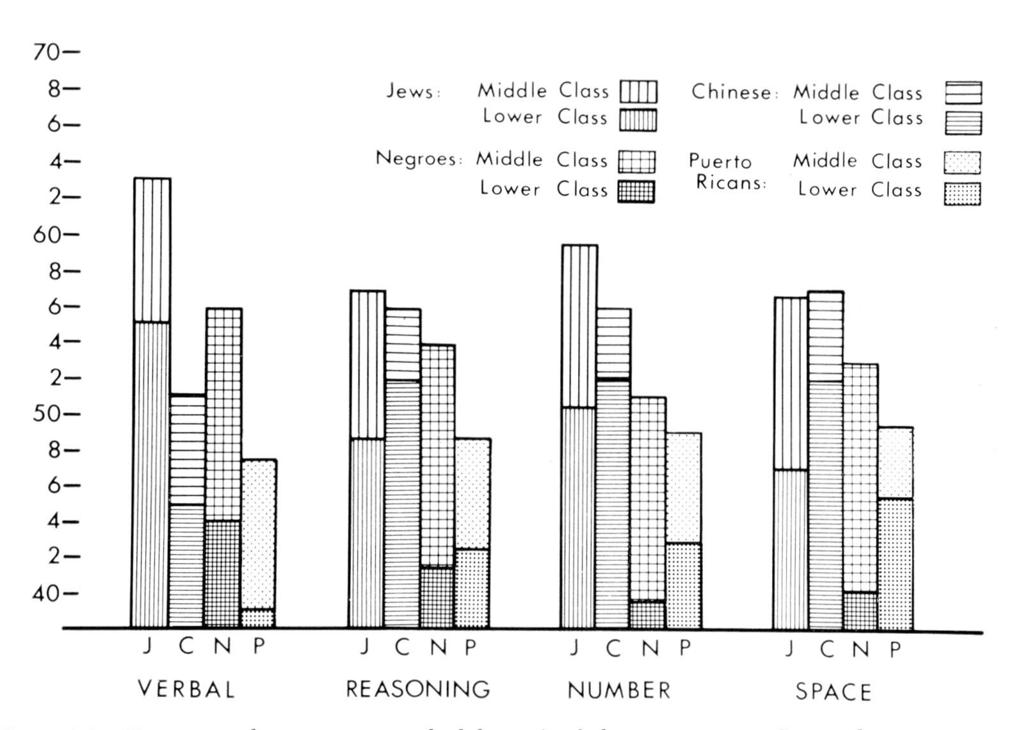

Figure 9-1. Test scores for various mental abilities (verbal, reasoning, number, and space conceptualization) for various cultural groups (Chinese, Jewish, Negro, and Puerto Rican) and socioeconomic levels in New York City school children.[8]

performance is poorer, but the degree of depression is different for different ethnic groups, and for different abilities.

In summary, then, one cannot exclude the possibility that the difference in IQ between blacks and whites has a genetic component, but the available data are inadequate to resolve the question one way or the other.[3,5] Furthermore, it would be extraordinarily difficult to obtain satisfactory data. We believe that the important thing is not whether races differ in the arrays of genes influencing intelligence, but whether human beings share an environment that permits each of us to develop his innate abilities to the maximum, keeping in mind that one cannot predict a man's intelligence, honesty, or wisdom by the color of his skin.

Because the race question causes so much irrational emotionalism, we would like to conclude this chapter with the following excerpts from a report prepared by a group of biologists who met in 1964, under the auspices of UNESCO, to consider, in the light of the latest scientific knowledge available, what could be said about the concept of race.[12]

- All men living today belong to a single species, Homo sapiens, and are derived from a common stock.
- Pure races do not exist in the human species.
- Differences between individuals within a race are often greater than the average differences between races.
- From the biological point of view it is not possible to speak of a general inferiority or superiority of this or that race.
- Human races cannot be compared at all to races of domestic animals.
- It has never been proved that interbreeding between races has biological disadvantages for mankind as a whole.
- The biological consequences of a marriage depend only on the individual genetic make-up of the couple and not on their race.
- No biological justification exists for prohibiting intermarriage between persons of different races.
- There is no national, religious, geographic, linguistic or cultural group which constitutes a race *ipso facto;* the concept of race is purely biological.
- The peoples of the world today appear to possess equal biological potentialities for attaining any civilizational level.
- Concerning the overall intelligence and the capacity for cultural development, there is no justification for the concept of "Inferior" and "Superior" races.
- The biological data stand in open contradiction to the tenets of racism. Racist theories can in no way pretend to have any scientific foundation and the anthropologists should endeavour to prevent the results of their researches from being used in such a biased way that they would serve non-scientific ends.

SUMMARY

Race may be defined as "a geographically or culturally more or less isolated division of mankind whose corporate genic content (gene pool) differs from that of all other similar isolates." Races originate through mechanisms that affect gene frequencies and the consequent genetic makeup of groups of individuals (as discussed in Chapter 8). There are a number of alternative ways used by geneticists and anthropologists to classify races, including arrays of physical features, blood groups, and serum proteins.

The genetic basis of racial differences is still poorly understood. The current controversy regarding genetic differences in intelligence has generated more heat than it has thrown light on the highly complex genetic and environmental variables and their interactions.

A concept of race that acknowledges that men belong to the single species Homo sapiens, that pure races do not exist, and that there is no justification for ideas of "inferior" and "superior" races has biologic validity, as well as being an essential sociologic proposition.

REFERENCES

1. Boyd, W. C.: Modern ideas on race in the light of our knowledge of blood groups and other characteristics with known mode of inheritance, p. 119. *In* C. A. Leone (ed.),

Taxonomic Biochemistry and Serology. New York, Ronald Press, 1964.
2. Burt, C.: The genetic determination of differences in intelligence: A study of monozygotic twins reared together and apart. Br. J. Psychol. 57:137, 1966.
3. Cavalli-Sforza, L. L., and Bodmer, W. F.: The Genetics of Human Populations. San Francisco, W. H. Freeman, 1971.
4. Garn, S. M.: Human Races. Springfield, Ill., Charles C Thomas, 1961.
5. Goldsby, R. A.: Race and Races. New York, Macmillan, 1971.
6. Jensen, A. R.: A two-factor theory of familial mental retardation. *In* Human Genetics. Proceedings of the Fourth International Congress on Human Genetics. Amsterdam, Excerpta Medica, 1972.
7. Lerner, I. M.: Heredity, Evolution and Society. San Francisco, W. H. Freeman, 1968.
8. Lesser, G. S., Fifer, G., and Clark, D. H.: Mental abilities of children from different social class and cultural groups. Monogr. Soc. Res. Child. Dev. 30:1, 1965.
9. Reed, T. E.: Caucasian genes in American Negroes. Science 165:762, 1969.
10. Stern, C.: Principles of Human Genetics. (3rd ed.). San Francisco, W. H. Freeman, 1973.
11. Turleau, C., de Grouchy, J., and Klein, M.: Phylogénie chromosomique de l'homme et des primates hominiens. Ann. Génét. 15:225, 1972.
12. United Nations Educational, Scientific and Cultural Organization. The Race Concept. Paris, 1964.
13. Willerman, L., Naylor, A. F., and Myrianthopoulos, N. C.: Intellectual development of children from interracial matings. Science 170:1329, 1970.

Chapter 10

THE GENETICS OF DEVELOPMENT AND MALDEVELOPMENT

Because the human embryo is to a large extent hidden from the investigator and because man is not subject to controlled matings, little is known of human developmental genetics, and much must be extrapolated from lower organisms. This chapter will touch briefly on what is known of how development is controlled by genes, both normal and abnormal.

Development of an organism has been compared to the building of a ship. The ship starts as a set of plans—specific instructions as to what is to go where and in what order. These are analogous to the organism's genes, which are specific instructions as to how the amino acids are put into sequences of particular polypeptides at specific places and times. The ship's plans are written on pages of blueprints. The organism's blueprints are its chromosomes—twenty-three pages, in duplicate, for man. The building materials of the ship are put together according to the instructions, and similarly the biological substrates are put together to form the developing embryo, according to the genetic instructions. Then the ship is launched, but development of the ship is not finished. Fitting of the ship continues after launching, and development of the human organism after birth. How well the ship—or baby—does on the sea of life will depend on the accuracy of the instructions and integrity of the blueprints, the accuracy with which they are translated into structure, the quality of the building materials, the quality of the shelter in which it is built, and finally, the rigor of the environmental hazards it meets as it sails, drifts, churns, or is towed through the sea of life.

Errors in development resulting in structural defects, or malformations, can occur at several levels. There can be errors in the instructions (mutations), errors in the way the instructions are carried out (translational defects), extra or missing or transposed pages of blueprints (chromosomal aberrations), defect in quantity or quality of building blocks (vitamin deficiencies, antimetabolites), or damage from termites, rust, and rot (environmental teratogens). Finally, the ship can be defective not for any such major reason, but by the interaction of "a lot of little things"—poor workmanship, inferior materials—no one of which would have caused a major defect, but which do so in combination (the multifactorial group, see Chapter 11). This analogy provides a useful way of thinking about the etiologies of various groups of diseases ranging from mental retardation to congenital malformations, in four major categories—diseases resulting from *mutant genes,* from *chromosomal aberrations,* from *environmental pathogens,* and from *multifactorial causes.*

But the analogy breaks down in a number of important ways when one thinks about the developmental process. Developing organisms show the phenomenon of *differentiation*—cells with the same genetic information may turn out quite differently. Second, development involves the process of *induction,* whereby a signal from one tissue or organ induces another tissue to begin developing along a new developmental pathway. Third, there is *morphogenesis,* emergence of formed structures from relatively unformed ones, and the synchronized integration of various tissues into structured organs.

DIFFERENTIATION

The human organism begins as a fertilized egg. Repeated cell divisions render it multicellular, each cell receiving a replica of the original nucleus and thus an identical complement of genes. Yet many hundreds of different cell types occur in the adult organism, migrate, become organs, and interact, virtually all with the same complement of genes. How can the same genome give rise to many different cell types? Much has been written but little is known, at least in higher organisms.[10]

A logical model can be developed from the assumptions that the same genotype will respond differently to cytoplasms of different compositions and that the cytoplasm of the original egg is not homogeneous; there are regional concentrations of yolk granules, mitochondria, and other components, as well as gradients of oxygen, glucose, and other ingredients. If so, then cleavage, which divides the egg cytoplasm into compartments, will result in identical nuclei operating in cytoplasms that differ in the concentration of various components. If the nuclei respond differently to the cytoplasm differences, they will create new ones, and so differentiation will proceed by a process of progressively more specialized nucleocytoplasmic reactions.

Differentiation, then, is the process by which genetically identical cells become functionally different. (*Determination* refers to the point where the progress of a cell line to its differentiated state becomes irreversible.) Differentiation appears to result largely from differential gene activity. All nucleated cells produce certain enzymes necessary for the maintenance of the cell (the "housekeeping enzymes"); a differentiated cell in addition has the enzymes to produce one or more special proteins relevant to its particular function, e.g., a reticulocyte produces hemoglobin, and a lymphocyte does not. Thus the genetic system for making hemoglobin molecules is "switched on" in the reticulocyte but "switched off" in the lymphocyte.

An example of gene activation during development is provided by the isozymes of lactic dehydrogenase (LDH). Isozymes are different forms of the enzyme, similar but not identical in structure and function. In the case of LDH there are five electrophoretically different isozymes. These are made up of varying combinations of two polypeptide chains, alpha and beta. Thus there must be a gene coding for each of these chains. The five isozymes represent varying combinations of the two chains. There may be four alpha, three alpha and one beta, and so on, to four beta chains—five types in all. The proportions of the five isozymes are different in different tissues and in the same tissue at different stages, indicating that the relative activities of the two genes differ from time to time and from place to place. The isozymes vary somewhat in their properties, such as degree of inhibition by lactate, and their varying proportion in different cells presumably has some functional significance. For instance, in striated muscle, lactate resulting from strenuous exercise inhibits the "muscle" type LDH, which weakens the muscle. This would be perilous for heart muscle, which conveniently has more of the isozyme not inhibited by lactate.[12]

Regulation of Transcription

It has been amply demonstrated that regulation of gene activity can occur at the level of transcription—the synthesis of the messenger RNA molecule from the DNA template. The array of messenger RNA molecules produced by one type of cell can be shown, by DNA-hybridization techniques, to differ from the array produced by another cell type. Furthermore, the same tissue may produce different arrays of mRNA's at different stages of development.

Little is known so far about the ways in which the onset and rate of transcription are controlled. It is clear from nuclear transplantation and cell hybridization experiments that the initiation of both DNA replication and transcription may depend on signals (DNA polymerases?) from the cytoplasm. In the amoeba, the use of radioactive tracers has demonstrated a class of proteins that migrate rapidly from nucleus to cytoplasm and from cytoplasm to nucleus that seem likely candidates for the conveyors of information that will initiate or suppress gene activity. Recent evidence shows that an adenine ribonucleotide known as cyclic AMP plays a fundamental role as a "second messenger" in regulating gene activity in many systems. For instance, when a hormone reaches its target cell, it appears to activate an enzyme, adenyl cyclase, in the cell membrane, and this leads to increased levels of cyclic AMP in the cell, and to activation of the appropriate genes. The details are still unclear, particularly in eukaryotes.[14]

The Operon. In bacteria, the activity of a certain group of genes (an *operon*) may be regulated by another gene (the *regulator*) by means of a cytoplasmically transmitted *repressor*, probably a protein, which binds to a locus (the *operator*) at one end of the operon. The RNA polymerase that synthesizes the messenger RNA from the DNA template binds to a site, distal to the operator, called the *promoter*. If the repressor is bound to the operator, the RNA polymerase cannot function. A small molecule in the cytoplasm, such as a metabolic substrate (the *inducer*) may combine with the suppressor and prevent its binding with the operator. Then the RNA polymerase can begin synthesizing messenger, and the genes of the operon are "turned on" when the appropriate substrate appears. A mutation in the promoter or the operator or regulator gene could prevent the operon from being "turned on" when the inducer appears, so the corresponding enzymes would not appear when needed.[12]

No well-substantiated example of an operon is known in mammals, but there are examples of genes that appear to regulate the activity of other genes. The switch from the production of fetal to adult hemoglobin is an example. Normal adult human hemoglobin consists largely of hemoglobin A, comprising two alpha and two beta polypeptide chains ($\alpha_2^A\ \beta_2^A$). Fetal hemoglobin (F) consists of two alpha and two gamma chains ($\alpha_2^A\ \gamma_2^A$). The human fetus makes mostly fetal hemoglobin, but well before birth the production of hemoglobin A in the blood begins to increase, and of F to decrease. This "switch" from hemoglobin F to A continues after birth and is virtually complete by the age of one year.

Thus in the fetal red cell precursor, the gene for gamma chains is active and the gene for beta chains is suppressed, while in the adult the beta chain is being transcribed and the gamma chain gene is inactive. The nature of the factor responsible for the switch is unknown, but the switch appears to be under genetic control. A mutant gene, "high F," closely adjacent to the beta chain, prevents the switch, so that homozygotes for this gene continue to produce hemoglobin F and not A. Heterozygotes produce intermediate amounts of F and A, showing that the high-F gene suppresses only the beta chain gene on the same chromosome, and not that on the homologous chromosome. Thus the normal allele at the high-F locus behaves like an operator. Furthermore, the gene for the delta chain, a component of the minor hemoglobin component A2 ($\alpha_2{}^A$ $\delta_2{}^A$), is closely linked to the beta chain gene, and its activity is also suppressed by the high-F gene. Perhaps in the normal switch-over from hemoglobin F to A, the production of beta or delta messenger suppresses the activity of the gamma chain gene, but so far this is just conjecture.

Learning how to "operate the switch" could have practical benefits. For instance, in a person homozygous for the sickle-cell gene ($\alpha_2{}^A$ $\beta_2{}^s$), if the switch from gamma to beta chain production could be prevented, the patient would continue to make hemoglobin F (which has no beta chains) and would not suffer from the effects of the mutation.

Heterochromatin. Another example of differential gene activation is provided by the heterochromatin—stretches of chromosome that remain tightly contracted, deeply staining, and genetically inactive during interphase, unlike euchromatin, which becomes dispersed. The outstanding example is the inactivated X chromosome of the mammalian female, most of which is heterochromatic, but there is heterochromatin in other chromosomes as well. Presumably, heterochromatic regions are so compacted that their DNA is not amenable to transcription, but what determines where and when the euchromatin will become heterochromatic remains a mystery.

Chromosomal RNA and Proteins. The activity, or inactivity, of the DNA in being transcribed by messenger RNA seems to depend on the chromosomal proteins. Histones repress messenger-RNA synthesis but lack sufficient specificity to account for regulation of individual loci. However, the specificity may be added by a special type of RNA, with a high content of 5-methyl dihydrocytidylic acid, which may bind to specific regions of the DNA and is also bound to chromosomal protein. This histone-RNA complex could bind to specific regions of the chromosome, thus achieving differential gene inactivation.

Hormones. Ample evidence demonstrates that activity of certain genes is regulated by hormones, but it is still not clear how this is done. There are semispecific proteins in the plasma that bind certain hormones, e.g., for estrogen and testosterone. In the cell cytoplasm there are very specific binding proteins, e.g., an estrogen cytosol receptor complex that somehow facilitates entry of estrogen into the nuclei of the endometrial cells. There, in combination with a specific acceptor protein, the hormone is bound to the chromosome, suggesting that it regulates gene activity at the transcriptional level. But the hormones appear to be organ-specific, and they may affect groups of seemingly unrelated enzymes. The mechanism of this coordinate control is still obscure.

Gene Amplification. A special example of differential gene activity has been found in the amphibian oocyte. The embryo, when it begins to develop, will need to have large numbers of ribosomes for the rapid synthesis of proteins. In the maturing oocyte, the genes that code for ribosomal RNA are replicated many times. The copies are released from the chromosome, and form hundreds of nucleoli, which then synthesize ribosomal RNA. Thus the activity of the gene for ribosomal RNA is amplified several hundredfold in apparent anticipation of the need for rapid protein synthesis.

Redundant RNA. Another way in which the activity of a particular kind of gene could be amplified would be to have it exist in many copies, in linear sequence. The amount of DNA in a mammalian cell is at least 1,000 times as much as needed to code for the actual number of proteins.

DNA-annealing techniques have shown that about 70% of the DNA of a haploid genome is present in single copies, and that 30% recurs in copies from 1,000 to 1,000,000 times per cell. Much of the redundancy is accounted for by the ribosomal RNA genes; the function of the remainder of the redundant DNA remains unknown but may well have something to do with gene regulation.

Polyploidy. One way of achieving cell differentiation would be to alter the number of chromosomes, and there are a number of tissues that are normally polyploid (bladder epithelium) or have a number of polyploid cells (liver), but in these cases it is not clear whether the polyploidy is a cause of the differentiation or merely associated with it. In any case, polyploidy does not seem to be important as a cause of differentiation in man.

Regulation of Translation

We have considered a number of examples of differential gene activity through changes in transcription. It appears that control at the level of translation (synthesis of polypeptide on the ribosome) may also be an important means of regulating gene activity.

It has been known for some time that a messenger-RNA molecule may be transcribed from the DNA long before it begins to synthesize its polypeptide. During oogenesis of the sea urchin egg, for example, messenger is synthesized and stored, inactive (the "masked messenger"), until fertilization, when there is an abrupt increase in protein synthesis without a concomitant increase in RNA synthesis. In the mouse oocyte, the ribosome (and messenger) are held in a protein lattice that presumably keeps them inactive and that disappears shortly after fertilization.[1]

Similarly, the messenger RNA synthesized in the erythroblast continues to be translated into hemoglobin long after the red blood cell has lost its nucleus. In the interval, it becomes associated with groups of four ribosomes (polysomes), where it appears to remain stable. Globin synthesis is regulated by the amount of heme present, and this appears to occur at the translational level.

Epigenetic Control

Brief mention should be made of regulation beyond the level of translation, the so-called epigenetic level. Enzyme activity can be regulated, for instance, by controlling the rate of degradation of the enzyme, rather than its synthesis. Or the way the molecule is folded may differ in different cytoplasmic states, resulting in changed activity. Polymerization and addition or deletion of part of a peptide chain are other ways of achieving epigenetic control.

Diseases Due to Defective Differentiation

Many genetic defects can be attributed to errors of differentiation, in which a specific cell type does not appear, or takes some abnormal form.[7] For instance, in the pituitary dwarf mouse, the eosinophils fail to differentiate, resulting in a specific growth hormone deficiency, and various hereditary chondrodystrophies result from failure of specific aspects of cartilage differentiation.

MORPHOGENESIS

Morphogenesis, or the emergence of form in the developing organism, is much more complicated than the activation or inactivation of genes. To account for the migration of cells, their aggregation into tissues, the synchronized spreading, bending, and thickening of tissues, the transfer of development instructions from one tissue to another (induction), and, in short, the whole complexly integrated series of interactions that eventually result in the adult organism seems a superhuman task. Yet a beginning has been made.

We cannot cover the whole subject of morphogenesis and its genetic control in this chapter, but we will refer to a number of representative examples.

There is no doubt that morphogenetic

processes are influenced by genes, since there are large numbers of mutant genes that alter the shapes of organs. Many of these are well described in structural terms, but little is known of their precise modes of action. Mutant genes can be useful in revealing the normal, and there are a larger number of mutant genes in the mouse and other animals that produce phenotypes resembling human diseases and malformations.[7,8,15,17]

Induction

Induction, to the embryologist, is the process by which a signal from one tissue initiates a change in the developmental fate of another. For instance, the optic cup, growing out from the brain, induces the ectoderm that lies over it to form a lens, and the two structures integrate with one another to form the eye. Recently, the use of mutant genes has shown not only that the induction is under genetic control but that inductive relations are more complicated than previously suspected. For example, the very early chick limb consists essentially of an ectodermal jacket surrounding an apparently undifferentiated mesoderm. Inductive interactions have been analyzed by the use of mutant genes in the chick, causing such things as extra digits or winglessness. By combining mutant ectoderm with normal mesoderm, and vice versa, and seeing how the resulting limb develops, it has been shown that the overlying ectoderm induces the mesoderm to organize digits. But the number of digits depends on an inductive stimulus to the ectoderm from the mesoderm. Thus morphogenesis of the hand depends upon a series of genetically controlled reciprocal inductive interactions between ectoderm and mesoderm. It is likely that some of the malformations of hands and feet seen in human babies result from disturbances in inductive relationships.

In some cases, abnormal development of an organ results from *failure of induction due to asynchrony* rather than an abnormal inductive mechanism. There is a gene that causes absent or small kidneys in the mouse, for example. Embryological studies show that the migration of the ureter is delayed so that it is late in reaching the kidney-precursor tissue. This suggested that the kidney tissue required an inductive stimulus from the ureter bud to initiate its differentiation and that the abnormal kidney resulted from a diminished or absent stimulus. In culture, when mutant ureter and mutant kidney precursor were put together, kidney differentiation occurred. Thus the ureter could induce, and the kidney precursor could respond—abnormality resulted from failure to bring the two together at the right time.

These examples show how a mutant gene may reveal normal developmental mechanisms, as well as the means by which they go wrong. Such studies can contribute to the understanding of human malformations.

Shape and Pattern

The most complex developmental problem of all is the means by which genes control the shape of organs and the patterns seen in such profusion and with such beauty wherever one looks in nature.

The influence of genes on pattern is beginning to be analyzed in higher organisms. The mouse mutant gene "reeler," for example, deranges the form of the cerebellum and cerebrum. The various organized layers are unrecognizable, the various cell types being intermixed instead of sorted out into their normal layers, and they lack their vertical orientation. Experiments have shown that dissociated isocortical cells from mutant day 18 embryos will form aggregates normally but do not organize themselves into an external molecular layer and an internal nerve-fiber zone as do aggregates of normal cells of this age (but not a day earlier or later). Thus the mutant produces a defect in self-organization of the mutant brain cells at a particular stage of development, showing that this property is under genetic control.[3]

In another example, a morphogenetic change due to a mutant gene has been traced to a property of the cell membrane. The mutant gene talpid,[3] in the chick, causes midline facial defects, fusion of mesenchymal precartilage condensations, and polydactyly. Cell aggregation experiments demonstrate that these result from increased adhesiveness and decreased motility of the mesenchymal cells.[4]

Much of the work on pattern comes from Drosophila. For instance, the bristles are distributed on the fly in a very regular pattern. Many mutant genes are known that change the number or position of specific bristles. How does a cell in a certain place on the Drosophila thorax "know" that it is to form a bristle, while the cell right next to it forms cuticle? Presumably, the genetic machinery of the Drosophila cells is programmed so that any cell finding itself in that particular environment and with that particular history will respond by, so to speak, turning on its bristle-making genes. If so, there must be an underlying system of gradients of chemical and physical change making that particular spot different from any other—a "prepattern." Ingenious experiments making use of bristle mutants and a technique for producing mosaic flies, containing a mixture of normal and mutant tissues, were able to show that the mutant bristle appears in a different place because its genes have changed its program rather than the distribution of the underlying pattern. Thus we return to the idea advanced at the beginning of the chapter—that differential gene activity may result from differences in the cytoplasm in which the genotype resides and that development proceeds by the emergence of more and more specialized kinds of cytoplasm.

GENES AND MALDEVELOPMENT

The Modes of Action of Mutant Genes

Theoretically, for every process involved in normal development, one might expect malformations resulting from mutations of each gene affecting that process. Thus one might have malformations resulting from a mutant structural protein (as in hydrotic ectodermal dysplasia); absent or abnormal enzymes (homocystinuria); defective properties of cell adhesiveness (the "talpid" chick) or defective capacity of cells to migrate or orient themselves (the "reeler" mouse); or failure to die at the proper time (syndactyly); or excessive cell death (the rumpless chicken); or failure to respond to signals from other tissues either by contact (anophthalmia resulting from failure of the ectoderm to respond to induction by the optic cup) or a humoral inducer (testicular feminization); or asynchronies in growth resulting in inductive failure (anophthalmia resulting from delayed growth of the optic cup; the kidneys of the Danforth short mouse), and no doubt many other causes. The wealth of mutant genes in the mouse, and other mammals, provides a fruitful field for research into the developmental links between mutant gene and phenotype, with the possibility of extrapolation to analogous human syndromes.

Gene-Environment Interactions

Environmental teratogens may also strike at various points in the developmental network, or may interact with mutant genes affecting the same developmental processes.

A particularly instructive example is the interaction of 5-fluorouracil (FUDR) and the mutant gene "luxoid" (lu) in the mouse.[2] A low dose of the teratogen or the mutant gene in the heterozygous condition produces only a minor defect, polydactyly of the hind foot. The homozygous mutant, or a high dose of the teratogen, produces polydactyly and tibial hemimelia. The combination of a low dose of teratogen and the heterozygous gene will produce polydactyly and tibial hemimelia even though neither would individually. One wonders whether an analogous situation may exist in man. Why, for instance, does a synthetic progestin produce a masculinization of the genitalia in only a minority of babies exposed to it at the appropriate gestational age? Could they be heterozygous for a rare recessive gene—for instance, one that causes the adrenogenital syndrome in the homozygote?

Our second example illustrates the possibility that malformations can be prevented by prenatal measures. The mutant gene "pallid" in the mouse causes ataxia, resulting from failure of the otoliths of the inner ear to form. Maternal manganese deficiency causes a similar condition in the rat. Putting these facts together, it was possible for Hurley and associates[5] to correct the ataxia in pallid mice by giving their mothers large doses of manganese during pregnancy!

Finally, we must mention the numerous examples of gene-environment interactions involving multifactorial threshold characters (Chapter 11). An embryo's genes may place a particular developmental variable near a threshold of abnormality, so that a relatively small additional environmental influence may place that organ beyond the threshold, and a malformation results. In another embryo, in which the organ is not near the threshold, the same environmental insult will have no effect. An example described in detail in Chapter 11 is cleft of the secondary palate, where the variable is the time at which the embryonic palatal shelves reach the horizontal, so they may fuse, and the threshold is the latest stage of development when they can reach each other when they do become horizontal.

The Developmental Basis of Pleiotropy, Penetrance, and Expressivity

Pleiotropy. Pleiotropy refers to the fact that a single mutant gene may have several end effects, as in numerous inherited syndromes. The multiple effects of single genes can be explained in several ways.

First, the several end effects may be secondary, tertiary, or even more removed results of the initial defect. Gruneberg[9] coined the term "pedigree of causes" to refer to this phenomenon. Thus, in sickle-cell disease, the basic molecular defect leads to *hemolysis,* which leads to anemia, pallor, and fatigability; to *intravascular sickling,* which causes leg ulcers, infarcts of various organs, and splenic rupture; and to *marrow hypertrophy,* which causes bone pain and the "tower skull." In phenylketonuria the mental defect, growth retardation, hypopigmentation, skin rash, and seizures are all, no doubt, results of the basic enzyme defect, though some of the steps are not yet clear. And in many syndromes the developmental connections are entirely obscure. What biochemical defect, for instance, is common to the retinitis pigmentosa and polydactyly of the Laurence-Moon-Biedl syndrome?

Pleiotropy can also occur if the same polypeptide is common to more than one protein. A mutant polypeptide would then result in more than one mutant protein, and more than one end-effect. We do not know of any relevant example. A great challenge for students of developmental mammalian genetics will be to trace the developmental connections revealed by pleiotropic mutant genes.

Penetrance and Expressivity. Little is known about the developmental basis of penetrance and expressivity. A convenient, if oversimplified, model is based on the developmental variable-threshold concept. If a group of individuals (such as a particular strain of mice) had a genotype and environment that placed it near the threshold, the effect of a major mutant gene might be to throw all the individuals beyond the threshold, and one would say that the gene had full penetrance (Fig. 10-1). In another group, who were far from the threshold, the effect of a mutant gene might place only a few individuals beyond the threshold, and the gene would be said to have low penetrance. Similarly, if individuals near the threshold were mildly affected and those far beyond the threshold were severely affected, it is clear that there would be a correlation between penetrance and expressivity, as there often

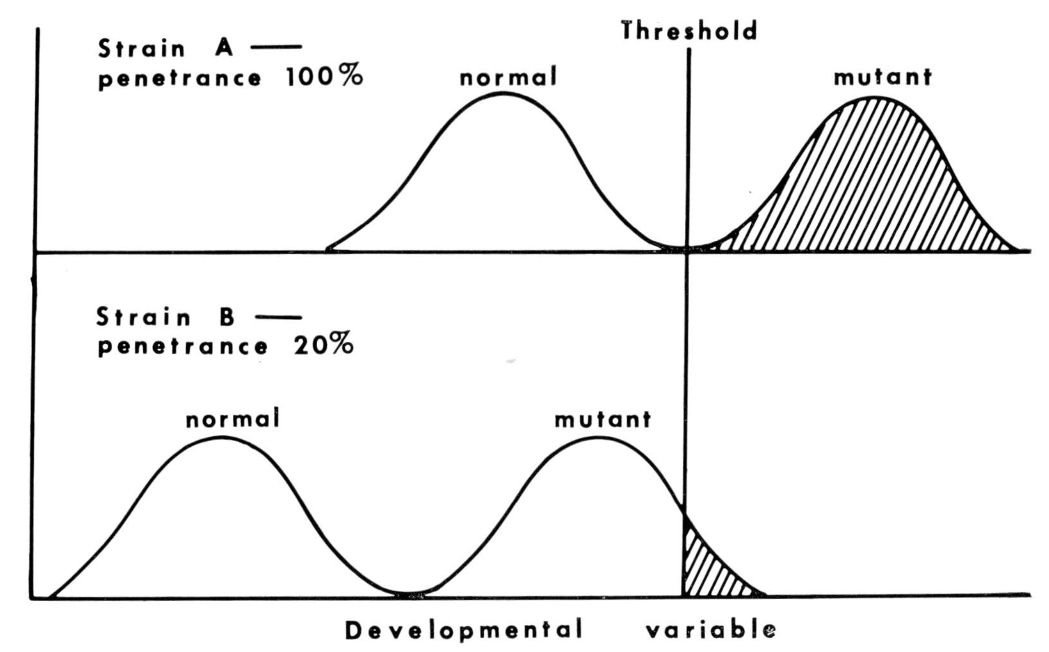

Figure 10-1. Hypothetical model showing penetrance of gene on two different genetic backgrounds. Note that the same model will account for strain differences in response to a teratogen—for instance cleft palate caused by cortisone in the A/J and C57BL inbred mouse lines—see Chapter 13.

seems to be in experimental animals where this can be adequately tested.

Phenocopies

Phenocopies are individuals with a mutant phenotype but a nonmutant genotype —that is, some environmental factor has simulated the effects of a mutant gene. Much has been deduced about the mode and time of action of mutant genes by analysis of their phenocopies, but this approach is only just beginning to be applied in man. An interesting example is given by Lenz,[11] who showed that, depending on the stage of exposure, the teratogenic effects of thalidomide on the limb could resemble those of the mutant genes for the Holt-Oram syndrome, the Fanconi anemia syndrome, or the radial aplasia with amegakaryocytosis syndrome.

The Genetics of Human Male Sexual Development

An interesting example, illustrating many of the above concepts, is provided by recent studies on the various types of male pseudohermaphrodite.

The gonad begins in the early embryo as an indifferent primordium, into which germ cells migrate from the yolk sac. Presence of a Y chromosome in the germ cells determines that the gonad differentiates into a testis. Otherwise it forms an ovary. Somewhat later the embryo lays down two sets of ducts: In female embryos, the Müllerian ducts are the precursors of the Fallopian tubes, uterus, and upper part of the vagina; and in male embryos, the Wolffian ducts will form the vas deferens, seminal vesicles, and epididymis, the ducts leading from the testis to the penis. Development of these ducts depends on a stimulus provided by the male hormone, testosterone, produced by the embryonic testis (induction). The testis also produces a hormone (not testosterone) that inhibits further differentiation of the adjacent Müllerian ducts. In the absence of a testis (i.e., in the normal female or castrated male embryo) the Müllerian ducts will continue to differentiate and the

Wolffian ducts will regress. Later on, the external genitalia form. In the presence of male hormone, the genital folds and tubercle form a penis; in its absence, they form the labia and clitoris (morphogenesis). Finally, it has recently been discovered in rodents that various target organs are "imprinted" by male hormone at various critical periods of development so that at puberty they will respond in a male pattern (determination). These include the secretion of gonadotrophic hormone by the hypothalamus, the way the liver metabolizes steroid hormones, the maturation of the prostate gland and seminal vesicles at puberty, and various behavioral characteristics. Furthermore (there are exceptions to every rule) there is some evidence that autosomal genes may play a role in testis differentiation; there are mutant autosomal genes in the mouse, goat, pig, and perhaps man that cause testis differentiation in XX individuals.

Male sexual development can be interrupted or diverted at various steps of the process, leading to various kinds of male pseudohermaphroditism.

Failure of Testis Differentiation. In familial XY gonadal dysgenesis, a mutant gene (X-linked or autosomal, sex-limited) prevents formation of the testis. The affected individuals have an XY karyotype, undifferentiated gonads, and a female anatomy.

In some families, some affected individuals do manage to achieve some degree of testis differentiation, so that although the internal anatomy is female there are varying degrees of masculinization of the external genitalia (variable expressivity). These may represent what the microbial geneticists would call "leaky" mutants.

Absence of the Müllerian-Duct-Inhibiting Substance. In the rare uterine hernia syndrome, the affected males are normal except that they have Müllerian ducts and an infantile uterus that may present in an inguinal hernia. The inheritance is probably autosomal recessive, and it seems likely that the mutant gene results in failure to produce the testis hormone that normally inhibits formation of the Müllerian duct derivatives.

Defects in Testosterone Synthesis. There are several recessively inherited forms of male pseudohermaphroditism in which the mutant gene causes a deficiency of one of the enzymes involved in testosterone synthesis. Most of them also involve synthesis of steroids by the adrenal cortex (pleiotropy) and are referred to as the adrenogenital syndromes. In males there are varying degrees of undermasculinization of the external genitalia, which may extend to a fully female anatomy, and varying degrees of breast enlargement (gynecomastia) at puberty. It is not clear why the Wolffian ducts manage to differentiate successfully. Perhaps the defect in testosterone synthesis appears after Wolffian duct differentiation has been induced, or the defect is incomplete and the Wolffian duct needs less of a stimulus than the genital tubercles in order to proceed with development in the male pathway. The gynecomastia may result from a failure of "imprinting" of the breast tissue at a critical period, that normally prevents the pubertal male breast from responding to estrogenic hormones from the adrenal cortex.

Failure of Target Tissue Response to Androgen. Perhaps the most interesting type of male pseudohermaphroditism is testicular feminization (TF). X-linked forms exist in the rat and mouse, and the human type fits this familial pattern too, though autosomal sex-limited inheritance has not yet been ruled out. The affected individuals have an XY karyotype, testes, no uterus or Fallopian tubes, and female external genitalia. At puberty, feminization occurs. Although the complete answer is not yet available, and there are significant differences between the rodent types and the human syndrome (which may itself be heterogeneous), it seems clear that the basic defect is in the response of the target organs. The testis in TF males produces testosterone. There is Müllerian duct inhibition, so the testis must also produce the inhibitory hormone. Fibroblasts from the genital skin of some TF individuals metabolize testosterone more slowly than normal fibroblasts, and the primary defect may turn out to be in the ability of the cytosol receptor in the cells of the target organs to bind with testosterone or its immediate derivative dihydrotestosterone.

This catalogue of gene-determined classes of male pseudohermaphroditism is far from complete, but it serves to demonstrate that mutant genes may interfere with male sex differentiation at many points and to show how study of their effects is contributing to our understanding of the normal process.

SUMMARY

Development from a single cell to an organism containing many cells and organ systems of diverse function is a complex process that cannot be studied in the human at many critical stages. Cells containing the same genetic information, through the process of differentiation (resulting largely from differential gene activity) become specialized to develop along different lines, to form different structures, and to perform different functions.

The regulation of gene activity can occur at the level of transcription. Differential gene activity appears to be influenced by heterochromatin, chromosomal RNA and histone, hormones, gene amplification, and redundant RNA. Gene regulation may also occur at the level of translation and beyond at levels of epigenetic control.

Morphogenesis (the emergence of form in the developing organism) is even more complicated than gene regulation. Cell migration and aggregation and inductive interactions (under genetic influence) participate in the morphogenetic process leading to the development of shape and pattern.

Maldevelopment may result from failures at the genetic level (mutations) or chromosomal level (aberration), from potent environmental pathogens and from interactions between polygenic diathesis and less potent environmental triggers.

REFERENCES

1. Burkholder, G., Comings, D. E., and Okada, T. A.: A storage form of ribosomes in mouse oocytes. Exp. Cell Res. 69:361, 1971.
2. Dagg, C. P.: Combined action of fluorouracil and two mutant genes on limb development in the mouse. J. Exp. Zool. 164:479, 1967.
3. DeLong, G. R., and Sidman, R. L.: Alignment defect of reaggregating cells in cultures of developing brains of reeler mutant mice. Dev. Biol. 22:584, 1970.
4. Ede, D. A., and Agerback, G. S.: Cell adhesion and movement in relation to the developing limb pattern in normal and talpid[3] chick embryos. J. Embryol. Exp. Morphol. 20:81, 1968.
5. Erway, L., Hurley, L. S., and Fraser, A. S.: Neurological defect: manganese in phenocopy and prevention of a genetic abnormality of inner ear. Science 152:1766, 1966.
6. Fraser, F. C., and McKusick, V. A. (ed.): Congenital Malformations. New York, Excerpta Medica, 1970.
7. Green, E. L. (ed.): Biology of the Laboratory Mouse. 2nd ed. New York, McGraw-Hill, 1966.
8. Green, Margaret C.: Genes and development, Ch. 15, p. 329. *In* E. L. Green (ed.), Biology of the Laboratory Mouse. 2nd ed. New York, McGraw-Hill, 1966.
9. Gruneberg, H.: An analysis of the "pleiotropic" effects of a new lethal mutation in the rat (Mus norvegicus). Proc. Roy. Soc. Lond. Ser. B. 125:123, 1938.
10. Hamburgh, M.: Theories of Differentiation. New York, American Elsevier, 1971.
11. Lenz, W.: Genetic diagnosis: molecular diseases and others, p. 1. *In* J. de Grouchy, F. J. G. Ebling, and I. W. Henderson (eds.), Human Genetics. Amsterdam, Excerpta Medica, 1972.
12. Markert, C. L., and Ursprung, H.: Developmental Genetics. Englewood Cliffs, N.J., Prentice-Hall, 1971.
13. Motulsky, A. G.: Biochemical genetics of hemoglobins and enzymes as a model for birth defects research, p. 199. *In* F. C. Fraser and V. A. McKusick (ed.), Congenital Malformations. New York, Excerpta Medica, 1970.
14. Pastan, I.: Cyclic AMP. Sci. Am. 227:97, 1972.
15. Pinkerton, P. H., and Bannerman, R. M.: The hereditary anemias of mice. Hemat. Rev. 1:119, 1968.
16. Pinsky, L.: Human male sexual maldevelopment. Teratology 9:193, 1974.
17. Sidman, R. L., Green, M. C., and Appel, S. H.: Catalogue of the Neurological Mutants of the Mouse. Cambridge, Harvard University Press, 1965.

Chapter 11

MULTIFACTORIAL INHERITANCE

METRICAL TRAITS

Everyone knows that close relatives tend to resemble each other with respect to a number of quantitative, or metrical, characters such as height, weight, size of nose, "intelligence," and so on. The question of how closely relatives resemble each other and how much of the familial tendency is due to genes shared in common is one that has received a good deal of attention from mathematical geneticists, and there is an extensive literature on the subject.[1] We will review only a few basic principles here.

For any particular metrical character, a first approach to the question of the genetic basis is to see whether the frequency distribution of the trait has a single mode, or peak, or more than one. A bimodal frequency distribution strongly suggests that a major genetic difference is segregating in the population, as in the case of isoniazid degradation[10] (Fig. 11-1). A unimodal curve (as in the case of blood pressure, or intelligence) suggests that no single factor is making a major contribution to the variation in the trait.

Heritability

Many quantitative traits have a distribution that fits the familiar bell-shaped curve known as the normal curve. This is compatible with the assumption that the magnitude of the trait is determined by a number of genes, each adding a small amount to the quantity of the trait, or subtracting a small amount from it, and each acting independently of the others (i.e., acting additively, with no dominance or epistasis). This is known as *polygenic* inheritance. There are a few individuals at the extreme of the distribution and many in the middle because it is unlikely for an individual to inherit a large number of factors all acting in the same direction.

For instance, if height were determined by one gene locus with three alleles, one adding two inches to the height (h^+), one subtracting two inches (h^-), and one neutral (h), and if h were twice as frequent as the other two alleles, the expected distribution of heights could be found by calculating the frequencies of pairs of gametes from the available pool, as in Table 11-1 and Figure 11-2. Thus, if the mean height was 68 inches, $\frac{1}{16}$ of the population would be h^-h^- and 64 inches tall, $\frac{1}{16}$ would be h^+h^+ and 72 inches tall, $\frac{4}{16}$ would be hh and 68 inches tall, $\frac{2}{16}$ would be h^+h^- and also 68 inches tall, and so on.

If we add another locus, also with three

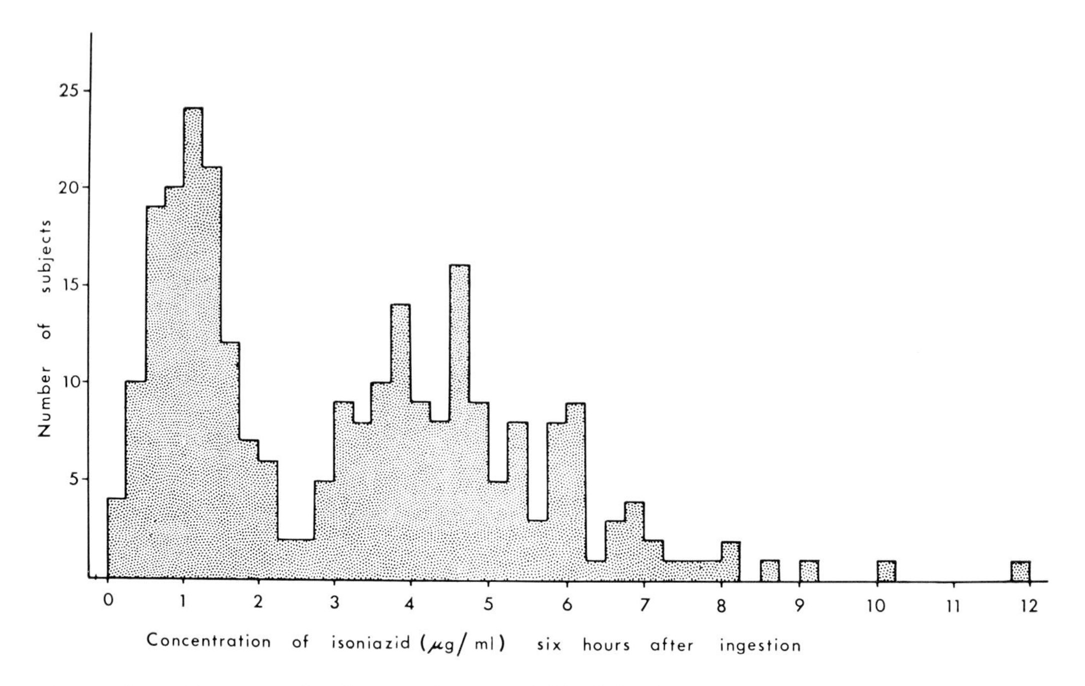

Figure 11–1. Frequency distribution of isoniazid blood levels 6 hours after a standard dose. The distribution is bimodal, illustrating the presence of two phenotypes—fast and slow inactivators.

TABLE 11-1. Frequencies of Genotypes for Height If Determined by Three Alleles at a Single Locus[a]

		Sperm		
		1 h+	2 h	1 h−
Eggs	1 h+	h+h+ 1 72″	h h+ 2 70″	h+h− 1 68″
	2 h	h+h 2 70″	h h 4 68″	h h− 2 66″
	1 h−	h+h− 1 68″	h h− 2 66″	h−h− 1 64″

h = average h− = minus 2 inches h+ = plus 2 inches

[a] After Carter, C. O.: Human Heredity. Baltimore, Penguin Books, 1970.

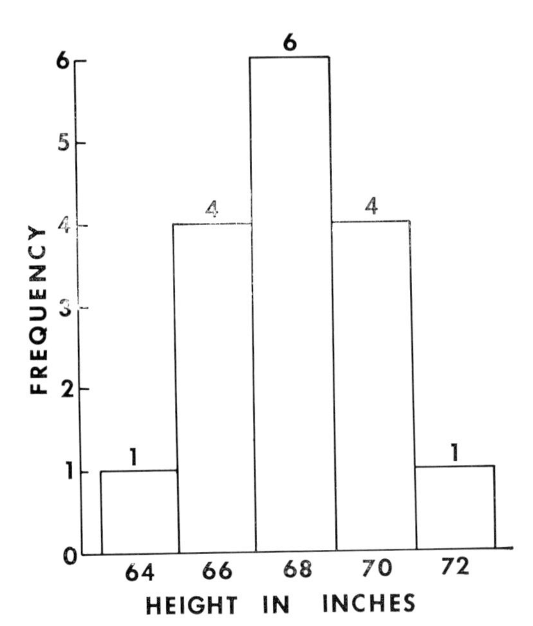

Figure 11–2. Frequency distribution of heights from Table 11–1.

alleles in the same proportions, the distribution of heights begins to look like the normal curve (Fig. 11-3). Thus a relatively small amount of genetic variation can produce a distribution that is fairly normal. In this case only 1/256 individuals would inherit all four minus or plus alleles and be at the extremes of the distribution.

Harris[7] has shown, for example, that the enzyme red cell acid phosphatase exists in three electrophoretically different forms, varying in their enzymatic activities, and that the apparently normal distribution of enzyme activities in the general population comes from various combinations of these alleles, very much as in the theoretical height model.

Of course, a number of environmental factors, each adding or subtracting a small amount to the final result, will also result in a normal distribution, even without any

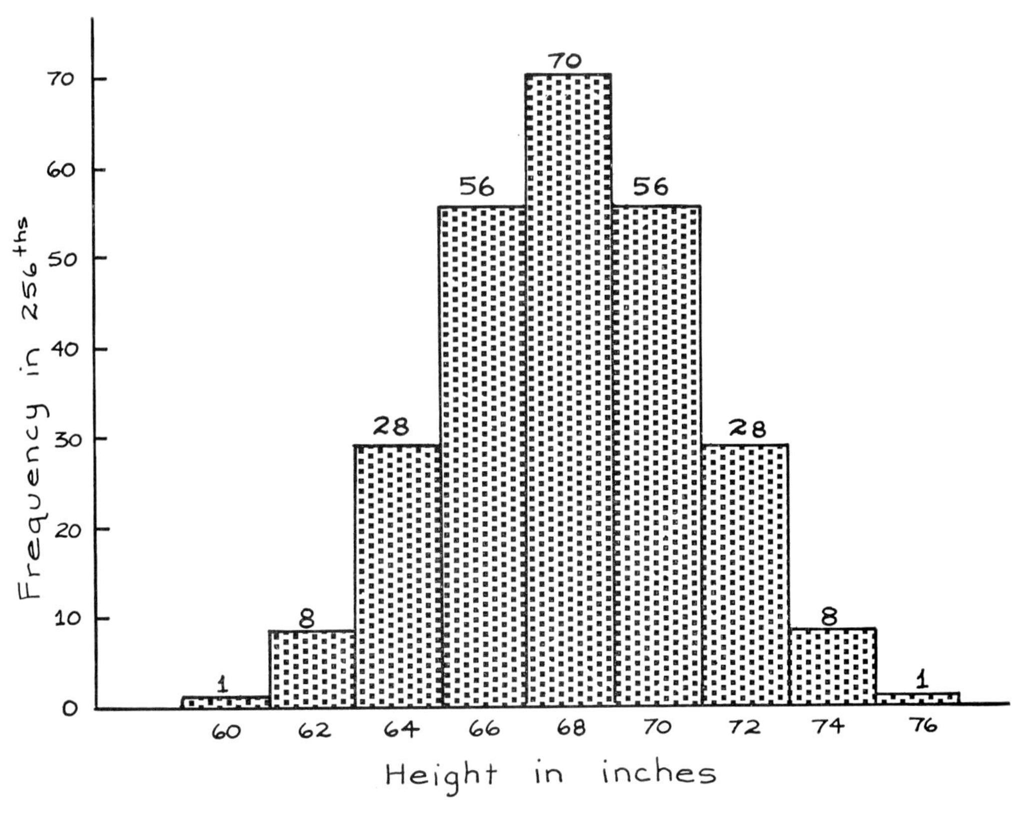

Figure 11–3. Frequency diistribution of heights assuming 2 loci each with 3 alleles.

genetic variation. In most cases, the variation in the population results from a number of genes and environmental factors acting together to determine the final quantity. This can be termed *multifactorial* inheritance. The proportion of the total variation in the trait that results from genetic variation is the *heritability* of the trait.

The problem then is to determine how much of the variation in the multifactorial trait is due to segregating genes and how much to environmental factors. One can reason as follows: If all the variation were due to environmental factors (which did not themselves show a familial tendency), there would be no tendency for relatives to resemble one another—i.e., the correlation between relatives would be 0. What would it be if the inheritance was polygenic, with no environmental variation? Consider, for instance, the correlation between father and son. The son gets half of his father's genes, so if the father's genes for the trait in question are such that he deviates from the mean, the son should, on the average, deviate by half as much. For a series of such pairs, this would lead to a father-son correlation (and regression of son on father) of 0.5. A similar situation would exist for pairs of brothers, who have half of their father's genes in common (Fig. 11-4).

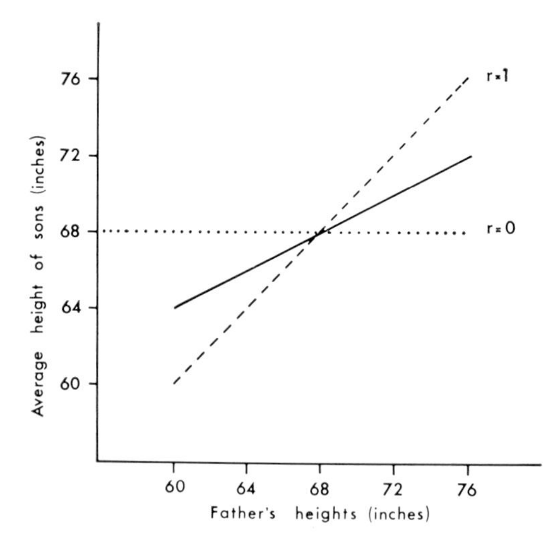

Figure 11–4. Regression toward the mean of sons' on father's phenotype for a polygenic character. For any father's value, the mean value for sons is halfway between father's value and the mean of the population (assuming no assortive mating).

This relationship was first formulated by Sir Francis Galton as the Law of Filial Regression—the mean value of the sons would be halfway between the mean of the fathers and the mean of the population (*assuming there was no assortive mating*). In other words, the mean of the sons "regressed" from the mean of the fathers toward the mean of the population. Obviously, the more environmental, nonfamilial variation there is, the lower the correlation will be. Conversely, the less important environmental factors are (i.e., the higher the heritability), the closer to 0.5 will be the correlation between first-degree relatives.

Note, however, that familial environmental factors will also increase the correlation between relatives. Some attempt to measure these can be made by comparing, for instance, similarities between pairs of unrelated children raised in institutions and raised in foster homes, respectively, but the practical difficulties are great. Furthermore, it was postulated that the genes act additively; nonadditive interactions such as dominance or epistasis would modify the correlations in a complicated way. They will, for instance, lead to parent-child correlations being lower than sib-sib correlations.

Finally, it should be emphasized that heritability estimates are made on specific populations in a specific range of environments and should not be extrapolated uncritically to other populations and environments. An estimate of heritability of skin color, for instance, based on a Swedish population would be misleadingly low if applied to the population of the United States, where there is much more genetic variation.

A variety of statistical techniques have been developed to estimate the various components of the variation in a trait from, for instance, comparisons between monozygotic and dizygotic twin pairs, twins reared together and reared apart, and correlation between child and biological parents and between child and foster-parents (see, for instance, Jensen[8]). Estimates of heritability have been made for many quantitative human traits, but they should in most cases be regarded as no more than indications of whether the role of genes in determining the trait is relatively large or small.[1] Chapter 16 provides further discussion.

THRESHOLD CHARACTERS

A number of relatively common defects and diseases that are clearly familial cannot be made to fit all the expectations for Mendelian inheritance, in spite of enthusiastic attempts to do so, sometimes by statistical methods more vigorous than rigorous. It was first recognized by Sewall Wright in 1934[13] that the inheritance of a discontinuous character (polydactyly in the guinea pig) could be accounted for by multifactorial inheritance of a continuously distributed variable, with a *developmental threshold* separating the continuous distribution into two discontinuous segments—polydactylous and not-polydactylous. Gruneberg[6] showed that a number of discontinuous but seemingly non-Mendelian traits in mice fitted this model and called them "quasi-continuous variants."

Developmental Thresholds

A well-documented example of a developmental threshold is cleft of the secondary palate. In order to close, the palatal shelves must reorient themselves from a vertical position, on either side of the tongue, to a horizontal plane above the tongue, where their medial edges meet and fuse. During the time of reorientation, the head continues to grow, carrying the base of the shelves farther apart. If the shelves become horizontal late enough, the head will be so big that they will be unable to meet, and a cleft palate will result. The latest point in development when the shelves can reach the horizontal and still meet can be considered a *threshold*—all embryos in which the shelves become horizontal later than this will have cleft palates. Other thresholds may involve other developmental asynchronies, such as neural tube closure, or physiological characteristics, such as renal tubular reabsorption, or mechanical relationship—e.g., the degree of occlusion of the pyloric canal necessary to cause the signs of pyloric stenosis, or the pressure at which a blood vessel ruptures. To return to the palate, the important point is that a continuous, multifactorial variable—stage at which shelf movement occurs—is separated into discontinuous parts—normal and cleft palate—by a threshold. If the continuous variable involves a postnatal process, e.g., blood pressure, it is possible to place any given individual on the distribution, but for a prenatal process, it is possible to tell only whether or not the individual fell beyond the threshold.

However, it is possible to make some deductions about how such a trait will be distributed in the population and in the relatives of affected persons. Furthermore, for traits that fit the predictions, one can then estimate the heritability of the underlying variable from the observed frequencies in families.

Models of Quasicontinuous Inheritance

Several mathematical models have been proposed from which estimates of heritability of liability to the disease can be calculated.

According to Falconer's model,[4] the term "liability" represents the sum total of genetic and environmental influences that make an individual more or less likely to develop a disease. The liabilities of individuals in a population form a continuous, normally distributed variable. A person develops overt disease if his liability reaches a certain threshold.

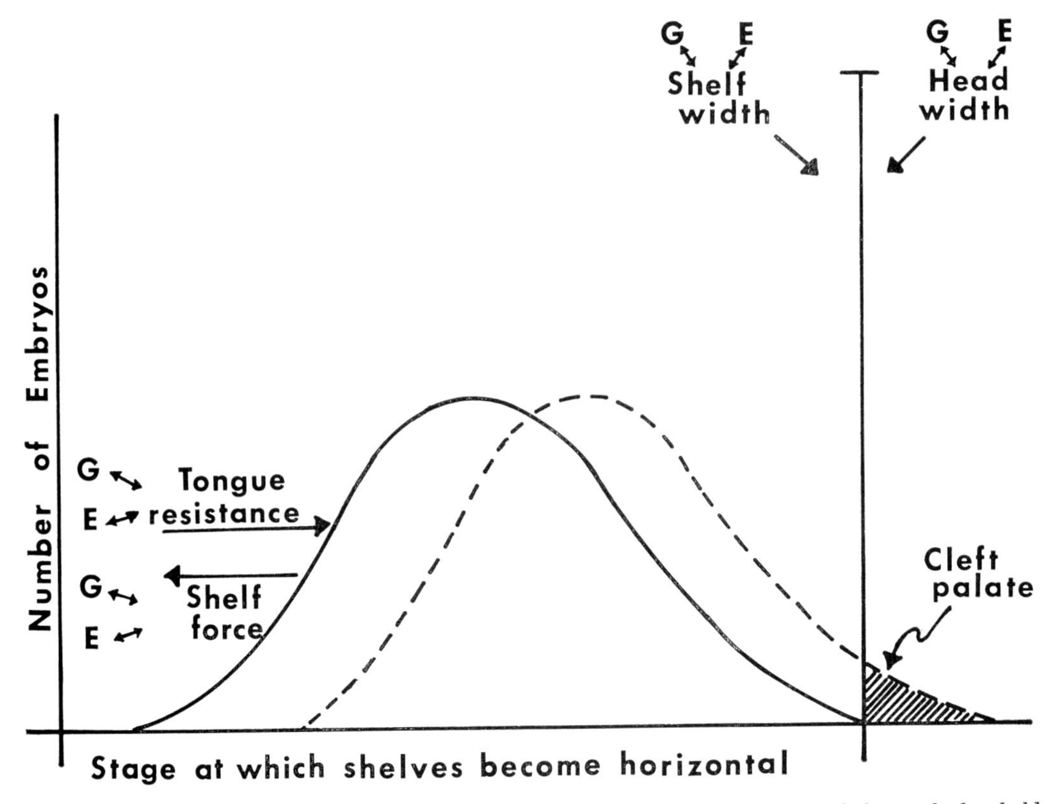

Figure 11–5. Diagram of factors influencing palate closure, illustrating its multifactorial threshold nature.

In the case of cleft palate, for instance, many things influence the stage at which the shelves move to the horizontal—the intrinsic shelf force, tongue size, forward growth of the mandible, and so on—and each of these is influenced by genetic and environmental factors. This, then, is a multifactorial model. The stage at which the shelves reach the horizontal would represent the liability to cleft palate—the later, the more liable—and the latest stage at which they can still bridge the gap would be the threshold. Note that genes and environmental factors can also influence the threshold—in this case by altering the size of the gap through changes in shelf width and head width. Figure 11-5 represents palate closure as a multifactorial threshold character.

Figure 11-6 illustrates a population in which liability for a given disease is normally distributed (solid curve), and all individuals beyond a certain threshold (T) actually have the disease (diagonal hatching). Thus the affected individuals have a mean liability near the tail of the distribution. The usual family study ascertains a series of such individuals, as probands, and measures the frequency of affected individuals in the near relatives. What will the frequency be in the proband's sibs and children?

The dotted line in Figure 11-6 illustrates the distribution of liabilities for first-degree relatives, assuming that all the variation is genetic (a heritability of 1). By the law of filial regression, it will have a mean halfway between the mean of the probands and the mean of the population, and the frequency of the disease will be higher than that in the general population (horizontal hatching). How much higher?

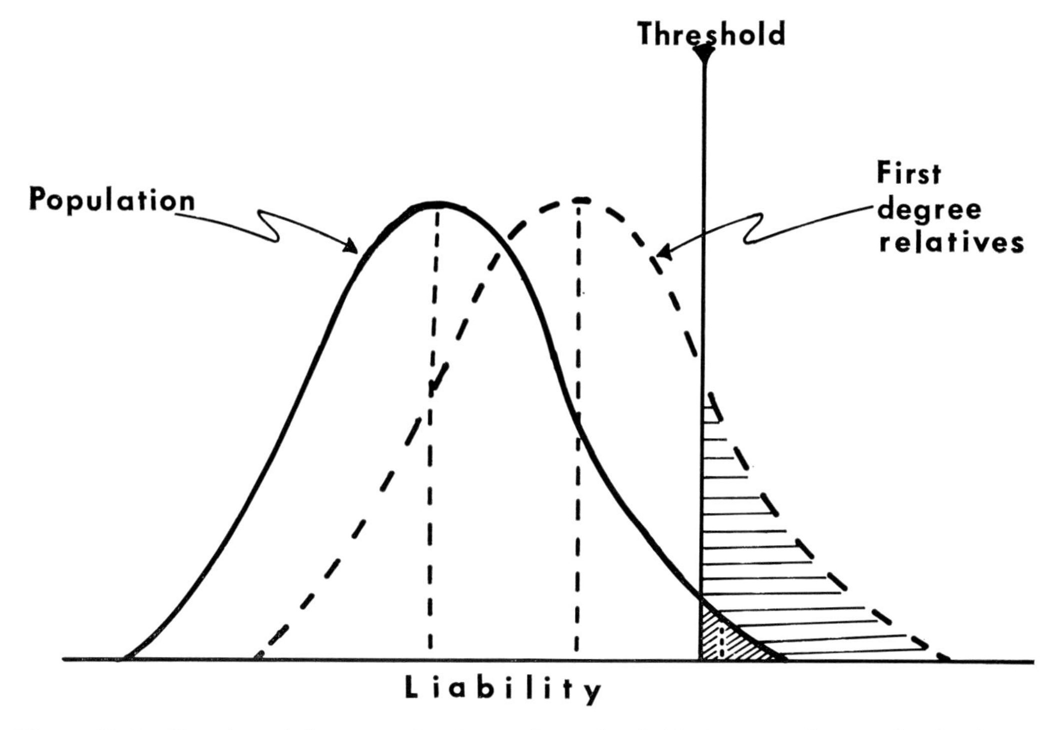

Figure 11–6. Hypothetical frequency distribution for a threshold character, showing the distribution of the population (solid line) and that of first-degree relatives (dotted line).

If liability is normally distributed, an estimate can be deduced from a property of the normal curve, which is that there is a fixed relationship between the distance from the mean (measured in standard deviations) and the number of individuals that lie under the curve beyond that distance. For instance, by consulting a normal curve area table, we can see that if a threshold were set two standard deviations from the mean, 2.27% of individuals would lie beyond the threshold and be affected, and at three standard deviations from the mean, 0.13% would be affected. Thus we can estimate, from the frequency in the population, how far the threshold is from the mean. For a frequency of 1 per 1,000, for instance, the table tells us that the threshold is 3.1 standard deviations from the mean. If we assume that all the variation is genetic, the mean of the distribution of liabilities for first-degree relatives should be intermediate between that of the affected probands (say 3.3 standard deviations) and that of the general population, or at about 1.65 standard deviations. If so we would expect (from the normal curve) that about 5% of the first-degree relatives would be affected. If the heritability is less than 1, the observed frequency in the relatives would be correspondingly lower, so it is possible to estimate the heritability by the difference between the observed figure and that expected if the heritability were 1. Mathematical details and appropriate tables can be found elsewhere.[4,11] It turns out that even for defects with a rather low recurrence risk the heritability can be quite high.

Edwards' model[3] assumes that the liability is continuously distributed and that the probability of being affected increases exponentially as the liability increases. This has some advantages, both conceptual and mathematical, over Falconer's model, but for practical purposes, such as predicting

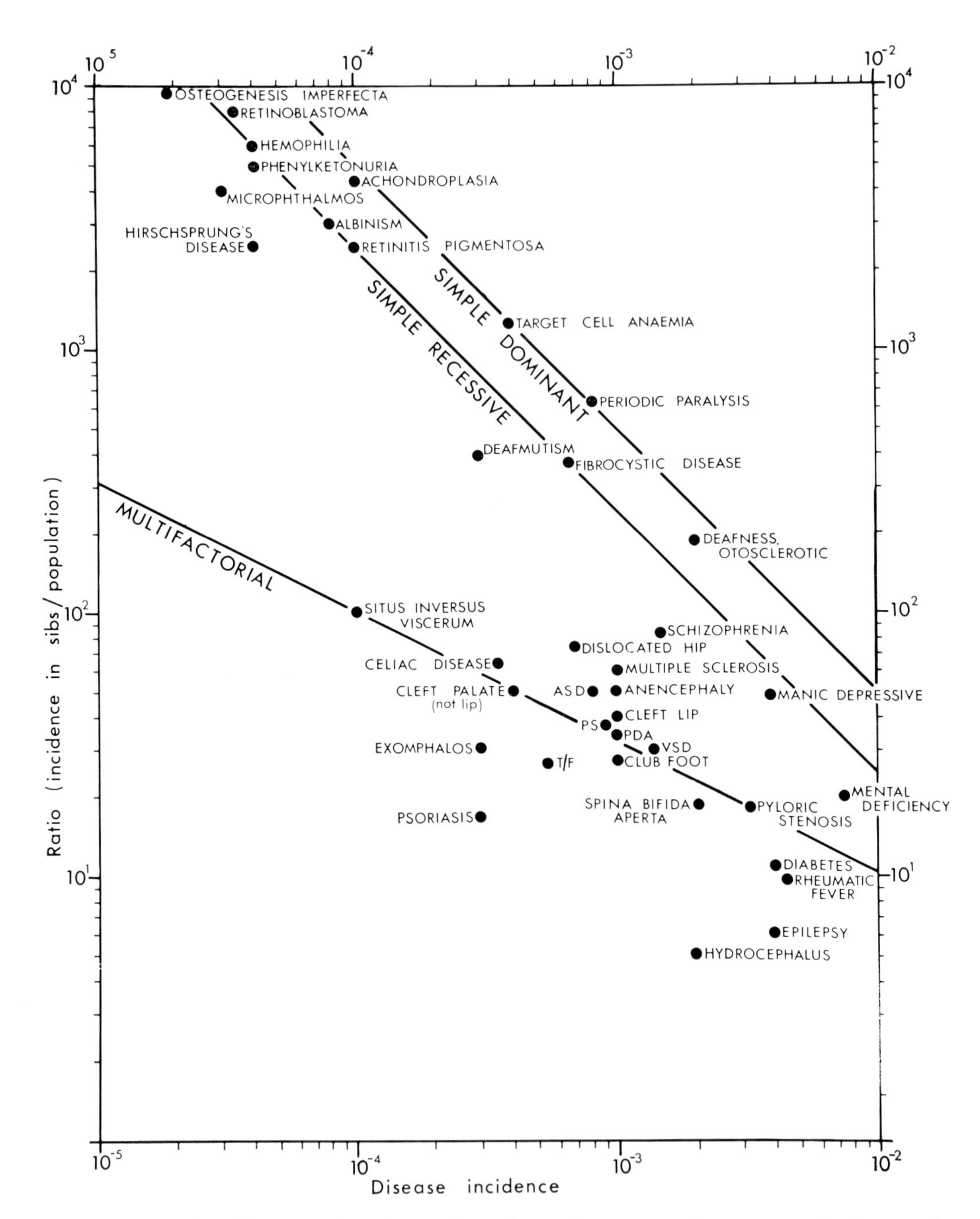

Figure 11–7. Chart illustrating the relation of population frequency to frequency in first-degree relatives for a multifactorial threshold character.

recurrence risks, it does not seem any more useful. Finally, Morton[9] proposes a model in which the disease is determined by rare genes in a small number of cases and small effects of many genes in others and shows that the three models predict about the same recurrence risks in conditions for which there is no evidence of recessive genes with major effects.

Family Features of Multifactorial Threshold Diseases

Nevertheless, there are features of the family distributions of certain common congenital malformations that are most easily explained by the multifactorial threshold model, as first pointed out by Carter[1] for pyloric stenosis. Some of the features are explained equally well by other models, but some are not.

Relation of Recurrence Risk to Population Frequency. With the aid of a number of reasonable mathematical approximations, it has been shown that for a threshold character with high heritability, the frequency of occurrence of the trait in first-degree relatives of affected individuals is approximately the square root of the frequency in the population.[3] This relationship holds for a number of common congenital malformations and not for diseases known to show Mendelian inheritance or diseases with a known major environmental component (Fig. 11-7), which suggests that the former are, indeed, multifactorially determined threshold characters.

Nonlinear Decrease in Frequency with Decreasing Relationship. We have already seen that the distribution of the underlying variable in first-degree relatives of affected individuals will have a mean halfway between the mean of the affected relatives and the mean of the population. For second-degree relatives, the mean will be between the mean for the first-degree relatives and that of the population, and so on. Thus if the distance between the curve for the first-degree relatives and the curve for the population is 1, the distance for the

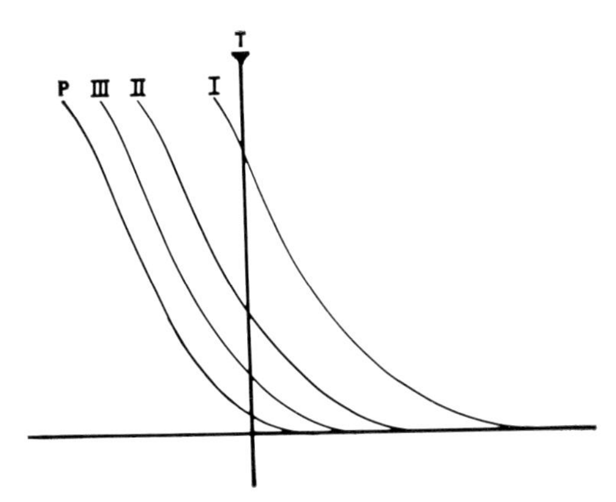

Figure 11–8. Relation of the tail of the frequency distribution to the threshold for the population (P) and the first (I), second (II), and third (III) degree relatives of affected individuals.

second-degree relatives will be ½, and for the third degree relatives ¼, and so on. However, the proportion of affected relatives will be represented not by these ratios but by the area under the curve beyond the threshold for the respective distributions (Fig. 11-8). Since the tail of the curve becomes progressively flatter, the drop in frequency should be greater between first- and second-degree relatives (on the steep part of the curve) than that between second- and third-degree relatives (on the flatter part of the curve). This has been shown for several common congenital malformations, including pyloric stenosis, dislocation of the hip, clubfoot and cleft lip with or without cleft palate.[2]

In the case of cleft lip, for instance, the frequency of the defect is about 40 per 1,000 for first-degree relatives, 7 for second-degree relatives, and only 3 for third-degree relatives.

Increased Risk of Recurrence after Two Affected Children. As we have said, with threshold characters one cannot tell where, on the distribution of liability, a given individual is. However, parents who have an affected child must have contributed a relatively large number of genes for liability to the child and are therefore likely to be carrying more than the average number of such genes themselves. That is, they will tend to lie between the population mean and the threshold. Thus their future children will have a greater than average risk of being affected. If they do have a second affected child, they can be assumed to carry still more predisposing genes, so the recurrence risk will be even greater than for parents after one affected and will lie still closer to the threshold. This has been shown to be true for cleft lip and palate and for spina bifida aperta, where the recurrence risk after one affected child is about 4% and after two affected children about 10%.

Increased Recurrence Risk with Increased Severity of Defect. It is reasonable to suppose that, for a threshold character, a person with a severe form of the defect would be nearer the tail of the distribution of liability than a person with a mild one. If so, the risk of recurrence should be higher in patients with the more severe defects. In the case of cleft lip, for instance, the recurrence risk is about 2.6% for probands with unilateral cleft lip and 5.6% for bilateral cleft lip and palate,[5] and for Hirschsprung's disease the risk of recurrence varies with the length of aganglionic segment.

Recurrence Risk and Sex of Proband. In defects that occur more frequently in one sex than the other, it must be assumed that the threshold is nearer the tail of the distribution in the sex less often affected. If, for instance, the defect appears less often in females than males, females must have more genes for liability in order to fall beyond the threshold, and be affected. If so, the recurrence risk should be higher for the relatives of female patients. This was first shown by Carter for congenital hypertrophic pyloric stenosis, which affects about five times as many males as females.[2] For instance, the risk is about 20% for sons of affected females as compared to 5% for sons of affected males. Similar though less striking differences occur in the case of cleft lip and palate and isolated cleft palate.[5] In each case, *the recurrence risk is higher when the proband is of the less frequently affected sex.*

CONCLUSION

These features suggest that a number of relatively common familial diseases and defects fit the multifactorial threshold model. Admittedly, it is difficult to distinguish critically between this model and that of Morton (described above), in which there is a mixture of "sporadic" cases with low recurrence risk and a smaller number of cases with a strong genetic component.[9] The nonlinear decrease in risk with decreasing relationship and the variation in recurrence risk with sex of proband (if there is a sex difference in frequency) are difficult to account for by Morton's model, but appropriate data are often not available. The difference is important. If Morton's model is correct, one should concentrate on attempting to identify the genetic cases by looking for biochemical differences. If the multifactorial threshold model is correct, there will be no major identifiable biochemical factor, and one might be better off to concentrate on identifying the underlying variable and its threshold. In any case, one should resist the temptation to invoke the concept, as some have done, for familial conditions that have not been tested by the above criteria. Neither should the term be used to refer to etiologic heterogeneity—that is, when different cases of the disease have different causes.

In counseling, it may be useful to develop with the parents of an affected child the idea that susceptibility is the result of the adding up of "a lot of little things," none of which is in itself abnormal, so that they should not feel guilty about having "bad genes" or having unknowingly exposed the baby to a prenatal insult.

What can be done to reduce the frequency of such conditions? One approach is to try to identify the individual components contributing to susceptibility—for instance, familial joint laxity with congenital hip dislocation, differences in face shape with cleft lip, or blood group O secretor status with duodenal ulcer. The more components that can be identified, the better the identification of susceptible individuals, though being a near relative of an affected person will probably be the best indicator for some time to come. The other approach stems from the observation that multifactorial threshold characters often vary in frequency with season of birth, socioeconomic class, geographic region, and other environmental differences. This shows that extrinsic factors may shift the relationship of underlying distribution and threshold; we must now learn to identify these factors. The preventive approach would then be to protect genetically susceptible individuals from all possible precipitating factors.

SUMMARY

Many quantitative or metrical traits such as height and intelligence fit a unimodal curve (normal distribution), suggesting that no major genetic factor is segregating but rather that the trait is determined by many genes, each contributing a small amount to the final result (polygenic inheritance). In most cases the trait (or variation) is produced by an interaction of polygenic and environmental factors. This system of genetic-environmental interaction is termed multifactorial inheritance. For a given population, the proportion of the variation in the trait that results from the genetic variation is the heritability of the trait.

Certain discontinuous characters may be produced by the imposition of a developmental threshold on a continuously distributed variable (quasicontinuous variation). A number of models of quasicontinuous inheritance have been proposed. For common traits it may be difficult, if not impossible, to distinguish the multifactorial threshold model from one that is based on a major gene with reduced penetrance in some cases and on many genes, each with a small effect, in the majority of cases, or from other similar models.

A number of features characterize multifactorial threshold diseases, including the relation of recurrence in first-degree relatives to the population frequencies, nonlinear decrease in frequency with decreasing relationship to the proband, increased risk after two affected first-degree relatives, increased risk with increased severity of the defect, and a higher recurrence risk when the proband is of the less frequently affected sex. These variations in recurrence risk are, in general, not large enough to make much difference to genetic counseling.

REFERENCES

1. Cavalli-Sforza, L. L., and Bodmer, W. F.: The Genetics of Human Populations. San Francisco, W. H. Freeman, 1971.
2. Carter, C. O.: Genetics of common disorders. Br. Med. Bull. 25:52, 1969.
3. Edwards, J. H.: Familial predispositions in man. Br. Med. Bull. 25:58, 1969.
4. Falconer, D. S.: The inheritance of liability to diseases with variable age of onset, with particular reference to diabetes mellitus. Ann. Hum. Genet. 31:1, 1967.
5. Fraser, F. C.: The genetics of cleft lip and palate. Am. J. Hum. Genet. 22:336, 1970.
6. Gruneberg, H.: Genetical studies on the skeleton of the mouse. IV. Quasi-continuous variations. J. Genet. 51:95, 1952.
7. Harris, H.: The Principles of Human Biochemical Genetics. Amsterdam, North Holland, 1970.
8. Jensen, A. R.: How much can we boost IQ and scholastic achievement? Harvard Educ. Rev. 39:1, 1969.
9. Morton, N. E., Yee, S., Elston, R. C., and Lew, R.: Discontinuity and quasi-continuity: Alternative hypotheses of multifactorial inheritance. Clin. Genet. 1:81, 1970.
10. Price Evans, D. A., Manley, K. A., and McKusick, V. A.: Genetic control of isoniazid metabolism in man. Br. Med. J. 2:485, 1960.
11. Smith, C.: Discriminating between modes of inheritance in genetic disease. Clin. Genet. 2:303, 1971.
12. Smith, C.: Recurrence risks for multifactorial inheritance. Am. J. Hum. Genet. 23:578, 1971.
13. Wright, S.: An analysis of variability in number of digits in an inbred strain of guinea pigs. Genetics 19:506, 1934.

Chapter 12

SOME MALFORMATIONS AND DISEASES SHOWING MULTIFACTORIAL INHERITANCE

Among the genetically determined disorders, a frequently accepted point of division between common and uncommon diseases is a population frequency of 1 in 1000. Multifactorial diseases are, in general, common, and diseases caused by mutant genes are uncommon. Diseases in the third major category of genetic diseases, chromosomal anomalies, fall on either side of this division. There is an increasing number of disease entities for which there are data consistent with multifactorial inheritance. A selection of these will be presented in this chapter.

The genetic counseling of families having diseases determined by multifactorial inheritance may be more difficult than for single mutant gene disorders, since the risks given are usually average risks, rather than precise probabilities. Empiric recurrence risks are becoming increasingly available but are often incomplete for a given disease, even if the multifactorial etiology is reasonably well established. For example, although recurrence risks after one affected first-degree relative may be known, the recurrence risk after two or three affected first-degree relatives may not be established. Under these circumstances, it is reasonable to apply generalizations gained from experience with multifactorial diseases in which such empiric risks have been established to multifactorial diseases for which there are no established risks. Smith[17] provides a useful theoretical model for calculating such risks, examples of which are given in Table 12-1.

From the **theoretical** model, and from experience with cleft lip and palate, spina bifida/anencephaly, and some congenital heart lesions, the recurrence following the birth of two affected first-degree relatives would be 2 to 3 times greater than that after one. Beyond two affected first-degree relatives, there are little data for any diseases, but the experience with some congenital heart lesions suggests that the risk becomes quite high, as predicted in the table.

Clinical Example

This couple appeared for genetic counseling because their second-born child, an 8-month-old boy, had spina bifida, menin-

TABLE 12–1. Recurrence Risk (%) for Multifactorial Diseases According to Number of Affected First-Degree Relatives and Heritability[a]

Population Frequency (%)	*Affected Parents*	0			1			2		
	Heritability (%)	*Affected Sibs* 0	1	2	*Affected Sibs* 0	1	2	*Affected Sibs* 0	1	2
1.0	100	1	7	14	11	24	34	63	65	67
	80	1	6	14	8	18	28	41	47	52
	50	1	4	8	4	9	15	15	21	26
0.1	100	0.1	4	11	5	16	26	62	63	64
	80	0.1	3	10	4	14	23	60	61	62
	50	0.1	1	3	1	3	7	7	11	15

[a] This table, adapted from Smith,[17] provides theoretic recurrence risks for a multifactorial threshold character. It can be used as a guide where there are no empiric figures available. For instance, to estimate the risk for the next child of diabetic parents who have a diabetic child, the population frequency is close to 1% and heritability is, say, 80%, so the risk would be about 4%. For a parent with cleft lip and palate who has two affected children, the population frequency is about 0.1, the heritability is about 80%, and the risk would be about 23%. The risk decreases with each unaffected child, but not very much.

gomyelocele, and hydrocephalus. The baby had paraplegia and an expanding head size despite surgical intervention with a shunting procedure. The couple also had a normal 3-year-old daughter. They were aware that their infant son would not survive much longer and hoped to have another child, and yet they felt they could not readily withstand the emotional and financial stress of having another child with spina bifida.

No other child on either side of the family was known to have spina bifida. There was a first cousin with cleft lip and palate and an aunt who was a "blue baby" and was considered to have had a congenital heart lesion, but there was no autopsy performed to confirm this clinical suspicion. This family, which had moved to our city recently from Boston, was third-generation Irish-American.

The mother had had a "virus" during the second month of her pregnancy and had taken dextroamphetamine to "initiate a diet," also during the second month. This young woman was slightly overweight and admitted that 2 or 3 times a year she had to diet for a week or so and always got her diet off to a successful start with 3 or 4 days of dextroamphetamine.

In counseling, we usually discuss in some detail the background information we have on diseases like the spina bifida/anencephaly complex—the hereditary predisposition and the environmental triggers. We began by discussing the hereditary predisposition in this family of Irish extraction, in which there had been no outbreeding with other Boston populations. There is evidence to suggest increased genetic liability in those of wholly Irish ancestry, whether they remain in Ireland or move elsewhere.[11] With outbreeding, the predisposition seems to diminish. This is not to cast aspersions on the Irish; it is only a way of making the point about the genetic predisposition to the lesion.

We then discussed the possible environmental triggers and mentioned that "clusters" of spina bifida cases have been reported on a number of occasions. This may mean, although there is no proof of it, that certain viruses may act as environmental triggers in individuals with hereditary predisposition to spina bifida. We also discussed what is known or suspected about dextroamphetamine–the suspicion of some investigators that dextroamphetamine may be implicated in some cases of spina bifida in the human and our own experience with this drug in the production of exencephaly in the mouse. Again we emphasized that we were discussing suspicions, not proof. At that time, blighted potatoes had not come under suspicion.[5]

We told the parents that the risk of having another child with spina bifida was of the order of 5%. Some families consider this a prohibitive risk. Other families think of this as quite a small risk, as did this family. The couple expressed a desire to consider a future pregnancy. Because they had been reading about amniocentesis, they wondered if this could be used to identify spina bifida *in utero*. We explained that it could not, although radiologic or ultrasound visualization might detect the more severe forms of the condition, and perhaps direct visualization of the fetus would be feasible within a few years. The significance of alpha-fetoprotein had not yet been demonstrated (see Chapter 22). We had no trouble in convincing the mother of the inadvisability of dextroamphetamine ingestion during pregnancy (or, for that matter, at any other time). We told her that although we could not assure her that avoidance of dextroamphetamine would reduce the recurrence risk below 5%, we felt that it would be prudent to eliminate as many potential environmental triggers as possible.

The affected child in this family died of sepsis at 14 months of age, when the mother was 3 months pregnant. She eventually delivered a normal 8-pound baby boy.

ANENCEPHALY/SPINA BIFIDA

These disorders are generally accepted as being related. Spina bifida is a failure of fusion of spinal lamina, most often in the lumbar region. In *spina bifida occulta* the defect is limited to the bony arch. In *spina bifida aperta* the bony defect is accompanied by a protrusion of the meninges covering the spinal cord (meningocele), or a protrusion of meninges containing tissue from the cord (meningomyelocele); hydrocephalus may also occur. This has led to some confusion in family studies reported in the literature because of a tendency to combine family data from cases of isolated hydrocephalus and hydrocephalus with spina bifida. Hydrocephalus associated with spina bifida should be considered a secondary manifestation of the spinal defect; hydrocephalus without spina bifida should be considered separately. *Anencephaly* is an absence of skin, skull, overlying membranes, forebrain, and midbrain that produces stillbirth or death shortly after birth. Embryologically, and genetically, spina bifida aperta and anencephaly appear to be variations of the same basic defect, a failure of the embryonic neural tube to close.

The usual findings in patients with spina bifida with meningomyelocele are paralysis of the lower extremities and urinary and fecal incontinence. Progressive hydrocephalus may be arrested by neurosurgical intervention, but the course has usually been one of progressive deterioration with an infection often being the terminal event. Recent improvements in surgical management are improving the outlook, however. Paradoxically, this may lead to an increase in the number of living but crippled children, since improved surgery may save the lives of children who would otherwise have died but now remain crippled, more often than it converts crippled to noncrippled children.

The population frequency of this group of disorders is variable.[11] It is, for instance, high in Ireland, Wales, Alexandria, and Bombay, and low in Orientals and black Africans. The population frequency for all lesions in the spina bifida/anencephaly group ranges from a low of about 1 per 1000 in several Oriental and African populations to a high of over 1% in Ireland. An intermediate figure of 3 per 1000 is the approximation we currently use for North America.

TABLE 12–2. Recurrence Risks for Sibs[a] of Children with Various Multifactorially Inherited Malformations

	Sex of Proband	Sex of Sib	*Proband has* Anencephaly[b]	Spina Bifida[b]	Cleft Lip ± Cleft Palate	Cleft Palate	Pyloric Stenosis
Population frequency/1000			0.2–5	0.2–4	1	0.4	3
Sex ratio M:F			0.6	0.8	1.6	0.7	5
% risk for sib after one affected sib—parents normal	male	male					5
			1.9	5.1	3.9	6.3	
	male	female					2.5
	female	male					18.0
			6.5	7.2	5.0	2.3	
	female	female					8.0
	either	either	4.1	6.1	4.3	2.9	6.0
% risk for sib after two affected sibs—parents normal				10	9		

[a] Risks for offspring are expected to be and (where measured) are found to be of similar magnitude.
[b] Risk for either anencephaly or spina bifida.

A recurrence risk of 3–6% in first-degree relatives has been derived from a number of European and North American studies.[10] Whether the proband has anencephaly or spina bifida aperta, the sibs are at risk for either one or the other or both. The frequency in first-degree relatives is about seven times the population frequency, and the recurrence risk after two affected first-degree relatives is about 10% (Table 12-2).

The sex ratio in this group of lesions reveals an excess of females, the male/female ratio being 0.8 for spina bifida and 0.6 for anencephaly according to one large study. There is also a small excess of first-born infants and infants for whom the maternal age exceeds 40 years.

That environmental triggers are in all probability involved in the production of these lesions was suggested in the preceding case presentation. The striking variations with season of birth, socioeconomic class, and geographic region, as well as variations from year to year, certainly suggest environmental factors.[11] However, no specific teratogen has yet been identified as playing a major role in this group of malformations, although blighted potatoes (or some associated factor) has recently been suggested as a possibility.[5]

CLEFT LIP AND CLEFT PALATE (FIG. 12-1)

The evidence that congenital clefts of the primary and secondary palate are multifactorially determined threshold characters has been discussed in Chapter 11. The secondary palate closes later in development

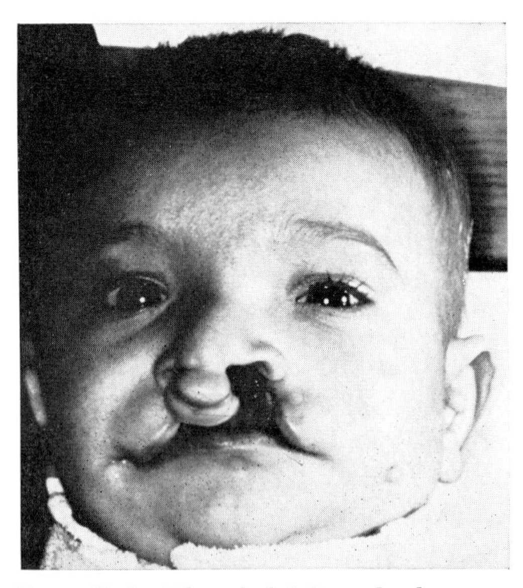

Figure 12–1. Bilateral cleft lip and palate.

than the primary palate, which forms the gum and lip, and the genetic and environmental factors that influence its closure may differ from those that influence closure of the primary palate. On the other hand, abnormal development of the primary palate leading to a cleft lip may interfere secondarily with secondary palate closure. Thus, on both embryologic and genetic grounds, congenital cleft lip (CL) and cleft lip with cleft palate (CLP) appear to be etiologically related [in data combining the two, they may be designated CL(P)]. Cleft of the secondary palate (CP) is an etiologically independent entity. Oddly enough, however, some mutant genes will cause cleft lip and palate in some individuals and cleft palate in others—the lip-pit syndrome, for instance (Chapter 5, p. 68). The embryologic basis for this remains obscure.

There are a large number of syndromes in which CL(P) or CP may be one of the features.[9] For most of these the cause is unidentified; a few are associated with recognizable chromosomal aberrations; and about a third are caused by major mutant genes. Each of these syndromes is rare, and together they may account for perhaps 5% of all cases, most of the rest being multifactorially determined. For counseling, it is important to distinguish cases associated with syndromes from the multifactorial type, as the recurrence risks are different. Inclusion of unrecognized cases of syndromes in family studies may account for the fact that the recurrence risk for sibs of patients with CL(P) is reduced from 4% to 2% if the proband has an additional malformation. Furthermore, we reiterate that it is difficult, on the basis of the present evidence, to distinguish between a multifactorial, additive model and one involving a major gene and several minor genes, but the recurrence risks are about the same.[4]

Cleft Lip with or without Cleft Palate—CL(P)

Most cases of CL(P) without associated malformations appear to be multifactorially determined, although even in this group there is almost certainly genetic heterogeneity.[5] More males than females are affected with CL(P). There are striking differences in frequency among races: Orientals have relatively high frequencies (1.7 per 1000 births), Caucasians are intermediate (1/1000), and Negroes tend to have low frequencies (0.4/1000). These differences persist in different geographic regions, suggesting that they do not result from environmental alterations, and it is tempting to think that they may be associated with differences in face shape.[8]

As expected for multifactorial traits, the risk for relatives of affected persons drops off sharply from first- to second-degree relatives and much less sharply from second- to third-degree relatives. Thus it is about 4% for sibs and offspring, 0.7% for uncles, aunts, nephews, and nieces, and 0.4% for first cousins. Occasionally, these figures are useful for counseling—for instance, the sib of a person with CLP can be advised that the chance of having an affected child is about 0.7%, or 1 in 140.

The recurrence risk is somewhat higher in the sibs of female probands (5.0%) than in sibs of male probands (3.9%), in the sibs of patients with a severe defect than in sibs of those with a mild form (5.6% when the proband has a bilateral cleft lip and palate and 2.6% when the proband has a unilateral cleft lip only), and after two affected sibs (9.0%) than after one (4%) (Table 12-2).[7] These findings are also consistent with the multifactorial-threshold concept.

There are few data on the risk for sibs of an affected child with an affected parent. The available figures suggest a figure of about 15%, but the numbers are small.

Experimental findings in the mouse predict that shape of the embryonic face influences the predisposition to cleft lip. If so, one should be able to detect indications of these differences in the near relatives. Preliminary evidence suggests that this is so; parents of children with CLP tend to have less prominent maxillae, an increased bizygomatic diameter, and more rectangular or trapezoid-shaped faces than controls.[8]

In spite of many attempts to demonstrate environmental factors associated with CL(P) no association has been convincingly demonstrated between CL(P) incidence and such things as seasonal trend, geographic location (except when there are differences in racial groups), social class, maternal age or parity, and paternal age.[11] A number of prenatal factors have been tentatively implicated, such as pernicious vomiting of pregnancy, antiemetics and other drugs (anti-epileptic drugs are currently under suspicion), maternal bleeding, toxemia of pregnancy, and Toxoplasma, but none is firmly supported.

Cleft Palate (CP)

Isolated cleft palate is rarer than CL(P), with an average frequency in Caucasians of about 0.45 per 1000 births. More females than males are affected. There is little, if any, racial variation. The frequency appears somewhat higher than average in children of older mothers of high parity and (for those with associated malformations) seems to increase with parental age.[11]

Where available, the recurrence risks for CP are similar to those for CL(P), except that the risk is higher for sibs of male probands (6.3%) than for sibs of female probands (2.3%), as expected for a multifactorial threshold character (Table 12-2). The rather low rate in MZ co-twins (23.5% vs. 10% in DZ co-twins) suggests that the environmental contribution is larger than for CL(P).[7]

DIABETES MELLITUS

It is becoming increasingly apparent that diabetes mellitus fulfills the expectations for multifactorial inheritance.[14] A development of this point of view and equal time for opposing opinions will not be provided here. The condition is undoubtedly heterogeneous, and recurrence risk estimates vary widely from study to study, varying, no doubt, with the degree of bias in ascertaining families, the diagnostic criteria used, and the statistical methodology. Since the most relevant event in counseling situations is usually the advent of juvenile diabetes, figures will be given for onset before age 20. It is clear that the risk in near relatives is greater if the proband has juvenile, rather than late-onset, diabetes and that it increases with the number of affected relatives. No doubt the risk will also vary with the diet and other environmental factors. The following estimates are presented as rough guides for counseling.

The frequency of diabetes in the general population is about 0.15% for individuals less than 20 years old, 0.6% for those under 40, 2.25% for those under 60 and 6% for those under 80. For normal parents, the risk for sibs of an early-onset (under 20) diabetic themselves having early-onset diabetes is about 2–5% and would be expected to double for sibs of two affected children.

The risk of a child having early-onset diabetes is about 10% if one parent has early-onset diabetes and about 6% if a parent has late-onset diabetes. For an affected parent and child, the risk for subsequent sibs should be higher than this—perhaps double, but there are very few data.

Information on offspring of two diabetic parents is also sparse, and estimates range from 20% to 70% or more. In any case, it seems well below the figure of close to 100% expected on the basis of recessive inheritance postulated by some workers.

So far, there is no reliable test for detecting those individuals who will later develop diabetes, though the presence of an insulin antagonist in the plasma albumen and/or impaired glucose tolerance curve with or without previous glucocorticoid "stress" can be demonstrated in near relatives more often than in the general population.

THE EPILEPSIES

Most workers put epilepsy in the multifactorial inheritance category, although a minority of investigators favor mendelian inheritance. One must be cautious about discussing epilepsy in broad terms and should speak of specific types of epilepsy.

Recent twin studies with careful documentation of zygosity have not been found, but a high concordance rate among presumed monozygotic twins and a low concordance rate among dizygotic twins have been reported by several workers.

Apart from convulsions associated with diseases of known mendelian causation, the only category of the epilepsies for which there are reasonably reliable figures on recurrence risk is that of subcortical or centrencephalic epilepsy.[1,13] Even here, the problem is complicated by variable age of onset and the fact that the characteristic "spike-wave" EEG abnormality, which may be the only indication of predisposition, may disappear with increasing age. The following figures, taken from the studies of Metrakos and Metrakos,[13] may provide a rough guide for counseling. A child who has a parent or sib with centrencephalic epilepsy has a basic risk of about 8% of being similarly affected. This is increased to 10% if the onset in the relative occurs before 2½ years of age. If both a parent and sib are affected, the risk for a sib is about 13%. The risk decreases the longer the child remains free of seizures and if the EEG is still negative after 5 years of age.

MENTAL RETARDATION

The distribution of "intelligence" in the general population is more or less normally distributed, as one would expect for a character with a polygenic hereditary component influenced by a number of subtle environmental factors. Thus, if one arbitrarily decides that anyone with an IQ of less than, say, 70, is retarded, some individuals will be retarded simply because they received an assortment of genetic and environmental factors, each detracting a small amount from the level of intelligence, that placed them in the lower tail of the normal distribution without any one of these factors being in itself abnormal. In addition, a child may be retarded because of a major insult to the brain, which may result from a chromosomal anomaly, an inborn metabolic error, or an environmental cause such as birth trauma, prenatal viral damage, or postnatal meningitis. These cases, which are relatively rare, form a small hump near the lower tail of the distribution of intelligence—i.e., the curve is "skewed to the left."

Thus the causes of mental retardation fall into the same four categories as do congenital cleft lip, heart malformations, and other common, familial disorders: those due to major mutant genes, to chromosomal aberrations, to major environmental insults, and to a multifactorial etiology.

As one might expect, children with specific causes of mental retardation, as in the first three groups, tend to be more severely retarded than those in the multifactorial group, since the damage in the former groups results from an insult that is likely to be major. Also consistent with this concept is the following, seemingly paradoxical, fact. The distribution of intelligence for the sibs of children with specific, and therefore severe, types of retardation is bimodal—there is a peak in the low-IQ range, representing similarly affected sibs, and another peak corresponding to the normal population. In other words, if a child is severely retarded his sibs are likely to be severely retarded or normal. On the other hand, in the sibs of children with nonspecific, and therefore milder, mental retardation, the distribution of intelligence is unimodal, but the mean is lower than that of the normal population. So the sibs of a child with severe retardation will have a higher IQ (excluding those similarly affected) than sibs of a child with mild retardation.

When dealing with a case of mental retardation, the genetic counselor first strives to place the child in one of the four etiological categories, by a thorough family prenatal and perinatal history, physical examination, and appropriate chromosomal and biochemical screening. If a specific cause is found, counseling is based on the appropriate estimate of recurrence risk. If no specific cause can be found, the child is assumed, by exclusion, to fall in the multifactorial group. On the average, then, the intelligence of the sibs will be midway between that of the two parents (averaged) and that of the general population. The empirical risks presented in Table 12-3, taken from the large study of Reed and Reed,[15] will serve as a rough guide in the appropriate family situation.

The growing evidence for an X-linked, nonspecific type of mental retardation, based on an excess of brother–brother pairs in sibs with mental retardation, would suggest a somewhat higher risk if the proband is a boy, particularly if the family history suggests a sex-linked recessive pattern.[18]

THE PSYCHOSES

The genetics of the major psychoses is too vast a subject to be dealt with adequately here. There is strong evidence for a genetic basis for susceptibility, particularly in the case of schizophrenia; concordance rates are much higher for monozygotic than for dizygotic twins, and there is an increased frequency in children of schizophrenic parents, even when reared in adoptive or foster homes. The population frequency of schizophrenia is about 1%, and the risk for the sib or child of a schizophrenic developing the disease is estimated in various studies as about 10%. For the offspring of two affected parents, estimates vary from 30 to 70%, in the same range as the risk for monozygotic co-twins. The figures are considerably higher if "schizoid" relatives are included. Whether these figures are interpreted as reflecting polygenic inheritance or an autosomal dominant gene with modifiers is a question that may be solved only when (and if) the genetic predisposition is defined in biochemical terms. Childhood autism seems to have a low recurrence risk and does not behave genetically as a form of schizophrenia.

For the affective psychoses the picture is also unclear, though most studies report higher recurrence risks than for schizophrenia, approaching the expectation for a dominant major gene in some studies. On the average, for probands with cyclic depression, the risk for parents and sibs has been estimated as about 20%, and for recurrent depression about 13%. For recent reviews of this confused and confusing subject, consult Erlenmeyer-Kimling,[6] and Rosenthal.[16]

TABLE 12-3. Risk of Mental Retardation in Children and Sibs of Retardates (IQ <70)[a]

Number of Retarded			
Parents	*Children*	*Risk for*	*Risk (%)*
0	—	child	1
1	—	child	11
2	—	child	40
0	1	sib	6
1	1	sib	20
2	1	sib	42

[a] Data from Reed and Reed.[15] Probands with profound retardation are excluded.

PYLORIC STENOSIS

This, the first of the multifactorial threshold characters, is a disorder in which projectile vomiting, undernutrition, dehydration, and electrolyte imbalance result from a muscular hypertrophy of the pylorus. A Rammstedt procedure, incising the hypertrophied pyloric muscle, permanently relieves the condition. The frequency in North American and European populations is about 3 per 1000.

The M : F sex ratio is 5 to 1. Carter's observation that the recurrence risks are much higher to the first-degree relatives of affected females than to the first-degree relatives of affected males was the clue to its multifactorial threshold nature.[3] As discussed in the preceding chapter, if a disease is found more frequently in one sex (in this case, males), then individuals of the other sex (females) require a greater genetic predisposition in order to develop the disease, since they are, on the average, farther from the threshold. If a greater number of "liability genes" are required for the disease to become manifest in the female, it follows that her first-degree relatives would also have greater genetic liability and a higher frequency of the disorder than the first-degree relatives of a male patient, with less genetic predisposition. Table 12-2 shows that this is so.

SUMMARY

A selection of relatively common disorders is discussed with respect to the evidence for their multifactorial (and often heterogeneous) nature, and figures are presented as guides for genetic counseling. In alphabetical order, the disorders discussed are anencephaly and spina bifida, cleft lip and palate, diabetes mellitus, the epilepsies, mental retardation, the psychoses, and pyloric stenosis.

REFERENCES

1. Beaussart, M., and Luiseau, P.: Hereditary factors in a random population of 5200 epileptics. Epilepsia 10:55, 1969.
2. Carter, C. O.: Multifactorial inheritance revisited, p. 227. *In* F. C. Fraser and V. A. McKusick (ed.), Congenital Malformations. New York, Excerpta Medica, 1970.
3. Carter, C. O., and Evans, K. A.: Inheritance of congenital pyloric stenosis. J. Med. Genet. 6:233, 1969.
4. Chung, C. S., Ching, G. H. S., and Morton, N. E.: A genetic study of cleft lip and palate in Hawaii. II. Complex segregation analysis and genetic risks. Am. J. Hum. Genet. 26:177, 1974.
5. Emanuel, I., and Sever, L. E.: Questions concerning the possible association of potatoes and neural-tube defects, and an alternative hypothesis relating to maternal growth and development. Teratology 8:325, 1973.
6. Erlenmeyer-Kimling, L. (ed.): Genetics and mental disorders. Int. J. Mental Health 1:8, 1972.
7. Fraser, F. C.: The genetics of cleft lip and cleft palate. Am. J. Hum. Genet. 22:336, 1970.
8. Fraser, F. C., and Pashayan, H.: Relation of face shape to susceptibility to congenital cleft lip. A preliminary report. J. Med. Genet. 7:112, 1970.
9. Gorlin, R. J., Cervenka, J., and Pruzansky, S.: Facial clefting and its syndromes. Birth Defects VII:3, 1971.
10. Laurence, K. M.: The recurrence risk in spina bifida cystica and anencephaly. Dev. Med. Child Neurol. 20(Suppl. 1):23, 1969.
11. Leck, I.: The etiology of human malformations: Insights from epidemiology. Teratology 5:303, 1972.
12. Lennox, W. G., and Jolly, D. H.: Seizures, brain waves and intelligence tests of epileptic twins. Proc. Assoc. Res. Nerv. Ment. Dis. 33:325, 1955.
13. Metrakos, J. D., and Metrakos, K.: Childhood epilepsy of subcortical ("centrencephalic") origin. Clin. Pediat. 5:536, 1966.
14. Neel, J. V.: Current concepts of the genetic basis of diabetes mellitus and the biological significance of the diabetic predisposition, p. 68. *In* Supplement to the Proceedings of the VI Congress of the International Diabetes Federation.
15. Reed, R. W., and Reed, S. C.: Mental Retardation: A Family Study. Philadelphia, W. B. Saunders, 1965.
16. Rosenthal, D.: Genetic Theory and Abnormal Behavior. New York, McGraw-Hill, 1970.
17. Smith, C.: Recurrence risks for multifactorial inheritance. Am. J. Hum. Genet. 23:578, 1971.
18. Turner, G., and Turner, B.: X-linked mental retardation. J. Med. Genet. 11:109, 1974.

Chapter 13

TERATOLOGY

PRINCIPLES OF TERATOLOGY

The developing embryo is not a "little adult." The requirements for making a heart or a brain or a bone or an eye are quite different from those for maintaining these structures. One dose of thalidomide, which most effectively accomplishes its purpose of sedating the mother, without any damaging effect to her, may horribly malform her unborn child. A corollary of this, first documented by the classical work of Warkany in the 1940's, is that the fetus is not as well protected by the uterus as had previously been assumed. Teratogens act by causing damage to the cells of the developing embryo, poisoning enzymes, or otherwise interrupting the normal developmental processes. Their results do not stem from changes in the genetic material. Thus teratogenic effects must be clearly distinguished from mutagenic effects. Some mutagens are teratogenic and vice versa, but the mechanisms are quite different.

PERIODS OF VULNERABILITY AND EXPOSURE

Teratogens act at vulnerable periods of embryogenesis and organogenesis. There is no demonstrable teratogenic effect prior to implantation (1 week in the human). After this, the nature of the teratogenic damage depends on what developmental processes are going on at the time of exposure.[3,13] However, the process deranged by the teratogen usually precedes the morphogenetic process that goes wrong as a result of the derangement. For example, exposure to thalidomide has its teratogenic effect on the heart about 2 weeks before the completion of the developmental event. Exposure to thalidomide in the 5th week may cause a ventricular septal defect, although the ventricular septum does not close until the seventh week.[6] From mouse studies, anti-heart antibody must be given later in gestation than dextroamphetamine ("speed") to cause ventricular septal defect[10]; presumably, it impairs some other step leading to closure. In general, the teratogenic exposure occurs a considerable time before the developmental event. This is what is treacherous about teratogenic insults; many organ systems, such as the heart, are vulnerable at a time when the mother is just becoming convinced she is truly pregnant and not "just a week or two late." She then decides to stop taking dextroamphetamine for her "diet," but this may already be too late, if the drug is teratogenic.

WHAT AGENTS ARE TERATOGENIC?

Although relatively few agents have been identified that are potent teratogens capable of causing malformations in a significant proportion of human embryos exposed at a vulnerable period of embryogenesis,[13] there are probably hundreds of agents that are teratogenic within a given set of circumstances, such as hereditary predisposition to a malformation and hereditary predisposition to respond adversely to a drug or virus, or some interacting environmental factor such as a nutritional deficiency. Rubella and thalidomide, which are discussed later in this chapter, may be less representative of teratogens than, say, coxsackievirus or dextroamphetamine.

Relationship to Dose[3,4]

In experimental animals, the teratogenic dose usually overlaps the dose that will kill some of the embryos. There are exceptions, however. Some agents that will kill embryos do not seem to be teratogenic at sublethal doses. Conversely, teratogenic effects may occur at doses well below the embryo-lethal dose, thalidomide being an outstanding example.

There is an additional dosage relationship. In experimental animals, the dosage of drugs on a milligram per kilogram basis required to produce a malformation is usually many times the normal dosage in the human. One assumption is that large doses in experimental animals may mimic the effect of "normal" doses in humans who have a predisposing condition (such as delayed clearance or abnormal metabolic response). An entire subdiscipline of genetics, pharmacogenetics, addresses itself to related problems (Chapter 14).

Hereditary Predisposition to the Effects of a Given Teratogen

It is clear that there are species and individual differences in response to the teratogenic effects of an agent. One of the first clearly demonstrated examples was that of cleft palate induced by cortisone in the mouse. The same dose produced a frequency of 100% in the A/J inbred strain and only 20% in the C57BL/6 strain.[3] Several genes are involved, and maternal uterine differences as well.

Genetic differences in susceptibility became particularly important when it was discovered that thalidomide, a sleeping pill used for nausea of pregnancy, was teratogenic in man. The malformations so readily produced by thalidomide in the human could not be reproduced in a number of traditional animal models. And not all pregnant women who ingested thalidomide during vulnerable periods of embryogenesis produced malformed infants. There is ample experimental evidence of genetic differences that influence the embryo's response to a teratogen, and these may be single gene or polygenic differences.[3] This makes it impossible to predict with confidence, from animal studies, whether a new drug or other agent will be teratogenic in man.[4,13]

Interaction of Teratogens and Other Agents

It should also be noted that drugs taken in combination with other drugs may produce malformations even though the drugs taken singly would have no teratogenic effect. It is often the case that a pregnant woman is treated for a disorder not with one but with several drugs. Furthermore, a given drug may be teratogenic only through interaction with another environmental factor, such as a virus or nutritional deficiency.[4,14] From the clinical point of view, this is another pitfall in the path of safe drug administration, and, from the research point of view, it is a further obstacle in defining specific teratogens.

Specificity of Teratogens

Finally, certain drugs and viruses cause malformations and patterns of malformation that are characteristic (e.g., phocomelia from thalidomide and patent ductus arteriosus, deafness, and cataracts from rubella virus). These patterns must be related to special properties of the teratogens: In the case of drugs, there are specific metabolic effects; and in the case of rubella virus, consequent continued proliferation interferes with the host's normal processes of growth and development.

Hereditary Predisposition to the Malformation

This brings us back to the concept of multifactorial inheritance and allows us to summarize the *three components of a typical teratogenic exposure* leading to a malformation: (1) hereditary predisposition to a malformation; (2) hereditary predisposition to the effects of a given teratogen; and (3) administration of the teratogen at a vulnerable period of embryogenesis. The latter two components have been discussed above.

To begin with animal models, about 1% of C57BL/6 mice "spontaneously" have a ventricular septal defect, but none has atrial septal defect. Exposure of the pregnant female at day 8 of gestation to a single large dose of dextroamphetamine produces ventricular septal defects in 11% of offspring. Exposure to antiheart antibody produces ventricular septal defects in 20% of C57BL/6 offspring. The A/J strain has a low spontaneous frequency of atrial septal defect but not ventricular septal defect. Similar exposures of dextroamphetamine and antiheart antibody to female A/J mice produce atrial septal defects in 13% and 22% of offspring, respectively, in this strain of mice. Thus, the C57 appears to have a hereditary predisposition to ventricular septal defect, and the A/J a predisposition to atrial septal defect. The same teratogens cause different lesions in the two strains of mice.[10] The predisposition to a rather specific type of heart defect is seen in human studies, in which affected members of a family have the same anomaly—ventricular septal defect in one family; atrial septal defect in another.

From experience with human beings and experimental animals, it would seem likely that in the majority of instances in which maldevelopment is produced by a teratogen, there is a hereditary predisposition to the specific malformation that results. The defect occurs when an exposure to one or more of any number of teratogens occurs at a vulnerable period of embryogenesis. There are probably few teratogens like thalidomide and rubella virus that are capable of producing such a high frequency of malformations following maternal exposures in the first trimester. Even in some of these cases, there is evidence of hereditary predisposition to the malformation, as will be discussed in the section on the rubella syndrome.

WHAT IS THE ROLE OF TERATOGENS IN HUMAN MALFORMATIONS?

Even before the thalidomide disaster and rubella pandemic of the past decade, parents of children with birth defects were deeply concerned about the effects of drugs, illnesses, radiation, and other environmental triggers on their unborn children. This concern has become intensified by continuing publicity about documented and potential environmental hazards.

The question of teratogens in human malformations may be looked at in two settings: that of the patient and his family and that of the research investigator. A clinical (and investigative) example will illustrate the problem.

Dextroamphetamine and Congenital Malformations

An unusually large and handsome newborn male infant was observed on the second day of life to have cyanosis. A cardiac catheterization revealed that the infant had transposition of the great vessels. The mother volunteered that she had been taking dextroamphetamine sporadically during the first 2 months of her pregnancy and regularly during the last trimester to curb her appetite and control her weight gain. She asked directly if this drug could have caused her baby's heart disease.

This was a more provocative question than the mother had anticipated, because she was the third mother in 2 weeks to present with this history of having a child with transposition of the great vessels and a first trimester exposure to dextroamphetamine. To the first mother in this series of cases, we had responded with the confident assurance that her baby's heart disease and the dextroamphetamine exposure were unrelated. To the second, we had said that there was no evidence in the literature implicating dextroamphetamine as a teratogen. But to this third mother, we had to confess that, while there was no evidence, our suspicions were now aroused and we would have to investigate the problem.

The parents of a child with a birth defect usually have some feelings of guilt. If the etiology is clearly genetic, they still feel some guilt regarding the contribution of "bad genes," however vague this concept may be. However, when a readily avoidable environmental exposure, such as an unnecessary drug, comes under suspicion, the cause-and-effect relationship becomes less vague to the parents, more easily understood, and more guilt-producing. To this mother, the sense of guilt became very great indeed. If only she had possessed more willpower, she could have dieted without dextroamphetamine, and her baby would have been normal, she said. We argued that there was still no evidence for an etiologic relationship, but the mother had already reached her own conclusions. In subsequent visits she began to seek for those who should share her guilt. The parents went so far as to retain legal counsel with the aim of bringing a suit against the physician who prescribed the dextroamphetamine and the company that manufactured it. We were eventually successful in dissuading the family from pursuing this course.

The next problem was to try to discover if dextroamphetamine really had the potential to produce human malformations—specifically, congenital heart diseases. The ultimate answer would be gained from studies in humans, so prospective and retrospective clinical studies were instituted. As a collateral investigation, experimental studies were undertaken in mice. This sequence, a clinical observation arousing suspicion and leading to investigations in humans and in animal models, is typical of studies in this area.

Retrospective studies take as the index case a patient with the disorder under investigation and search, retrospectively, for some cause. The retrospective protocol employed in the study of dextroamphetamine required an extensive teratogenic history, which explored exposure not just to dextroamphetamine but to over 100 teratogens and potential teratogens—viruses, radiation, and chemicals as well as drugs. A family history was also taken.

A single history-taking session was rarely sufficient for either the genetic or teratogenic history. Family members had to be contacted who were more familiar with the "cousin who died in infancy and may have been blue." Autopsy records had to be obtained. Physicians had to be contacted regarding prescriptions, and a search undertaken of medicine cabinets for nonprescription items.

Without elaborating on all the pitfalls, we should mention a few points. Maternal-memory bias is an enormous obstacle—the mother is more likely to remember an unusual incident if the baby was abnormal than if it was normal. This applies particularly to teratogenic exposure in which a physician has not been involved (mild infections, nonprescription drugs, toxic chemicals, and the like). One would think that prescription items, at least, would be an easy matter to verify. Some physicians are very cooperative, but, surprisingly, others are not. Fear of lawsuits may be playing a role here. A control group *matched* for as many factors as possible must be utilized. To reduce maternal-memory bias, the control proband should have some defect not environmentally induced.

When we chose infants 1 year old, or less, and a period of vulnerability based on recent evidence regarding the time of closure of the ventricular septum during embryonic development, a statistically significant difference was obtained. The congenital heart group had a higher frequency of maternal exposure to dextroamphetamine at the vulnerable stage of cardiac development than the control group and also had a higher frequency of positive family history for congenital heart lesions.[11] This suggested that exposure to dextroamphetamine may indeed interfere with normal development of the heart, particularly if there is a genetic predisposition to the resulting defect.

A *prospective* study (in which cases known to be exposed are identified, and the outcome then observed) on the teratogenic effects of maternal exposure to dextroamphetamine was initiated at the same time as the retrospective study. The obstetrical service of a large private multispecialty group from which a number of patients with congenital heart defects had recently originated allowed us to obtain teratogenic histories from their patients. This looked like a very promising arrangement because dextroamphetamine was very commonly prescribed by this group. Unfortunately, one of our medical students involved in chart review at the group clinic "leaked" the information that we were most interested in dextroamphetamine. The obstetricians immediately stopped prescribing the drug and lost their enthusiasm for our reviewing their charts. Although we obtained a small number of prospective patients and a larger number of retrospective patients, the opportunity to find the answers we sought from this group was lost.

We hope that the logistic difficulties we encountered in carrying out a prospective study on the relationship of teratogens to human malformations will not discourage the efforts of other teams. However, some insight may be gained from this review into why there is so little information on the role of teratogens in human maldevelopment.

By comparison, the studies in experimental animals of the teratogenic effects of dextroamphetamine were relatively easy. The pertinent results and inferences from these studies have been discussed previously. Although no "proof" regarding human teratogenic effects could be derived from the animal studies, our thinking was greatly influenced about how teratogens produce cardiac malformations and about predisposition to lesions and vulnerable periods of cardiogenesis. This is a great deal of information and is just about what one should hope to gain from studies in experimental teratogenesis.

SYNDROMES PRODUCED BY TERATOGENS

Only two teratogens have been established as being responsible for widespread disability: rubella virus and thalidomide. Other agents, including folic acid antagonists and androgenic steroids, have proved to be teratogenic to the human but on a much smaller scale; and still other drugs, including anticonvulsants, amphetamines, hypoglycemics, alkylating agents, and aspirin (large doses), are suspected of teratogenicity.[12] The possibility that hundreds of agents may be somewhat teratogenic has been suggested. That thalidomide and rubella virus *are* teratogenic has been documented beyond question.

Rubella Syndrome

The rubella pandemic of 1964–1965 had a medical, sociological, and economic impact that is still being measured. Estimates of the costs of medical care and rehabilitation have been in the billions of dollars, and far exceeding this is the loss and tragedy at the level of the individual family.

The rubella virus was first recognized as a cause of birth defects by the Australian ophthalmologist Gregg in 1941.[5] Cataracts, hearing loss, and cardiac disease (mainly patent ductus arteriosus) were recognized as the early triad of abnormalities resulting from maternal rubella infection during the first trimester of pregnancy. Following the pandemic of 1964–1965, additional features were recognized and referred to by some authors as the "expanded" rubella syndrome. These features include neonatal purpura, enlarged liver and spleen, hepatitis, bone lesions, psychomotor retardation, and anemia. Furthermore, it now seems that there is an appreciable risk of mental retardation, visual problems and deafness following maternal infection in the second trimester.[12]

Congenital Heart Disease. The most common manifestation of the rubella syndrome in most series is congenital cardiovascular disease. Personal experience with a series in Houston revealed that 71% of patients with congenital rubella had some form of cardiovascular disease. Pulmonary artery stenosis was present in 55% of patients, while patent ductus arteriosus was present in 43%. A variety of other lesions was found in 8% of patients, including ventricular septal defect, atrial septal defect, tetralogy of Fallot, and aortic stenosis. Some patients had more than one lesion.

Deafness. The next most common manifestation of the rubella syndrome is deafness (56% of patients). Hearing loss is permanent, may be unilateral or bilateral, and may be progressive. Interestingly, from the Houston series, among those rubella patients being followed in the heart clinic who also had deafness, there were a significant number who had a family history of deafness. A similar positive family history for deafness in patients with deafness associated with the rubella syndrome was found in Stockholm.[1] This would suggest that the rubella virus is capable of producing deafness in certain genotypes and not in others—another example of a genetic-environmental interaction.

Cataracts and Retinal Degeneration. Cataracts were the lesions that initially called attention to the entire rubella syndrome. Eye defects are present in 40% of patients with congenital rubella. The vast majority of lesions are cataracts and/or a patchy retinal degeneration.

Immunologic Considerations. During the height of the pandemic, a number of pregnant nurses caring for these sick infants contracted rubella and in turn delivered infants with the birth defects of the rubella syndrome. Through longitudinal observation, it became obvious that sick infants could shed virus for months and in some rare cases even for years.

This is an example of immunologic tolerance. The viral invasion takes place before the fetus is immunologically mature enough to recognize the virus as being "nonself," so for varying lengths of time, there is tolerance to the virus until the host mounts an immunologic attack of sufficient strength to destroy the invader.

Prevention. Rubella as a cause of birth defects is now a problem for which there is a reasonable solution. A vaccine that offers a seroconversion rate of 95% is available and is being used in women of the reproductive age (who have no serologic evidence of past infection and who are not pregnant at the time of immunization) and in schoolchildren.[2]

Thalidomide Syndrome

In late 1960, the first cases of a new syndrome were recognized in Australia[8] and West Germany.[7] Through 1961, amelia, phocomelia, and other anomalies became "epidemic" not only in Germany but in other western European countries, Canada, and Japan. Yet it was not until 1962 that Lenz and Knapp painstakingly developed the evidence that led to an irrefutable indictment of thalidomide.[7] But the damage was already widespread. Again, as in the rubella syndrome, the cost of medical care and rehabilitation of these patients, on a worldwide scale, has been enormous, and the tragedy at the level of the individual and family has been incalculable.

The thalidomide syndrome has been associated with certain characteristic malformations, but there has been involvement of many systems, depending on the stage in embryogenesis at which exposure occurred.

Limb anomalies are perhaps the most characteristic stigmata of thalidomide embryopathy. These malformations range from mild dysplasia of the thumbs to complete absence of all limbs. A typical "thalidomide baby" would have some combination of the following: absent or markedly underdeveloped and hypoplastic arms, ranging from phocomelia through absence of radius and thumb to triphalangeal thumb; capillary hemangioma of the midface; aplasia or hypoplasia of the external ear with atresia of the canal; a congenital heart defect; a stenotic lesion somewhere in the gastrointestinal tract; and malformed legs, ranging from phocomelia through tibial absence to clubfeet (Fig. 13-1).

Lenz, through careful analysis of almost 800 patients with thalidomide embryopathy, was able to devise a timetable, specifying at which stages of embryogenesis thalidomide exerted its teratogenic effect on which developing structures.[6] For example, no malformations occurred when thalidomide was taken before 34 days *after the last menstrual period.* This coincides with the general expectation of relative resistance to teratogenic effect during the first weeks of gestation, the period before implantation and shortly thereafter. The vulnerable period was essentially from the 34th to the 50th day after the last menstrual period, a 16-day period from 3 to 5 weeks of gestational age.

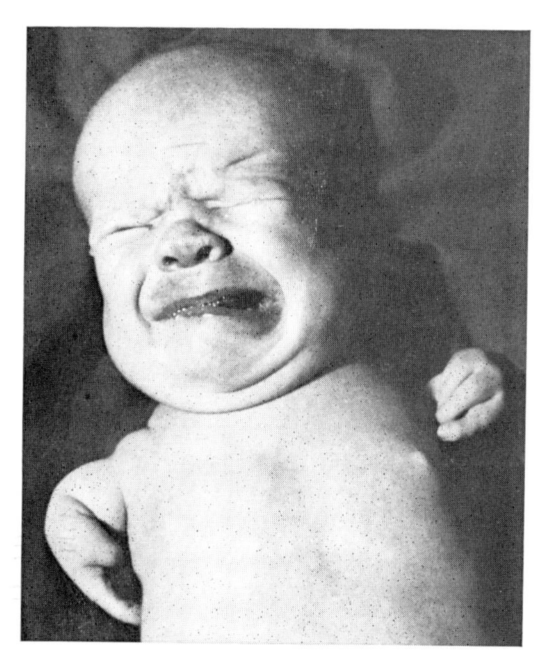

Figure 13–1. Phocomelia in infant with thalidomide syndrome. He also has facial paralysis on the right and a missing ear.

For hypoplasia or absence of the arms, thalidomide exposure must be between the 39th and 44th days; for similar malformations of the legs, from the 42nd to 48th days; for heart malformations, the 41st to 43rd days.

In general, the teratogenic exposure to thalidomide precedes the completion of the developmental event by about 2 weeks. Studies of comparable developmental stages in animal models suggest that many teratogens would have to be administered on about the same timetable as thalidomide to produce malformations (e.g., dextroamphetamine in the mouse), while other teratogens may have somewhat different timetables (e.g., antiheart antibody in the mouse).

These and many other insights derived from the tragic experience of thalidomide have contributed significantly to our understanding of human teratology.

SUMMARY

Teratogens act by causing damage to the cells, tissues, or interactions between tissues of developing embryos, in contrast to mutagens, which produce changes in the genetic material.

The three components of a typical teratogenic exposure are: (1) hereditary predisposition to a malformation; (2) hereditary predisposition to the effects of a given teratogen; and (3) administration of the teratogen at a vulnerable period of embryogenesis.

Although it is quite likely that there are a large number of environmental teratogens that operate within this three-component reaction, there are only a handful of agents whose teratogenicity in the human has been precisely documented. Only two teratogens have so far been identified as being responsible for widespread damage: rubella virus and thalidomide.

REFERENCES

1. Anderson, H., Bengt Barr, E. E., and Wedenberg, E.: Genetic disposition—a prerequisite for maternal rubella deafness. Arch. Otolaryngol. 91:141, 1970.
2. Cooper, L. Z.: Rubella: a preventable cause of birth defects. *In* Intrauterine Infections. Birth Defects IV:23, 1968.
3. Fraser, F. C.: The use of teratogens in the analysis of abnormal developmental mechanisms, p. 179. *In* First International Conference on Congenital Malformations. Philadelphia, J. B. Lippincott, 1961.
4. ———: Experimental teratogenesis in relation to congenital malformations in man, p. 277. *In* Second International Conference on Congenital Malformations. New York, International Medical Congress Ltd. 1964.
5. Gregg, N. M.: Congenital cataract following German measles in the mother. Trans. Ophthalmol. Soc. Aust. 3:35, 1941.
6. Lenz, W.: Chemicals and malformations in man, p. 263. *In* Second International Conference on Congenital Malformations. New York, International Medical Congress Ltd., 1964.
7. Lenz, W., and Knapp, K.: Thalidomide embryopathy. Arch. Environ. Health 5:100, 1962.
8. McBride, W. G.: Thalidomide and congenital abnormalities. Lancet ii:1358, 1961.
9. Menser, M. A., Forrest, J., Slink, R. F., Nowale, M., and Dorfman, D. S.: Rubella virus in 29-year-old woman with congenital rubella. Lancet ii:797, 1971.
10. Nora, J. J., Sommerville, R. J., and Fraser, F. C.: Homologies for congenital heart diseases: Murine models influenced by dextroamphetamine. Teratology 1:413, 1968.
11. Nora, J. J., Vargo, T. A., Nora, A. H., Love, K. E., and McNamara, D. G.: Dexamphetamine a possible environmental trigger in cardiovascular malformations. Lancet i:1290, 1970.
12. Peckham, C. S.: Clinical and laboratory study of children exposed *in utero* to maternal rubella. Arch. Dis. Child. 47:571, 1972.
13. Wilson, J. G.: Present status of drugs as teratogens in man. Teratology 7:3, 1973.
14. Wilson, J. G.: Environment and Birth Defects. New York, Academic Press, 1973.

Chapter 14

PHARMACOGENETICS

Physicians have long recognized that patients vary in their response to drugs. This is not surprising, since the fate of a drug in the body depends on its rate of absorption, whether and how much it is bound to the serum proteins, its distribution to organs and transfer across cell membranes, its interaction with cell receptors and organelles, and its metabolism and excretion. Not only are many of these processes modified by the environment (diet, other drugs, and so on), but it is reasonable to suppose that they will be modified by genes, since they involve many enzymes and other proteins.

Twin studies show that the heritability of drug clearance is very high for most of the drugs tested,[1] suggesting that the determination of dosage for chronically administered drugs should not be based on body weight but on systematic monitoring of blood level and adjustment of dose to the patient's individual response.

Occasionally, the response of an individual to a drug is dramatically different from the norm and perhaps life-threatening, and it was the discovery that some of these marked deviations in response showed simple Mendelian inheritance that led Vogel, in 1952, to coin the term "pharmacogenetics."

The pharmacogenetic disorders are a special type of inborn error of metabolism, involving those proteins that function in drug metabolism or drug action; they are considered a special subcategory because of their pharmacologic implications and because they represent an interesting kind of gene-environment interaction. We have presented below a number of conditions involving aberrant reactions to drugs and showing simple mendelian inheritance (Table 14-1). We have also included examples of gene-determined diseases that may be precipitated or exacerbated by certain drugs (Huntington's chorea, periodic paralysis, porphyria). Several recent reviews may be consulted for a more detailed treatment.[2,3,4]

For simplicity's sake, we have arranged the disorders alphabetically rather than attempting a classification by organ system or class of drug or according to whether the gene alters the way the body acts on the drug (e.g., acatalasia, pseudocholinesterase deficiency, isoniazid inactivation) or the way the drug acts on the body (e.g., G6PD

TABLE 14-1. Examples of Inherited Conditions with Altered Response to Drugs

Trait or Deficient Enzyme	*System Affected*	*Drug or Factor*	*Frequency of Trait*	*Clinical Effect*
Acatalasia	Tissues	H_2O_2	High in Japanese	No response to peroxide
Alcohol dehydrogenase atypical	Liver	Alcohol	?	Increased tolerance
Alpha-1-antitrypsin deficiency	Plasma	Smoking	Moderately rare	Emphysema (cirrhosis)
Dicumarol resistance	Clotting	Dicumarol	Rare	Decreased response
Glaucoma	Eye	Glucocorticoids	Frequent	Increased ocular pressure
Glucose-6-phosphate dehydrogenase	RBC	Fava beans, primaquine, others	High in Mediterraneans, Negroes	Hemolysis
Hemoglobins, unstable	RBC	Sulfonamides, oxidants	Very rare	Hemolysis
Huntington's chorea	Brain	Levodopa	Rare	Tremors in "gene carriers"
INH transacetylase	Liver	Isoniazid	Common	Polyneuritis
Malignant hyperthermia	? Neuromuscular	Anesthetics	Very rare	Rigor, hyperthermia
Methemoglobin reductase	RBC	Nitrites, oxidants	Variable	Methemoglobinemia
Periodic paralysis	? Cell membrane	Insulin, adrenalin, others	Rare	Paralysis
Porphyria (some kinds)	Liver	Barbiturates, sulfas, others	Variable	Acute "attacks"
Pseudocholinesterase deficiency	Plasma	Succinylcholine	Moderately rare	Apnea

deficiency, Dicumarol resistance, ocular response to steroids, malignant hyperthermia).

ACATALASIA

This odd example of a gene–drug interaction was discovered in 1959 by a Japanese otolaryngologist, Takahara, who removed some diseased tissue from the mouth of a patient and applied hydrogen peroxide to disinfect the wound. The usual bubbling of oxygen did not occur, and the tissue turned black, presumably because of oxidation of hemoglobin by the drug. Takahara, who must have been unusually well up in biochemistry, deduced that the tissue must lack the enzyme catalase and demonstrated that this enzyme deficiency showed autosomal inheritance, the heterozygote having intermediate enzyme levels. There is genetic heterogeneity, with five different types described so far, severity ranging from mild (ulcers of the tooth sockets) to severe (gangrene of the gums and recession of the tooth socket). The condition is not uncommon in Japan, but it has not yet been reported in North Americans. A very rare type of acatalasia has been described in Switzerland. In the Swiss type, the small amount of residual enzyme differs in its physicochemical properties from the normal, whereas the residual enzyme is normal in the Japanese type, suggesting that the Swiss type represents a structural gene mutation and the Japanese type a mutation of a controller gene.

ALCOHOL DEHYDROGENASE

Ethyl alcohol is metabolized in the liver by alcohol dehydrogenase. A variant has been found with increased activity in Swiss (20%) and English (4%) populations, which results in more rapid clearance of alcohol from the system. The genetic basis for this polymorphism and its possible relationship to alcoholism are unclear. Research is impeded by a certain reluctance of healthy individuals to provide liver biopsies.

Racial differences in alcohol metabolism have been demonstrated, but the relationship to alcohol dehydrogenase type or to alcoholism is not clear. White volunteers showed a more rapid fall in blood level after a standard dose of alcohol than did Eskimos or Canadian Indians.

Finally, it has been reported that Japanese, Taiwanese, and Koreans show facial flushing and signs of intoxication after drinking amounts of alcohol that produced no detectable effect in Caucasians. The genetics has not yet been worked out.

Thus we have no clear-cut Mendelian differences in reaction to alcohol and no evidence that alcoholism is related to a pharmacogenetic difference. However, there are enough suggestions that alcohol metabolism and the reactions to alcohol are genetically influenced to warrant their inclusion in this chapter.

ALPHA-1-ANTITRYPSIN ACTIVITY

This condition can be considered a pharmacogenetic disorder if cigarette smoking can be considered a form of drug-taking, although it is not known what constituent of cigarettes is the interacting factor in this case.

Alpha-1 antitrypsin is an enzyme that inhibits the proteolytic (protein breaking-down) action of trypsin and other enzymes. It is absent in homozygotes for a mutant gene Pi^{z} (Pi for protease inhibitor; normal allele Pi). Heterozygotes have intermediate levels. The homozygote has a frequency of about 1 in 1500. Several rarer alleles have been identified. Many persons suffering from lung emphysema (dilatation of the lung alveoli due to loss of elastic fibers) or childhood liver cirrhosis are found to be homozygous for the mutant gene. Presumably, inflammation, as from smoking, leads to release of trypsin, which is normally inhibited by alpha-1 antitrypsin, but which in deficient individuals remains active and destructive to the lung tissue. Smoking certainly exacerbates the lung disease, and there is evidence that heterozygotes are also predisposed to chronic obstructive lung disease if they smoke. If so, population screening for heterozygotes (about 5% in the general population) and warning them of their increased risk from smoking would be a worthwhile application of preventive medicine—if heterozygotes could be persuaded to stop smoking, which, in view of the experience with lung cancer, seems doubtful. (See also the aryl hydrocarbon hydroxylase polymorphism referred to in Chapter 19, p. 210.)

DICUMAROL (WARFARIN) RESISTANCE

Dicumarol is an anticoagulant used in the treatment of patients at risk for intravascular clotting, such as patients with coronary disease. It is also used as a rat poison, causing massive internal bleeding. The drug acts by competing with vitamin K for receptor sites in the liver that are involved in synthesizing various blood-clotting factors. When the drug, rather than vitamin K, occupies the sites, synthesis of the clotting factors decreases, hence the anticoagulant effect.

A rare dominant gene is known in which the effective anticoagulant dose of Dicumarol in carriers is about 21 times normal. The gene appears to act by increasing the affinity of the receptor site for vitamin K, so that the Dicumarol is no longer able to compete.

GLUCOCORTICOIDS AND INTRAOCULAR PRESSURE

Glaucoma is an eye disease resulting from increased intraocular pressure due to obstruction of the canal of Schlemm that drains the fluid from the inside of the eye back into the bloodstream. It was discovered that in some individuals local treatment of the eye with cortisone or other glucocorticoids (e.g., for inflammatory diseases) might precipitate an attack of glaucoma.

Further study showed that corticoid treatment causes minor increases in intraocular pressure in some persons, marked rises in others, and intermediate rises in a third group. Family studies showed that these three classes represented the two homozygous and the heterozygous genotypes for a locus with two alleles, P^L for low pressure and P^H for high pressure following glucocorticoid treatment. Overt glaucoma was much more likely to occur in the P^HP^H genotype, even without glucocorticoid treatment. About 5% of the population tested were P^HP^H, and their risk of developing glaucoma was about 5 times that of the heterozygote and 100 times that of the P^LP^L homozygote. Thus this genetic polymorphism determines not only a marked difference in response to a drug but predisposition to a disease.

GLUCOSE-6-PHOSPHATE DEHYDROGENASE (G6PD) DEFICIENCY (FAVISM, PRIMAQUINE SENSITIVITY)

The structural gene for G6PD is on the X chromosome, close to the locus for hemophilia A and for colorblindness. Over 90 mutant alleles exist, causing either enzyme deficiencies of varying degree or electrophoretic variants or, in at least two examples, an increased activity. Deficiency of the enzyme in the red blood cell renders the cell sensitive to certain drugs, for reasons not yet well understood. This is the cause of *favism*, long recognized as a hemolytic condition peculiar to Mediterranean peoples and related to the eating of uncooked fava beans, a Mediterranean delicacy. The enzymatic basis was not recognized until the antimalarial drug primaquine was issued to American troops in malarial regions during World War II, following which a number of hemolytic reactions were noted, particularly in Negroes; the affected men were found to be deficient in G6PD.

G6PD is involved in a minor pathway for red cell glycolysis, the hexose monophosphate shunt, and plays a role in maintaining the concentration of reduced glutathione, which, in turn, is necessary for stability of the red cell in the presence of certain drugs, although the mechanism for this is not clear.

There are two major electrophoretic variants, type B being the most frequent. Type A, faster-migrating, is present in about 20% of American Negro males.

The so-called African type of G6PD deficiency involves only type A (a curious association, not understood). The A^- mutant produces a relatively mild enzyme deficiency, with 8–20% of normal activity. Since enzyme levels decrease with increasing age of the red cell, the deficient cells do not become susceptible until they are about 50 days old. Thus, exposure of a mutant individual may cause a mild hemolytic response, with dark urine, and perhaps jaundice with some abdominal and back pain, but if the drug is still given, the episode passes and the patient recovers, since his red cell population now contains mostly young cells.

In the Mediterranean or B^- type, there is a severe deficiency, with 0–4% activity, and a much more severe response to precipitating drugs. Over 20 drugs are listed that will cause hemolytic crises, including acetanilide, sulfanilamide and other sulfa drugs, quinidine (a heart drug), primaquine, and naphthalene (mothballs). In some circumstances, e.g., in some newborns, hemolysis occurs spontaneously.

A variety of other mutant types have been described, particularly in East and Southeast Asia. In general, severity of hemolysis correlates with the degree of enzymatic deficiency. The population frequency correlates with malaria frequency, suggesting that the gene confers an advantage against malaria, as in the case of hemoglobin S. In some populations (e.g., Sardinia), the gene is so frequent that patients are screened routinely before being treated with sulfa drugs or other drugs known to cause hemolysis in mutant subjects.

Because the gene is X-linked, heterozygous females are mosaics for two populations—in fact, the demonstration of this mosaicism by cloning of fibroblasts was one of the first convincing demonstrations of the Lyon hypothesis in man. The ratio of the two populations of cells is about 50 : 50 on the average, but there is wide variation, from as high as 99% to as low as 1% in occasional heterozygotes. This would be expected if a fairly small number of cells made up the anlage of the blood-forming tissues. Since Lyonization is random, if there were n cells involved, there would be a 2^n chance that all would have either the X carrying the mutant gene, or the X carrying the normal gene, inactivated. If the latter, the female would be just as susceptible to hemolysis as a mutant male. About one third of heterozygous females have enough mutant cells to predispose them to clinically significant hemolysis on exposure.

By reversing the argument, one can calculate from the variation in proportions of mutant to normal red cells in various mutant females how many cells must have been present in the anlagen of the blood-forming tissues—probably about 5. The role of G6PD in supporting the somatic mutation hypothesis is mentioned in Chapter 19 (p. 206). Thus this pharmacogenetic trait has contributed notably to our knowledge of human genetics.

UNSTABLE HEMOGLOBIN MUTANTS

Several rare mutant hemoglobins are known in which the carrier is predisposed to hemolytic crisis when exposed to certain drugs, mainly of the same types that affect G6PD-deficient males. These include hemoglobin Zurich and Torino, as described in Chapter 6, page 91.

HUNTINGTON'S CHOREA

This autosomal dominant disorder causes progressive deterioration of the personality, dementia, and chorea (uncontrollable jerking and writhing); it may not appear until the 30's or 40's. There is degeneration of the basal ganglia of the brain. The drug levodopa causes chorea-like movements in patients with Parkinsonism (another disease involving the basal ganglia), and it was thought that small doses of this drug might cause chorea in still-asymptomatic carriers of Huntington's chorea. Preliminary evidence suggests that this may be so, but this has raised some controversy as to the wisdom of performing such a test unless there is some preventive measure to be taken. Persons at risk of developing the disease seem about equally divided in their opinions as to whether they would want to know if they were going to get it. Those who decided they would like to know would be relieved of unnecessary fear if the test were negative and could restrict their family size if it were positive, but whether to take such a test must be a terribly difficult decision.

ISONIAZID INACTIVATION

The observation that some patients excreted the antituberculosis drug isoniazid quite rapidly and others relatively slowly led to the discovery of a pharmacogenetic polymorphism (see Fig. 11-1), rapid inactivation of isoniazid being transmitted as an autosomal dominant trait. There are marked differences in frequency between populations, the proportion of rapid inactivators being about 45% among North American Caucasians and Negroes, 67% among Latin Americans, and 95% among Eskimos.

The responsible enzyme is an acetylase (N-acetyltransferase) in the liver that acetylates isoniazid and a few related drugs. The polymorphism is of some importance, as the drug interacts with pyridoxine (a B vitamin) and may cause a pyridoxine deficiency with peripheral neuritis, but only in slow inactivators. Treatment with pyridoxine is effective. Furthermore, where the drug must be given intermittently, as in some Eskimo and American Indian communities, rapid inactivators have a poorer therapeutic response. For these, a slow-release form of the drug has been developed.

A similar pharmacogenetic situation exists with the antidepressant phenelzine, which is acetylated by the same enzyme. Side effects such as blurred vision and psychosis occur, but mainly in slow inactivators.

In patients with both tuberculosis and epilepsy, a curious pharmacogenetic drug interaction may occur. When the epilepsy is treated with Phenytoin (diphenylhydantoin, DPH), high levels of isoniazid may inhibit the metabolism of DPH by the liver oxidases and the DPH may reach toxic levels, but this happens only in slow inactivators of isoniazid.

MALIGNANT HYPERTHERMIA

A recently discovered pharmacogenetic trait is a dominantly inherited tendency to react to certain anesthetics, including ether, nitrous oxide, and halothane, with a rapid rise in body temperature, progressive muscular rigidity, and often death from cardiac arrest. The population frequency is about 1 in 15,000 in Canada. Some cases also show an ill-defined muscular disease. The underlying defect is not known. The message here is that a history of anesthetic-related death in a relative should not be taken lightly when preparing to give (or take) an anesthetic.

METHEMOGLOBIN REDUCTASE

Methemoglobin, in which the iron is in the ferric state, is reduced to hemoglobin (ferrous iron) by the enzyme methemoglobin reductase. A recessive mutation causes absence of the enzyme, leading to congenital methemoglobinemia (not to be confused with the methemoglobinemia produced by structurally abnormal hemoglobin M molecules described in Chapter 6, p. 91). About 1% of the general population are heterozygotes. They have approximately 50% of the normal activity and are more likely than normal persons to develop methemoglobinemia and cyanosis when given methemoglobin-forming drugs, such as dapsone, primaquine, and chloroquine.

PERIODIC PARALYSIS

There are three types of periodic paralysis, all autosomal dominant, in which the plasma potassium rises, falls, or remains unchanged, respectively. *Low-potassium paralysis* may be precipitated, in gene carriers, by a variety of drugs, including insulin, adrenalin, ethanol, some mineral corticoids—and licorice! *High-potassium paralysis* may be brought on by potassium chloride and some kinds of anesthesia. The third type has no pharmacogenetic significance.

HEPATIC PORPHYRIAS

The porphyrins, formed by the union of four pyrrole rings, are a major constituent of the world's most important molecule (chlorophyll) as well as the second most important (hemoglobin). There are three genetically determined inborn errors of porphyria metabolism, all autosomal dominant. They qualify as pharmacogenetic disorders because the clinical disease may be precipitated by the administration of certain drugs, particularly barbiturates, sulfonamides, estrogens, and some anticonvulsants and tranquilizers.

In *porphyria variegata,* the "South African type," skin lesions may develop on exposure to the sun, and there are attacks of abdominal and muscle pain, with dark red, fluorescent urine—the symptoms of acute porphyria. About 9,000 white South Africans carry the gene, which also occurs in other parts of the world.

Intermittent acute porphyria, the "Swedish type" (but not restricted to Swedes) lacks the skin lesions.

Hereditary coproporphyria may develop similar symptoms in childhood, whereas the other two are usually postpubertal.

The characteristic feature of the hepatic porphyrias is an overproduction of the rate-limiting enzyme, delta-aminolevulinic acid (ALA) synthetase. Precipitating drugs, such as the barbiturates, induce ALA synthetase in the liver, thus precipitating the attacks. Personality changes, which may be labeled "hysteria" or "neurosis," may occur in patients with visceral attacks. Needless to say, the diagnosis must be made correctly to avoid the disaster of sedating such a "hysterical" patient with barbiturates.

PSEUDOCHOLINESTERASE DEFICIENCY

One of the most dramatic pharmacogenetic disorders involves a deficiency of serum cholinesterase, or "pseudocholinesterase," a defect that seems to have no harmful effect whatsoever unless the patient is given the drug succinylcholine or suxamethonium as a preoperative muscle relaxant or prior to electroconvulsive therapy. This drug causes paralysis of striated muscle, which, at the doses used, lasts only a minute or two, because it is rapidly broken down by the serum cholinesterase. In the absence of the enzyme, the paralysis may be greatly prolonged, and the patient may stop voluntary breathing for half an hour or more—an event that is likely to produce a certain amount of commotion in the operating room!

Genetic analysis shows a system of multiple alleles. The locus is called E_1 (E for esterase, and 1 because it is the first esterase to show genetic variation). The E_1 locus is linked to the transferrin locus. The normal allele is E_1^U (for "usual") and the mutant allele is E_1^A (for "atypical"). The enzyme in the mutant homozygote has a structurally abnormal enzyme, which is resistant to inhibition by dibucaine (Cinchocaine). A standard concentration of dibucaine causes 80% inhibition of the normal enzyme (the dibucaine number), 60% inhibition in the $E_1^UE_1^A$ heterozygote, and 22% inhibition in the E_1^A homozygote, who is sensitive to succinylcholine.

Another allele, E_1^S, is the "silent" allele, the homozygote having no activity and being very sensitive to succinylcholine. As one would expect, the $E_1^SE_1^A$ heterozygote is also sensitive to succinylcholine. The use of other inhibitors has led to the detection of other alleles, for example, E_1^f, in which the resulting enzyme is inhibited by fluoride, but not by dibucaine.

The "atypical" allele has a frequency of about 3% in the general U.S. population and 10% in Oriental Jews, with homozygote frequencies of 1 in 2500 and 1 in 400, respectively. The "silent" allele may be quite frequent in some Eskimo populations. Thus, although only about half the reported cases of succinylcholine sensitivity are due to detectable abnormalities in serum cholinesterase, the genetic trait is frequent enough to justify routine screening of patients to be exposed to this drug. A simple screening test exists.

The high frequency suggests that the gene has, or had, some selective advantage, but this is difficult to study, since we do not even know why the normal enzyme is there; presumably, it was not evolved simply to degrade succinylcholine! One theory for the high frequency of this mutant gene is that tomatoes, potatoes, and related plants may sometimes produce toxic amounts of an alkaloid, solanine, which is a potent cholinesterase inhibitor. The mutant enzyme is less sensitive to this inhibition, so mutant gene carriers would be more resistant to solanine poisoning.

CONCLUSION

It is likely that further genetic differences influencing responses to drugs will be found, thus permitting increasing accuracy in determining doses, choosing appropriate drugs, and avoiding undesirable side reactions. For example, of women taking oral contraceptives, those of blood group O are least likely to develop thromboembolism, perhaps because they have reduced levels of antihemophiliac globulin. (On the other hand, they are more prone to bleeding peptic ulcers.) Another example involves antidepressant drugs: Patients seem to fall into two classes, those who respond well to tricyclics and those that do best on monoamine oxidase inhibitors. This difference seems to be family-specific—probands and their near relatives respond best to the same group of drugs, suggesting that there are at least two genetically different kinds of depressive disorders. Thus pharmacogenetic principles may be an aid to disease classification.

In conclusion, the examples given in this chapter, besides demonstrating the truth of the adage that one man's medicine is another man's poison, remind us that our genes make us unique pharmacologically, as in so many other ways, that pharmacogenetic differences may make it desirable to screen high-risk families and high-risk populations before exposing them to certain agents, and that patients being treated with drugs should be regarded as individuals, not just as so many kilograms of body mass.

SUMMARY

Pharmacogenetics concerns itself with a special type of inborn error of metabolism, the response of individuals to drugs. Aberrant responses may be manifested only in a need for a dose different from that usually given to achieve a desired therapeutic effect or may be responsible for marked deviations from normal that may have mild, moderate, or serious untoward effects, including death. A number of Mendelian disorders have been identified and are discussed. One of the most informative loci in human genetics is the X-linked gene determining the red blood cell enzyme glucose-6-phosphate dehydrogenase. Over 90 mutant alleles are known, many of which determine enzyme deficiencies, resulting in a hemolytic response to more than 20 drugs, as well as infection. This locus has also been useful in the investigation of the Lyon hypothesis and in linkage studies.

REFERENCES

1. Eidus, L., and Schaefer, O.: Tuberculosis treatment for "rapid" metabolizers of isoniazid: a problem particular in Canadian Indians and Eskimos. Mod. Med. Can. 29:18, 1974.
2. La Du, B. N.: The genetics of drug reactions. *In* Medical Genetics. McKusick, V. A., and Claiborne, R. (eds.), New York, HP Publishing Co., 1973.
3. Vesell, E. S.: Advances in pharmacogenetics, p. 291. *In* A. G. Steinberg, and A. G. Bearn (eds.): Progress in Medical Genetics. Vol. 9. New York, Grune & Stratton, 1973.
4. World Health Organization Scientific Group: Pharmacogenetics. WHO Tech. Rep. Ser. No. 524. Geneva, 1973.

Chapter 15

DERMATOGLYPHICS

TERMINOLOGY

Dermatoglyphics are the dermal ridge configurations on the digits, palms, and soles. They begin to develop at about the thirteenth week of prenatal life[4] as the fetal mounds on the digit tips, the interdigital, thenar, and hypothenar areas of the hand, and the corresponding areas of the foot begin to regress. The pattern formation is complete by the nineteenth week.

Certain syndromes include unusual combinations of dermatoglyphic patterns. These can help to establish a probability index for a particular diagnosis, as first demonstrated by Norma Ford Walker[7] for Down syndrome, in which dermatoglyphic analysis is still a useful screening procedure. Even when a specific syndrome cannot be identified, abnormal dermatoglyphics are a sign of prenatal growth disturbance, which may have some diagnostic value—for instance, in cases of mental retardation where there is some question whether the cause was birth trauma or a prenatal factor.

Finger Patterns

The patterns on the finger tips are of three main types, classified by the number of triradii present (Fig. 15-1). The sim-

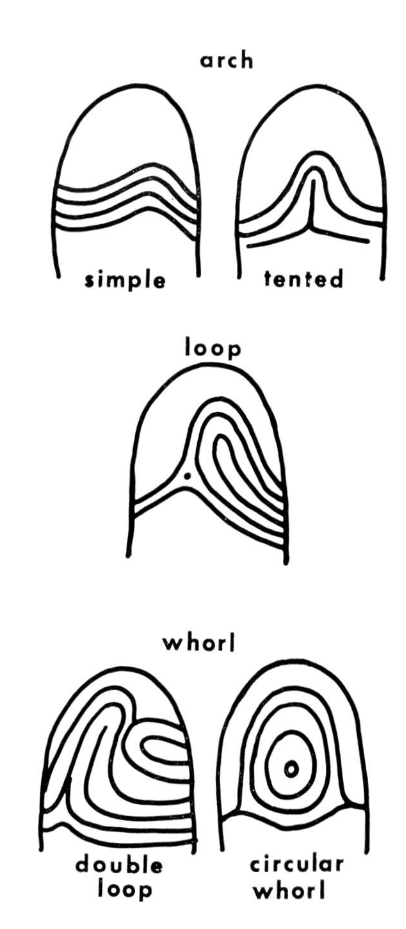

Figure 15–1. The three basic finger patterns (see text).

plest pattern is the arch (A). A simple arch has no triradius; a tented arch has a central triradius. The loop has a single triradius and is called ulnar (U) or radial (R), depending on the side to which it opens. The whorl (W) has two or more triradii and may be a double loop or, more commonly, a circular type of pattern.

The ridge-count is a quantitative way of measuring finger patterns. The number of ridges is counted between the center of the pattern and the farthest triradius. Thus arches have a count of 0, and whorls tend to have the highest values. The total ridge count is obtained by summing the counts for the 10 fingers.

TABLE 15-1. Frequency (%) of Pattern Types on the Fingers[a]

	Digit					
	1	*2*	*3*	*4*	*5*	*Total*
A	3	10	8	2	1	5
U	65	36	72	58	86	63
R	0	23	4	1	0	6
W	32	31	16	39	13	26

[a] The values for left and right and male and female do not differ appreciably and have been combined for simplicity. (Adapted from Holt.[1])

The frequency of these patterns for Caucasians is shown in Table 15-1. Conventionally, the digits are numbered from thumb to little finger. Arches and radial

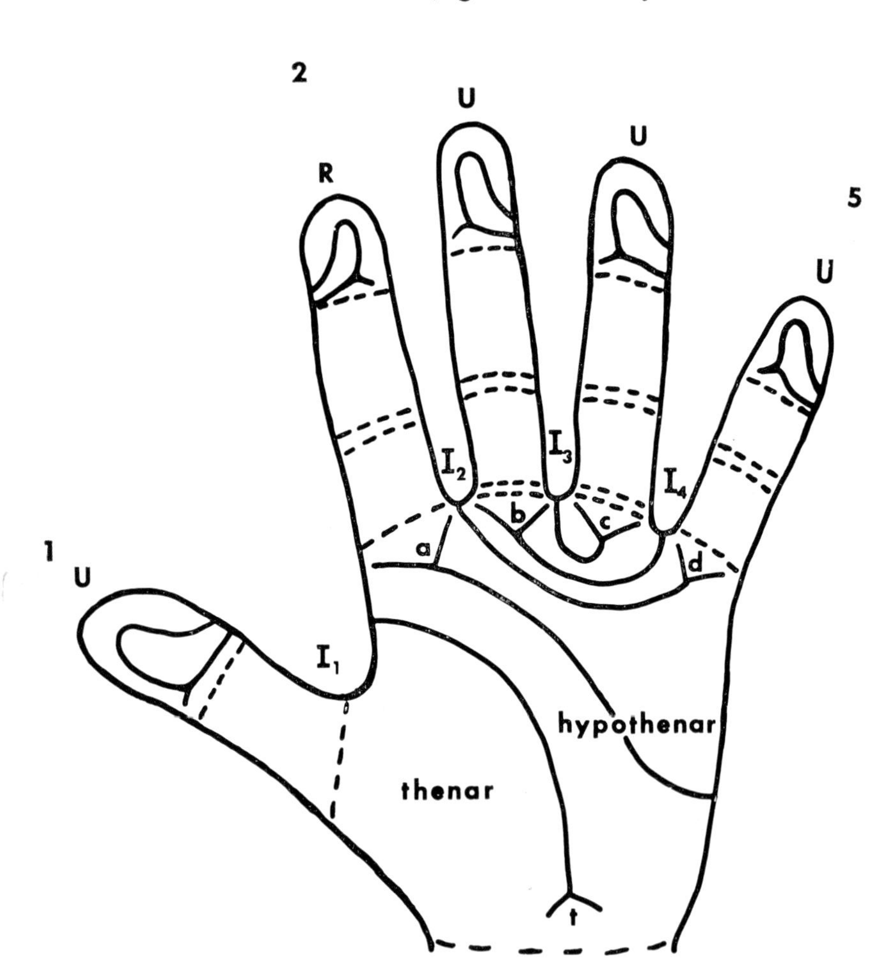

Figure 15–2. One of the more common finger and palm patterns.

loops have the lowest overall frequency; when present, they occur most often on digit 2, especially in the case of radial loops. Whorls occur most often on digits 4, 1, and 2. Ulnar loops occur more frequently than any other pattern type. A common distribution is shown in Figure 15-2.

The pattern frequencies vary somewhat with side and with sex, females having slightly more arches and fewer whorls than males. There are also racial differences in pattern frequencies. Orientals, for example, have a higher frequency of whorls than Europeans and white Americans.[1]

Palmar Patterns

The palm can be divided into the hypothenar, thenar, and four interdigital areas I_1, I_2, I_3, and I_4 (see Fig. 15-2). The normal palm has a triradius at the base of the palm between the thenar and hypothenar areas. This is the axial triradius (t). A variety of patterns (loops and whorls) which may be found in the hypothenar area are classified by the location and number of triradii. A pattern in either I_3 or I_4 is common, and patterns in the thenar/I_1 or I_2 area are less common. The mainline from the a triradius usually exits in the hypothenar area, that from b in I_4, that from c in I_3, and that from d in I_2. The axial triradius t is usually proximal but may be displaced distally. Its height from the base of the hand can be measured as a percentage of the height from the distal wrist crease to the proximal crease at the base of the third digit; t is defined as a height of 0–14%, t' as 15–39%, and t'' as greater than 40% of the total height. An alternative method is to measure the *atd* angle (the higher the axial triradius, the greater the angle), but this varies more with age than the first method. Since both methods are used in the litera-

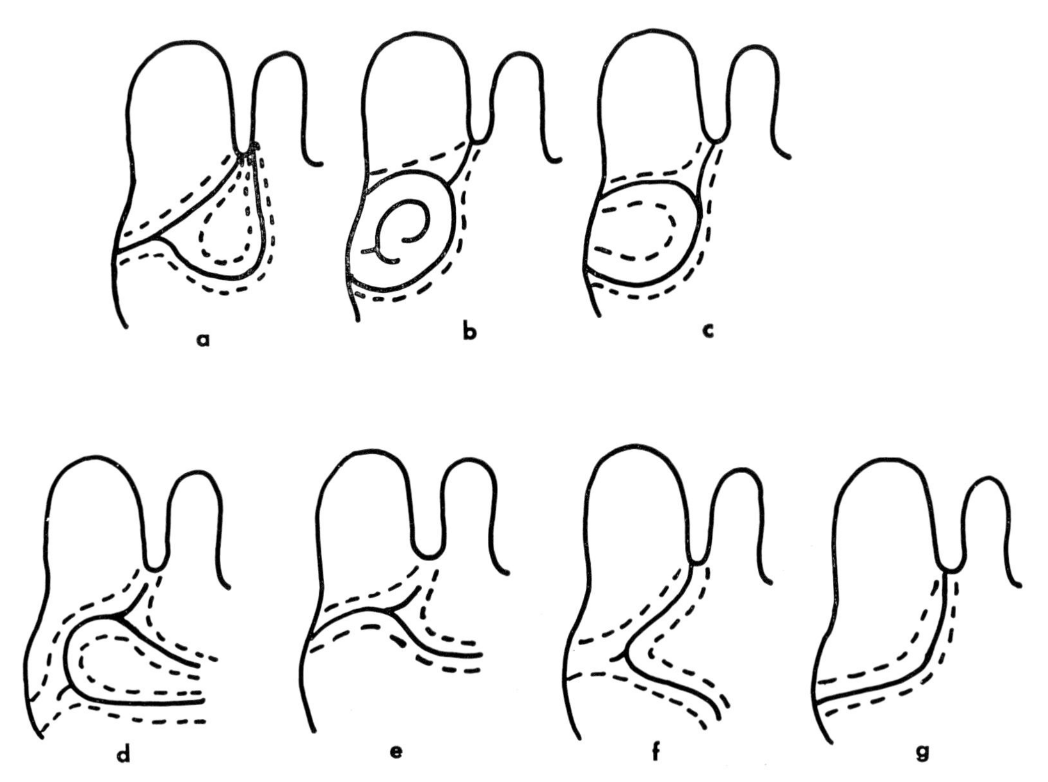

Figure 15–3. Hallucal patterns in order of frequency: (a) distal loop; (b) whorl; (c) tibial loop; (d) fibular loop; (e) proximal arch; (f) fibular arch; (g) tibial arch (no triradius).

ture, we have established approximate values from our data for converting the angle measurements to t, t', or t''. We define t as less than 46° and t'' as greater than 63°.

Soles

Because of the difficulties in printing the foot, observations are limited largely to the hallucal area (the "ball of the foot"), although other areas can also give valuable information. The patterns in the hallucal area are shown in Figure 15-3. The most frequent patterns in normal individuals are the whorl and the large distal loop (>21 ridges).

FLEXION CREASES

Strictly speaking, the flexion creases are not dermatoglyphics, but they have come to be included in dermatoglyphic analysis. They represent places of attachment of the skin to underlying structures and are formed between the seventh and fourteenth week of development.[4]

The palmar creases generally consist of a distal and proximal transverse crease and a thenar crease (Fig. 15-4). About 6% of normal individuals have at least one *simian crease*, a single crease extending across the entire palm, or a *transitional* simian crease —two transverse creases joined by an equally deep, short crease (type 1) or a single crease with a branch above and below the main crease (type 2). The considerable variation from study to study in reported frequencies results from the fact that some workers count a transitional form as a simian crease while others do not. About 11% of normal individuals have a *Sydney*

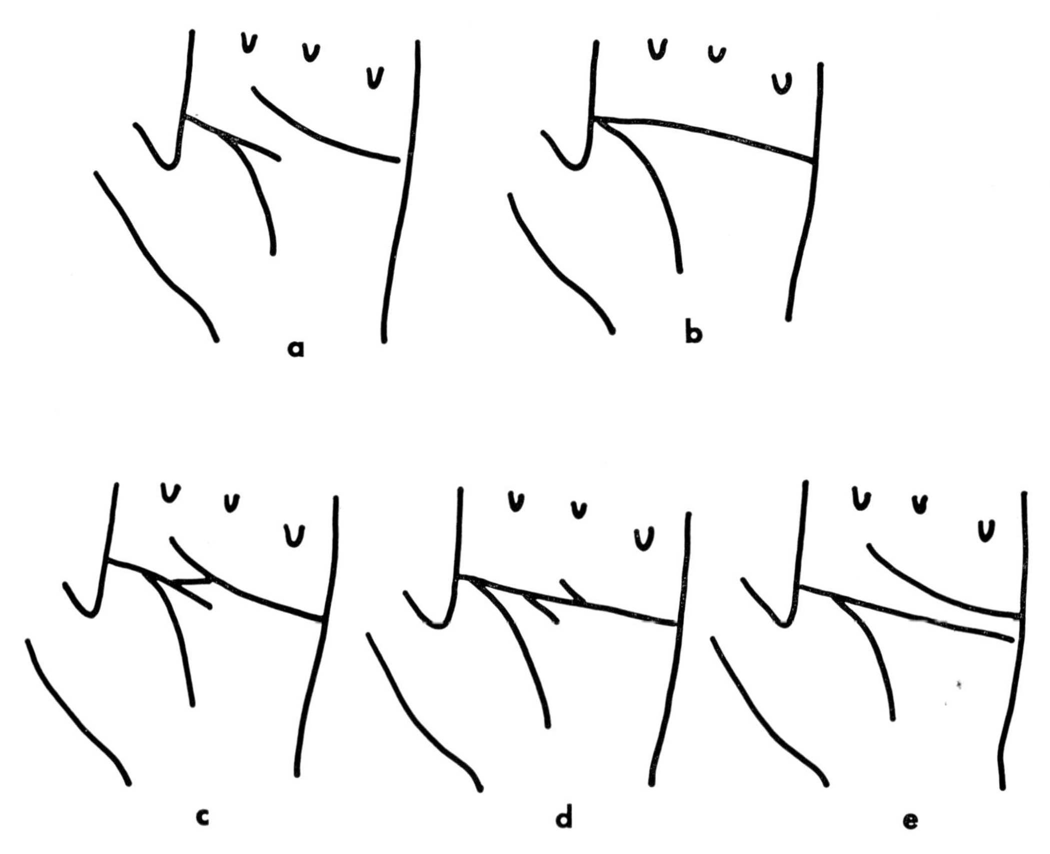

Figure 15–4. Palmar creases: (a) normal; (b) simian; (c) transitional, type 1; (d) transitional, type 2; (e) Sydney.

line, in which the proximal transverse crease extends across the entire palm, rather than stopping short of the ulnar border.

METHODS OF OBSERVATION AND PRINTING

If the dermal ridges are too small or poorly developed to observe directly, as in the newborn, an ordinary otoscope provides a satisfactory light and magnification for observation. The position of the hand or foot is adjusted to get the best interplay of available and direct light. Depending on the interests of the observer and the nature of the patient's defect, the patterns may be recorded directly or printed for a permanent record.[8]

USES AND LIMITATIONS OF DERMATOGLYPHIC ANALYSIS

The strong resemblance of dermatoglyphic patterns in pairs of monozygotic twins suggests that their determination has a major genetic component. This is well demonstrated in Table 15-2, in which observed correlations for various degrees of relationship are compared with those expected on the basis of determination by multiple additive genes, with a heritability of 1. One would therefore expect that, when a large number of genes were missing or present in excess, the dermatoglyphics would be altered. This appears to be so. In several chromosome syndromes, these alterations are consistent enough to be of diagnostic value.

TABLE 15-2. Correlations between Various Pairs of Relatives for Total Finger Ridge-Count[a]

Relationship	Correlation	
	Observed	*Expected*
MZ twins	0.95	1.0
DZ twins	0.49	0.5
Sibs	0.50	0.5
Father–child	0.49	0.5
Mother–child	0.48	0.5
Mother–father	0.05	0.0

[a] After Holt.[1]

Intuitively, one might expect that the dermal ridge patterns reflect the conformation of the fetal hand at the time of ridge development. If so, we would expect the dermal ridges to be altered when the limbs are deformed, and they are.[2] We would therefore not expect the dermatoglyphic patterns to be diagnostically useful if the patient has gross malformations of the limbs, as in Apert's syndrome (premature closure of the cranial sutures and webbing or bony fusion of digits)—the patterns reflect the obvious anatomical defect. They may be useful, however, as a reflection of more subtle morphological changes in the embryo. Cutaneous syndactyly, for instance, is a feature of a number of syndromes, but it is sometimes difficult to decide on gross examination if it is present in a minor degree. Fusion of the triradii at the base of the digits is good evidence that the basic defect resulting in syndactyly was present when the ridges formed, even if the syndactyly is not obvious at birth.

Examination of dermatoglyphics may help to decide whether a causative agent acted early or late. For example, the unusual longitudinal mainline configurations in a number of patients with arthrogryposis multiplex (a congenital condition where the joints are fixed in flexion) suggests that the defect was present in these patients at least as early as the 13th to the 19th week.

Finally, when there is no early anatomical effect, as in the simple biochemical disorders, there is also no change in the dermatoglyphics.

ZYGOSITY DIAGNOSIS IN TWINS

If dermatoglyphic patterns are in large part genetically determined, then the hands of monozygotic twins should resemble one another as closely as do the hands of a single individual, which are also genetically identical. This appears to be true, and the argument can be used in reverse to develop an aid to the diagnosis of zygosity. That is, the more closely the hands of twins resemble one another, the more likely it is that the twins are monozygotic (Chapter 16). Several methods have been developed to estimate the probability of monozygosity.[1]

DERMATOGLYPHIC FEATURES OF SYNDROMES

Trisomy 21 (Down Syndrome)

Long before the chromosomal basis of Down syndrome was established, Cummins demonstrated characteristic differences in dermal configurations between affected and normal children. In 1957, Walker[7] used these differences in frequency to derive an estimate of the probability that a child has Down syndrome on the basis of the dermatoglyphics alone. The principle is that, for any particular pattern present, the probability that the individual has Down syndrome varies directly with the ratio of the pattern's frequency in patients with Down syndrome to that in normal individuals. A probability index may be derived by multiplying the probabilities for each pattern or, after conversion to logarithms, by adding them. The more differences, and the greater their magnitude, the better the discrimination. The frequency distributions of the index for a group of patients with Down syndrome and controls overlap somewhat, but scores in the nonoverlapping range are strong evidence for or against the diagnosis of Down syndrome. It is assumed that the patterns in the various areas are independent of one another, and, although this is not strictly so, the method allows discrimination of about 70% of affected individuals from controls and 76% of controls from affected individuals. The remainder fall into the overlap zone.

As Walker suggested, this index could be improved by including other characteristics. Indeed, over 90% of patients with Down syndrome and over 80% of controls can be separated by including such stigmata as Brushfield spots, and simian crease.[5]

Other Syndromes

Dermatoglyphics are also of some help is distinguishing other syndromes.[6] For instance, almost all babies with E trisomy have 7 or more arches on their fingers, as compared to 1% of the general population. Diagnostic indices for a number of these syndromes have been worked out by Penrose and Loesch.[3]

SUMMARY

Dermatoglyphics, the study of dermal ridge configurations on the digits, palms, and soles, has proved useful in the investigation and identification of disorders and syndromes in which there is a prenatal influence, either genetic or teratogenic. In certain syndromes the dermatoglyphic analysis may provide findings that are almost diagnostic, whereas in other conditions of maldevelopment the most information one may derive is whether a causative agent was acting early or late. In simple biochemical disorders without structural abnormalities, there is no abnormality of dermatoglyphics.

References

1. Holt, S. B.: The Genetics of Dermal Ridges. Springfield, Ill., Charles C Thomas, 1968.
2. Mulvihill, J. J., and Smith, D. W.: The genesis of dermatoglyphics. J. Pediat. 75:579, 1969.
3. Penrose, L. S., and Loesch, D.: Diagnosis with dermatoglyphic discriminants. J. Ment. Defic. Res. 15:185, 1971.
4. Popich, G. A., and Smith, D. W.: The genesis and significance of digital and palmar hand creases: Preliminary report. J. Pediat. 77: 1017, 1970.
5. Preus, M.: Department of Medical Genetics, The Montreal Childrens Hospital. Unpublished observations.
6. Preus, M., and Fraser, F. C.: Dermatoglyphics and syndromes. Amer. J. Dis. Child. 124:933, 1972.
7. Walker, N. F.: The use of dermal configurations in the diagnosis of mongolism. J. Pediat. 50:19, 1957.
8. Walker, N. F.: Inkless methods of finger, palm, and sole printing. J. Pediat. 50:27, 1957.

Chapter 16

TWINS AND THEIR USE IN GENETICS

Twins have always been a subject of interest, particularly to their surprised parents, but also as mythical, historical, and literary figures. Yet it was only in 1875 that Galton pointed out their value in estimating the relative importance of heredity and environment—or, as he put it, of "nature and nurture." "Identical" or monozygotic (MZ) twins result from the splitting of a fertilized egg, giving rise to two genetically identical individuals. "Fraternal" or dizygotic (DZ) twins result from fertilization of two different eggs and are therefore no more similar genetically than sibs. Galton saw that this experiment of nature allowed comparison of genetically identical and genetically different individuals in similar environments.

A third type of twin has been postulated to arise from fertilization of an egg and its polar body by two sperm. Although this may happen, it apparently gives rise to mosaicism rather than to twins.[3]

In Caucasian populations, about 1 in every 87 deliveries is a twin birth, and the proportion of monozygotic to dizygotic pairs is about 30 : 70. This proportion was first estimated by Weinberg, who reasoned that for dizygotic twins there should be equal numbers of like-sex and unlike-sex pairs (1 boy/boy : 2 boy/girl : 1 girl/girl). Since MZ twins are always like-sexed, the excess of like-sexed twins in a representative series of twin pairs measures the proportion of MZ twins.

The frequency of MZ twins is remarkably constant, ranging from 3.5 to 4 per 1000 deliveries in various populations. Virtually nothing is known about the causes of MZ twinning.

The frequency of DZ twins is much more variable. It varies with maternal age from near 0 at puberty to 15/1000 at age 37 in Caucasians, and then falls sharply to near 0 again just before the menopause. The frequency is low in Mongoloid races (around 4/1000), higher in Caucasians, (8/1000), and higher still in Negroes (16/1000 or more).

Whereas MZ twinning does not show any familial tendency, DZ twinning does. As one would expect, the predisposition to have DZ twins is a maternal characteristic. The probability of recurrence of DZ twins in subsequent deliveries of mothers who have had a pair is about 3%—about a fourfold increase. After "trizygotic" triplets, the recurrence rate for DZ twins is increased about ninefold.

DETERMINATION OF ZYGOSITY

In any study in which twins are used to estimate heritability, it is necessary to determine whether each pair is monozygotic or dizygotic. This can be done in several ways, of varying degrees of reliability.

Fetal Membranes

The fetal membranes can sometimes be useful in the diagnosis of zygosity. In order to show how, a brief digression into embryology is necessary.[2]

At about the fourth day after fertilization, the ovum has developed into a 16-celled solid ball, the morula. By the sixth day this has progressed to a hollow sphere, the blastula, which contains an outer layer of cells, the trophoblast, and an inner cell mass. The trophoblast develops into the *chorion*, a heavy outer membrane which eventually lines the uterus and in one area forms the fetal component of the placenta. About 7 days after conception, the trophoblast begins to implant itself in the uterine wall. The inner cell mass forms the embryo proper and also, in the second week, develops the amnion, a thin membrane which forms a fluid-filled sac around the

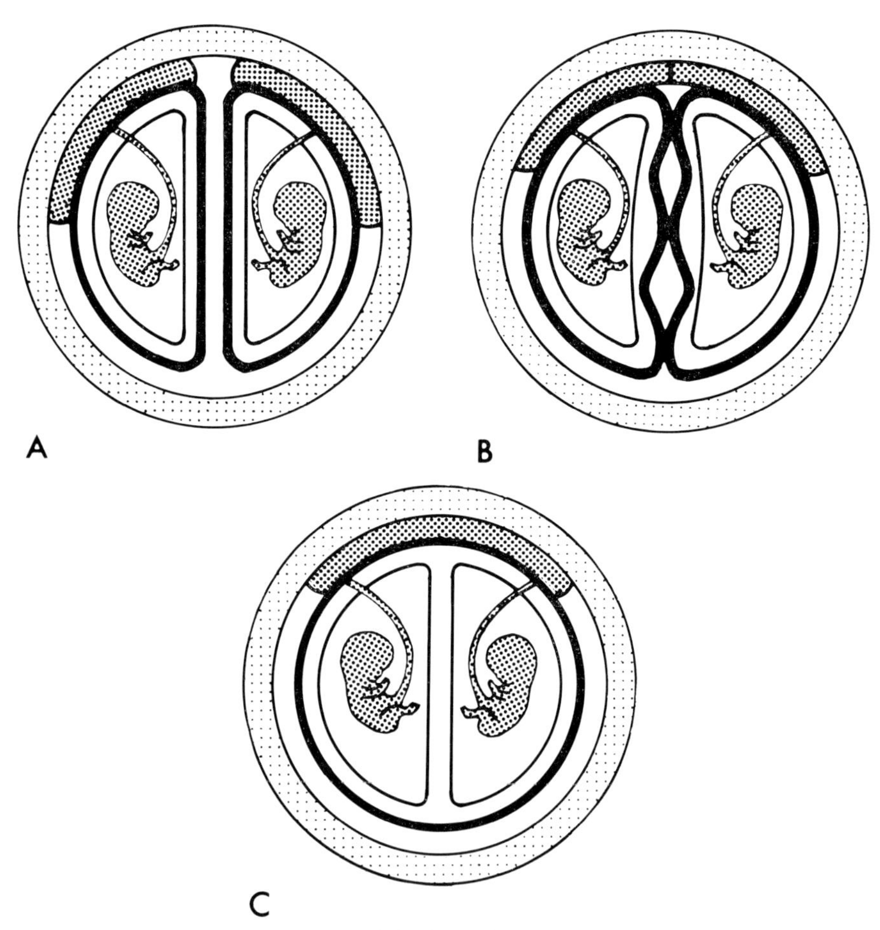

Figure 16–1. Arrangements of placentas and fetal membranes in twins (after Thompson and Thompson[10]). The placenta, umbilical cord, and fetus are heavily stippled; the thick line is the chorion, and the thin line enclosing the fetus is the amnion. *A*, 2 placentas, 2 chorions, 2 amnions. *B*, 2 placentas, fused, 2 chorions fused, 2 amnions. *C*, 1 placenta, 1 chorion, 2 amnions.

embryo, thus providing a protective barrier for the developing baby.

The splitting of the egg to form monozygotic twins may occur at one of several stages. If (as occasionally happens) it occurs at the two-celled stage, or at any point before the end of the morula stage, each resulting embryo will form a complete set of membranes. Thus there will be two amnions, two chorions, and two placentas, *just as there are in dizygotic twins* (Fig. 16-1A). If the inner cell mass splits after the trophoblast has formed but before the amnion appears, the twins will have a common chorion (and placenta), but separate amnions (Fig. 16-1B). Finally, if the embryonic disc splits after the amnion has developed (a rare event), the twins will share both placenta, chorion, and amnion (Fig. 16-1C). Table 16-1 shows the relative frequencies of the various types.

Dizygotic twins will always have separate chorions and amnions, though if they implant close together the placentas may fuse and appear to be single. Very occasionally, the membranes separating DZ twins may break down (due to mechanical pressure?), leaving the twins in a common sac.

Thus, examination of the fetal membranes will sometimes, but not always, be useful in establishing zygosity. If there is only one chorion, the twins are almost certainly monozygotic. If there are separate chorions and amnions, the twins may be either monozygotic or dizygotic. The distinction between a fused dichorial placenta (Fig. 16-1B) and a monochorial diamniotic placenta (Fig. 16-1C) requires careful examination, preferably histological.

Physical Similarity

In most cases the zygosity of pairs of twins can be estimated fairly reliably from their physical similarity alone. If the twins look so much alike that they are difficult to tell apart, they have about a 95% probability of being monozygotic. However, if they are discordant for a congenital malformation such as cleft lip or for a chromosomal anomaly, this criterion becomes unreliable. If twins differ in some trait known to be genetically determined, such as eye color, or webbed toes also occurring in other family members, they are almost certainly dizygotic.

Dermatoglyphics

The dermatoglyphic patterns (Chapter 15) are determined largely by the genetic constitution and can be used for zygosity determination. The corresponding hands of MZ twins should be just as similar, dermatoglyphically, as the right and left hands of each twin, whereas this is not so for DZ pairs. The degree of dissimilarity and the corresponding probability of dizygosity has been put on a statistical basis and presented in tabular form for convenient reference.[7] For instance, if the mean ridge count differs by 33 or more, the twins have a relative probability of 34:1 of being DZ.

Genetic Markers

A more reliable method than physical or dermatoglyphic similarity is the use of ge-

TABLE 16-1. Placentas and Fetal Membranes in Twins

			Percent of All Twins	
Placenta	*Chorion*	*Amnion*	*MZ*	*DZ*
1	1	1	rare	—
1	1	2	~22.5	—
1 (or fused)	2	2	~ 7.5	~35.0
2	2	2	rare	~35.0
			30.0	70.0

netic markers such as blood groups, serum proteins (Chapter 17) and, of course, sex. If the twins differ by even one such marker they must be dizygotic—ignoring the minute possibility of mutation. If they are similar for a particular marker, they may be MZ, or may have just happened to inherit the same marker. Obviously, the more markers are shown to be the same, the higher the probability that the twins are monozygotic.

The exact probability can be calculated if the genotype of twins and parents are known. Consider a pair of twin boys who are both blood group MM and whose parents are group MM and MN, respectively. To begin with, a pair of twins has an *a priori* chance of $7/10$ of being DZ, as this is the proportion of DZ to MZ twins in the population. If the twins are DZ, they have 1 chance in 4 of being boys, 2 in 4 of being a boy and a girl, and 1 in 4 of being girls, so the chance of being like-sexed is $1/2$. The possible genotypes of DZ twins would be 1 in 4 of being MM-MM, 2 in 4 of being MM–MN and 1 in 4 of being MN–MN, so the chance of being alike in their MN groups is also $1/2$. Thus the probability of the twins being DZ and alike for sex and the MN blood group is $7/10 \times 1/2 \times 1/2 = 7/40$.

If the twins are MZ, they must be like-sexed and have the same blood groups, so the probability of these twins being MZ is $3/10 \times 1 \times 1 = 3/10$ or $12/40$. So the relative probability of being MZ is $12/(7 + 12) = 12/19 = 63\%$.

Each additional marker for which the twins are identical (provided there is an opportunity for segregation) increases the probability of monozygosity. For instance, if both parents are group AB and the twins are alike for this group, one can calculate that DZ twins have 3 chances in 8 of being alike for this locus (work it out). The probability of being DZ then becomes $7/40 \times 3/8 = 21/320$, the probability of being MZ is $12/40 \times 1 = 96/320$, and the relative probability of being MZ is $96/(21 + 96) = 96/117 = 82\%$. Of course, if there is no opportunity for the twins to be different—e.g., if both twins and parents are group O—this marker will not contribute any information.

If the parental genotypes are not known, they can sometimes be deduced from those of their other children, or their parents. If not, they can be calculated from the known population frequencies of the genes concerned, but the arithmetic is complex. Fortunately, tables of probabilities for such situations have been prepared by Maynard-Smith *et al.*[8]

Skin Grafts

If a piece of grafted skin contains an antigen different from the host's, the host will make antibodies against it, and the graft will be rejected (Chapter 17). Dizygotic twins are almost certain to be antigenically different, since there are several histocompatibility loci, with many alleles, that will result in skin graft rejection if graft differs from host. Monozygotic twins, on the other hand, are antigenically alike and will therefore accept grafts from their co-twins. Thus, skin-grafting provides the ultimate test of zygosity and may be used when zygosity determination is important, as in the case of a contemplated kidney transplant.

A somewhat less sensitive, but easier and faster method is the mixed lymphocyte assay, which also depends on antigenic dissimilarity. The lymphocytes from one of a pair of MZ twins will not stimulate division in the lymphocytes from the other, whereas there is often a response if the twins are DZ (Chapter 17).

USE OF TWINS IN ESTIMATING HERITABILITY

Quantitative Characters

Twin studies have contributed a great deal to our knowledge of the genetic component in many quantitative characters, particularly intelligence. The statistical methodology is complicated, and its detailed discussion is not within the authors' competence or the scope of this book. We will present some of the concepts involved and refer the reader to textbooks of quantitative genetics for the technical aspects.[5,6] Let us begin with some definitions.

In twin studies designed to estimate heritability, the aim is basically to measure how much more closely members of MZ pairs resemble one another than do members of DZ pairs. This can be done in several ways.

Mean Pair Difference. The tendency of pairs (of twins, sibs, parent–child, and so on) to resemble each other can be measured by obtaining the difference between the members of each pair by subtraction and calculating the mean of these values. If y is the value of the measurement for one twin and y^* for the other, the mean pair difference for n pairs is $\Sigma(y-y^*)/n$. The smaller the mean pair difference, the greater the tendency for pairs to be similar.

Variance. Variance (V) is a measure of the variation, calculated by squaring the difference (d) of each value from the mean, summing the squared differences and dividing by the number of observations; $V = \Sigma d^2/n$. The greater the tendency of pairs to resemble each other, the smaller will be the mean pair difference and the variance of the mean pair differences.

Correlation. Correlation (r) is another measure of the tendency of pairs to resemble each other. Several methods of calculation exist. For twin pairs, it may be calculated from the variance of the mean pair difference, $V(y-y^*)$, as compared to the variance of the population, $V(y)$, by the formula

$$r = 1 - \frac{\frac{1}{2}V(y-y^*)}{V(y)}.$$

The formula shows that the smaller the variance of the mean pair difference, the greater the correlation. A perfect correlation (complete resemblance) has a value of 1. For a quantitative character that is completely determined by additive genes, MZ twin pairs would have a correlation of 1, and parent–child, sib–sib, and DZ twin pairs would have a correlation of 0.5.

Partitioning of the Variance

Suppose we want to determine the degree to which a given quantitative character is genetically determined. We will assume that the character is normally distributed. The questions can be framed in terms of the total variation shown by the character in the population and how much of this variation is due to genetic factors.

The observed value of the character (y) for a particular individual will deviate from the population mean ($\overline{y}$) by an amount determined by genetic factors (g) and an amount determined by environmental factors (e). That is,

$$y = \overline{y} + g + e.$$

For the population as a whole, the variance of y will similarly be determined by the variance of g and the variance of e. This can be symbolized as

$$V = V_G + V_E.$$

The degree of genetic determination can be estimated by the heritability (H), which was defined in Chapter 11 as the amount of variation resulting from genetic differences as a proportion of the total variation. Thus,

$$H = V_G/(V_G + V_E).$$

For a character determined completely by genes, the heritability would be 1, since V_E would be 0. Similarly, if genes played no part in determining the value of the character, H would be 0.

Assortative mating (the tendency for people to choose mates similar to themselves) increases the genetic variance by increasing the proportion of offspring falling in the tails (extremes) of the distribution. There is considerable assortative mating for several characteristics. For instance, the correlation between mates is about 0.3 for height and 0.4 or more for intelligence.[5]

The genetic variance can in turn be partitioned into V_A, that due to additive genes (where the heterozygote is intermediate between the two homozygotes) and V_D, that due to genes showing dominance. If there is dominance, the children will not be intermediate between the parents. The dominance variance tends to reduce parent–child correlation relative to sib–sib correlation and can be estimated from this comparison. Other deviations from additive gene action can also be considered as contributing to the variance. Epistasis, or the interaction of different loci in a nonadditive way, will also contribute to lack of resemblance between parents and offspring. Furthermore, there may be a correlation between genetic and environmental factors. For instance, children with a better than average genotype for intelligence have a better than average chance of having relatively intelligent parents, who will provide a relatively better environment for fostering the intellectual endowment. This is referred to as the covariance of heredity and environment. Distinct from this is the interaction of heredity and environment, which refers to the fact that different genotypes may respond differently to different environments. For instance, two genetically different individuals may have the same caloric intake and activity levels, but one may become fat and the other stay thin, because their different genotypes cause them to metabolize their food quite differently. In an "annotation," which anyone interested in the subject should study, Lewontin points out that the complexities of the interaction between genotype and environment may render conclusions about heritability based on the partitioning of variance virtually meaningless.[7]

Thus the estimation of how much of the variation in the population is due to genes is a complicated procedure and, in man, entails the use of many untested assumptions. (The procedure and the limitations are well described by Cavalli-Sforza and Bodmer.[5]) Nevertheless, the partitioning of variance can be a rough guide to the relative importance of genetic factors, which is necessary if we are to interpret what has happened to or predict what may happen to the genes influencing quantitative characters.

It should also be emphasized that any such estimate of heritability refers to a particular population and its particular range of environments. In the case of intelligence, for instance, a poor environment will prevent the expression of some favorable genetic potential, and a good environment will allow the full expression. Thus, for the same population, the heritability (proportion of variation due to genes) will be greater in the better environment.

Twin Studies

A major contribution of twin studies to genetics has been in the estimation of H for various quantitative characters. Since monozygotic twins are genetically identical, any differences between them must result from environmental variation. Thus the measurement of differences between MZ twins provides a direct estimate of V_E. Differences between DZ pairs represent $V_G + V_E$, though this will be a biased estimate, since V_G for DZ twins will be less than that for the general population. For any sex-influenced character, one should compare only like-sexed pairs, of course.

Another bias is introduced by the fact that the environment is more similar for twins than for unrelated pairs, and perhaps for MZ than for DZ pairs; for postnatal measurements, this can be partly compensated for by using twins reared apart, but these are relatively uncommon. Thus, in one study of intelligence quotient by Burt, the correlation coefficient was 0.88 for MZ twins reared together, but 0.78 for MZ twins reared apart. Similarly, the correlation coefficients for pairs of sibs reared together and apart were 0.55 and 0.45, respectively. Presumably, DZ pairs would have shown similar differences.

Using, in addition to correlation between twin pairs, the correlation between sibs and between parent and child, Burt did an analysis of variance and estimated that about 48% of the variation in IQ was due to additive genes, 18% to assortative mating, and 22% to dominance and epistasis: Thus, 88% of the variation resulted from what is commonly called genetic variance. It must be emphasized that the statistical models are necessarily oversimplifications. Table 16-2 gives some other examples of similarities between pairs of MZ and DZ twins for quantitative characters, along with estimates of heritability.

For characters that are classified as present or absent, rather than measured, heritability can be estimated from the frequency with which twin pairs are *concordant* (both affected) or *discordant* (only one affected). If the trait is determined in part by genes, the concordance rate will be higher in MZ than in DZ twins.

However, as with ordinary sib studies, the concordance rate should be corrected for the method of ascertainment. If the concordance rate is defined as the proportion of affected individuals among the cotwins of previously defined index cases (called the proband concordance rate), the corrected concordance rate, c', will be

$$c' = \frac{c + 2c^*}{c + 2c^* + d},$$

where c is the number of concordant pairs ascertained through only one of the affected twins, c^* is the number of concordant pairs in which both members were ascertained independently, and d is the number of discordant pairs.[1] In other words, concordant pairs ascertained twice (once for each twin) are counted twice.

Smith[9] provides a useful guide to the management and interpretation of data on concordance rates for discontinuous characters in twin pairs. Discontinuous familial traits not due to single mutant genes are likely to be multifactorial threshold characters, and it therefore seems appropriate to use the threshold model to estimate heritability from twin-concordance rates. It turns out that even when MZ concordance rates are relatively low, heritability can be quite high. In an individual whose genes place him near the threshold, relatively small environmental differences will place him on one side of the threshold or the other and determine whether he is affected or unaffected. Table 16-3 gives ex-

TABLE 16-2. Heritability Estimates for Some Quantitative Traits Based on MZ-DZ Twin Comparisons

	Sex		
Trait	*Male*	*Female*	*Both*
Height (child)			0.92
Height (adult)	0.79	0.92	
Weight (child)			0.83
Weight (adult)	0.05	0.42	
Arm length	0.80	0.87	
Age at menarche		0.93	
Alcohol clearance			0.88
Stanford-Binet IQ			0.83
Arithmetic			0.25
School achievement			0.09

TABLE 16-3. Twin Concordance and Heritability Estimates for Various Diseases (Assuming No Dominance)[a]

	% Concordance		
Disease	*MZ*	*DZ*	*H*
Manic depressive psychosis	67	5	1
Congenital hip dislocation	40	3	0.90
Club foot	33	3	0.88
Cleft lip ± cleft palate	38	8	0.87
Rheumatoid arthritis	34	7	0.74
Bronchial asthma	47	24	0.71
Tuberculosis	37	15	0.65
High blood pressure	25	7	0.62
Rheumatic fever	20	6	0.55
Cancer, same site	7	3	0.33
Death from acute infection	8	9	−0.06

[a] Estimate from Cavalli-Sforza and Bodmer[5] and Smith.[8]

amples of concordance and heritability estimates for a number of discontinuous traits (diseases and malformations).

Remember that these estimates are subject to sampling error, and that they are only approximations. For instance, it is most unlikely that the heritability of manic depressive psychosis is really 1—we should conclude only that it is high. Similarly, the heritability of death from acute infection is obviously not a negative value, but it must certainly be low.

Twin studies have their limitations. They cannot determine the mode of inheritance of a character. For prenatal traits, and particularly malformations, assumptions about similar environments are complicated by the mechanical effects of having two babies growing in the same confined space, the relations of the fetal membranes, and the fact that monozygotic twins may have vascular connections between the placentas that may favor one twin at the expense of the other. Other difficulties have been referred to previously. Nevertheless, careful studies of twins, their sibs, and their parents are the most valuable method available for demonstrating whether the familial tendencies observed in many quantitative traits and common diseases have a genetic basis and, if so, its relative magnitude.

SUMMARY

Twins have proved to be useful in weighing the relative importance of heredity and environment in normal variation and in disease. Traits or disorders having an important genetic component will be found in higher frequency in the co-twins of affected monozygotic (MZ, identical) twins than in the co-twins of affected dizygotic (DZ, fraternal) twins. The frequency of identical twins in various populations is about 3.5 per 1000 deliveries; the frequency of dizygotic twins varies greatly depending on race and maternal age.

The zygosity of twins may be determined by a number of methods, including examination of fetal membranes, physical similarity, dermatoglyphics, genetic markers, immunological reactions, and skin grafts.

Twin studies have been useful in the study of the genetic component of quantitative characters, such as intelligence, and threshold characters, such as common diseases. A number of statistical methods have been employed; the results must be evaluated within the context of their limitations.

REFERENCES

1. Allen, G., Harvald, B., and Shields, J.: Measures of twin concordance. Acta Genet. 17: 475, 1967.
2. Benirschke, K.: Origin and clinical significance of twinning. Clin. Obstet. Gynecol. 15: 220, 1972.
3. Bulmer, M. G.: The Biology of Twinning in Man. Oxford, Clarendon Press, 1970.
4. Burt, C.: The genetic differences in intelligence: A study of monozygotic twins reared together and apart. Brit. J. Psychol. 57:137, 1966.
5. Cavalli-Sforza, L. L., and Bodmer, W. F.: The Genetics of Human Populations. San Francisco, W. H. Freeman, 1971.
6. Falconer, D. S.: Introduction to Quantitative Genetics. Edinburgh, Oliver and Boyd, 1960.
7. Lewontin, R. C.: The analysis of variance and the analysis of causes. Am. J. Hum. Genet. 26:400, 1974.
8. Maynard-Smith, S., Penrose, L. S., and Smith, C. A. B.: Mathematical Tables for Research Workers in Human Genetics. London, Churchill, 1961.
9. Smith, C.: Concordance in twins: methods and interpretation. Am. J. Hum. Genet. 26: 454, 1974.
10. Thompson, J. S., and Thompson, M. W.: Genetics in Medicine. 2nd ed. Philadelphia, W. B. Saunders Co., 1972.

Chapter 17

IMMUNOGENETICS

The subject matter of modern immunology has its roots in clinical medicine, but it spreads its branches to shade a wide area of biology.[6] In this chapter, several topics will be presented: the development of immunity; histocompatibility; transplantation; maternal–fetal interaction; autoimmune diseases. Immunogenetics as it pertains to blood groups will be discussed in Chapter 18.

DEVELOPMENT OF IMMUNITY

The basis of immunity is the capacity within each individual to recognize what is "self" and what is "nonself".[2] This capacity is vital to survival: When bacteria or viruses or cancer cells appear, the body can recognize the invaders as being "nonself" and destroy them before being destroyed by the invading cells. The appearance of lymphoid tissue (in man at about 12 weeks *in utero*) coincides with and is directly related to the beginning of immune defense capability. However, there is some evidence that immunity may be induced in sheep and man before the appearance of lymphoid tissue.

At present, it is fairly widely accepted that there are two major immune systems—the bursa system and the thymus system—which originate and differentiate from the same stem cells[3] (Fig. 17-1).

The *bursa system* is responsible for humoral immunity—immunity carried by circulating *antibodies*, small globulin molecules that arise in response to stimulation from an antigen. An *antigen*, then, is a substance (a protein or related material) that stimulates the formation of an antibody. The antibody is able to recognize the antigen and combine specifically with it. The result depends on the nature of the antigen and antibody but may be, for instance, the destruction of the cell, agglutination of red blood cells, or the release of histamine, with its symptoms, so well known to hay-fever sufferers.

To illustrate how the *bursa system* works in the development of immunity, let us hypothesize that the body is invaded by bacteria, in this case, beta hemolytic streptococci group A, type 3. The first cells to try to halt this invasion are macrophages, which engulf the bacteria by a nonimmunological process. Following this initial contact, a series of transformations takes place in which antigens (parts of bacteria or products of bacteria) processed by the macrophage are taken up by small lymphocytes that become transformed to lympho-

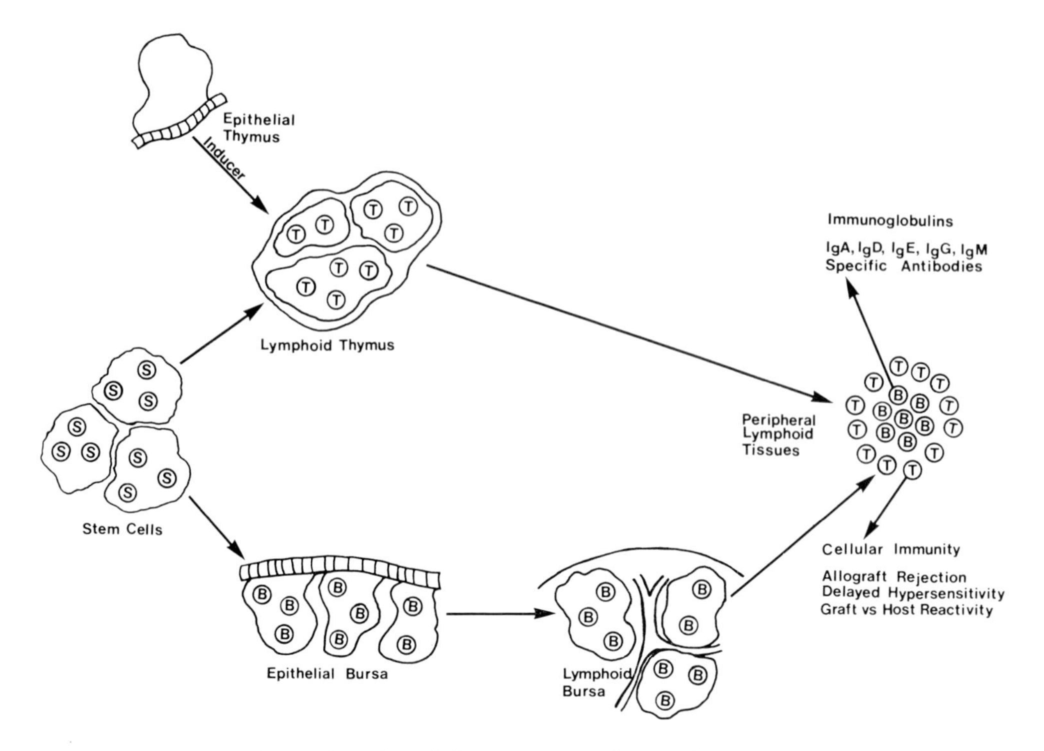

Figure 17-1. Origin of the two branches of the immune mechanism from the same lymphoid precursor.

blasts and then to plasma cells. It is in the plasma cells that the *immunoglobulins*, which constitute the antibodies, are manufactured. There are five types of immunoglobulins designated: IgG, IgM, IgA, IgD, and IgE, separable on the basis of their physicochemical properties. Each is called upon for certain functions.

These immunoglobulins are then released into the circulation as antibodies, capable of combining specifically with the corresponding antigens. In the case of the invasion of bacteria, the antibodies inactivate the antigens in collaboration with other constituents of the blood, such as complement and polymorphonuclear cells. In the example of the beta hemolytic streptococcus, group A, there are several antigens that stimulate antibody production, including erythrogenic toxin, streptolysin O, and M substance, a protein fraction of the cell.

The nature of the genetic mechanism by which the body can produce antibodies of a wide variety of specificities in response to exposure to a large number of antigens is a fascinating problem. Rapid progress is being made, but the answer is still not clear.[4] Among the theories advanced are

1. *Clonal selection theory*—the antigen selects from a library of cells with pre-existing antibody patterns a unit that will combine with itself. The selected cell then multiplies rapidly, and the resulting line makes the appropriate antibody.
2. *Direct template theory*—the antigen becomes part of the immunoglobulin-manufacturing unit.
3. *Indirect template theory*—the antigen alters an existing manufacturing unit to fit itself.

Some of the functions of the specific immunoglobulins have been defined. IgG accounts for the major portion of the immunoglobulins and takes part in reactions against a variety of bacteria, viruses, and toxins. It plays the central role in fighting the streptococcal invasion and is the best-suited immunoglobulin to neutralize toxins such as erythrogenic toxin. IgM, on the other hand, may be adapted to deal with particulate antigens, such as bacterial cells, and may combine with cell-membrane antigens, activate complement, and provoke immune lysis of the cell, as in the case of skin-graft rejection. IgA has the remarkable property of being secreted locally into saliva, respiratory secretions (where it protects mucous membranes), intestinal juice, and colostrum. It is now known that IgE is involved in allergy, such as hay fever, but a role for IgD has not been elucidated at the time of this writing.

Since our example represents a first exposure to this type of streptococcus, the patient would become clinically ill while developing antibodies to fight the infection. The response to the streptolysin O antigen is a rising antistreptolysin O (ASO) titer, which assists in the diagnosis of the streptococcal infection. The development of antibodies against the M substance confers a permanent immunity against the specific type 3 group A beta hemolytic streptococcus that was the infecting agent in our hypothetical case. However, the patient is still vulnerable to any of the other types of group A streptococci.

Now what would happen should the patient be exposed to another invasion of type 3 group A streptococci? The body would recognize this group of foreign proteins as a previous invader, and there would be a prompt and vigorous response that would eradicate the invader without allowing it a sufficient foothold to produce clinical illness. The patient is said to be "immune," and this is a manifestation of "immunologic memory." The weight of current evidence favors a small lymphocyte originating from the marrow (a "B cell") as the "memory cell."[4] On antigenic stimulation, "memory" is rapidly translated into the activity of antibody production. The "memory cell" immediately recognizes the M substance of the type 3 organism.

The second system of immunity, the *thymus system,* is mediated by entire cells, lymphocytes. Small lymphocytes derived from the thymus ("T cells"), which may live in the circulation for 10 years, are thought to be "memory cells." This system of cellular immunity has recently been emphasized because of its importance in organ transplantation. It also may be a major factor in the body's natural defense against cancer, as well as in many viral, bacterial, fungal, and protozoal diseases. The sequence of events following the introduction of foreign cells such as a kidney allograft is similar to that found in the invasion of bacteria discussed above. The antigens present in the cells of the kidney allograft are detected as being "nonself" by small thymus-dependent lymphocytes, probably after the antigen has been processed by the macrophages. These lymphocytes are now sensitized. They are transformed to lymphoblasts, which divide into many new lymphocytes, each one sensitized to the grafted kidney and each one bearing antibodies against the foreign kidney cells. Unlike humoral antibodies, which circulate freely in the blood, the antibodies in cellular immunity remain fixed to the lymphocytes. What happens next is not precisely understood, but the sensitized lymphocytes return to the kidney allograft, which has been recognized as "nonself," and initiate a rejection of the graft, possibly by enlisting the aid of macrophages.

The defense mounted against the invasion of streptococci, in our earlier example, may not always result in an unqualified victory. The immune response may turn against some patients and produce damage. Such is the case with rheumatic fever and glomerulonephritis. It is also true that the body may not respond to a foreign antigen by developing immunity. It may, instead, develop tolerance—accept the antigen as "self." This has been demonstrated to occur in the immunologically immature (or deficient) individual, for instance, in dizygotic twins that may exchange blood cell precursors *in utero* and thus become histocompatible. The concept of tolerance is particularly relevant to organ transplantation.

HISTOCOMPATIBILITY—THE GENETIC BASIS OF TRANSPLANTATION

That every human being differs genetically from every other human being (except for monozygotic twins) does not require elaboration. The fact that "nonself" is open to immunologic attack whether it is a virus, bacterium, cancer cell, or transplanted organ would appear to present a formidable barrier to transplantation. How, then, can one even consider transplanting an organ, which cannot avoid being genetically dissimilar, into a recipient? Fortunately, in the totality of genetic differences between the donor and recipient, only certain genetic differences play significant roles in whether a transplanted organ will be accepted as "self" or rejected as "nonself." The ABO blood groups and the histocompatibility antigens are of major importance.

Great progress has been made in elucidating the genetic basis of the histocompatibility (HL-A) antigens. The situation appears complicated, as there are now more than 40 histocompatibility antigens known. It appears that they are controlled by alleles at two closely linked loci, with less than 1% crossing-over between them. Thus there are several thousand possible phenotypes, which is why it is unlikely that one will find an unrelated donor who is histocompatible with a would-be recipient. However, siblings have a much higher chance of being histocompatible—specifically, 1 chance in 4, as illustrated in Figure 17-2.

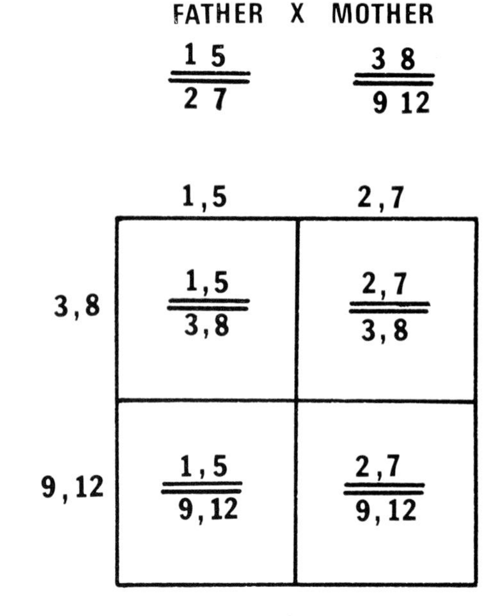

Figure 17-2. HL-A maternal and paternal specificities derived from each of the two loci yield four different genotypes, each with a 1-in-4 chance to recur.

This 1-in-4 chance of histocompatibility among siblings may be exploited to obtain a kidney that will not undergo rejection. The mixed leukocyte culture test consists of culturing leukocytes from donor and recipient together and adding mitomycin C after 7 days to stimulate division. If stimulation occurs, the cells are antigenically different. The basis of the test is that lymphocytes exposed to materials that are antigenically incompatible (including other lymphocytes) undergo a lymphoblastic transformation, which can be observed under the microscope. This test may be used in any situation where there is a living donor and at least 1 week available to complete the study, but cannot be applied to cadaver donors for heart and liver transplants, where there are only a few hours to prepare for the procedure.

Perhaps the clearest way to illustrate the features of histocompatibility in man is to begin with a clinical situation: A patient has irreparable brain injury and is about to die in the emergency room of a hospital that has an active transplantation service. The relatives are approached for possible consideration of donation of the patient's organs to the transplantation service at his death. They agree. What steps are now followed to prepare for organ transplantation and reduce the risk of incompatibility and rejection?

1. ABO Incompatibility

This is the first and strongest barrier to transplantation, so the donor's blood groups must be compared with the blood groups of potential recipients in the same way that blood groups are matched for transfusion. If the donor is type A, the only suitable recipients would be type A or AB. However, if the donor is type O, he may be considered a universal donor—that is, he would be compatible in this first step of matching with recipients of group O, A, B, and AB.

2. Lymphocyte Cross-match

The lymphocytes of the donor are presumed to carry the antigens present in other tissues and to reflect the antigen content of the donor organs. These donor lymphocytes are cross-matched with the blood serum of the recipient in an effort to detect in the recipient the presence of antibodies already formed against the donor antigens. Such preformed antibodies could be responsible for hyperacute rejection of the donor organ. This second step in deciding whether to proceed with transplantation cannot, of course, be taken until some potential recipients are selected for cross-matching on the basis of the information obtained in steps 1 and 3.

3. Histocompatibility Antigen Matching

The specific antigens possessed by the donor are determined by serologic means, using the lymphocytes as the source of antigen.[11] Let us assume that the donor turns out to have the antigens HL-A2, HL-A3, HL-A5, and HL-A12. The antigen profile of available recipients who have already been typed is now studied. First, the potential recipients are selected on the basis of ABO compatibility. The donor is blood type A_1, so only type A_1 recipients will be considered initially. Three renal

TABLE 17-1. Histocompatibility Typing in Clinical Example of Renal Transplantation

	ABO	*HL-A1*	*HL-A2*	*HL-A3*	*HL-A9*	*HL-A10*	*HL-A11*	*HL-A5*	*HL-A7*	*HL-A8*	*HL-A12*	*Cross-match*
Donor	A	+	−	+	−	−	−	+	−	−	+	Negative
Rec. 1	A	+	+	−	−	−	−	−	+	−	+	
Donor	A	+	−	+	−	−	−	+	−	−	+	Negative
Rec. 2	A	+	−	+	−	−	−	+	−	−	+	
Donor	A	+	−	+	−	−	−	+	−	−	+	Negative
Rec. 3	A	−	−	+	−	−	−	+	−	−	+	

patients whose life is being sustained by chronic hemodialysis are found who are type A_1 (Table 17-1). Recipient 1 has no HL-A3 or HL-A5 antigens. This is a two-group, two-locus mismatch of the specificities for which the laboratory is currently able to test (not to mention potential mismatches for which no test was performed due to the limits of the current state of the art). Recipient 3 has no HL-A1; this is a one-group mismatch of known specificities. Recipient 2 has all four antigens the donor has: HL-A2, HL-A3, HL-A5, and HL-A12. A completely compatible match would be judged confidently only in identical twins, but this match would be considered very acceptable for proceeding.

It has been calculated from data on ABO and HL-A antigen compatibility between donors and recipients that a minimum of 500 prospective recipients may be required to give a cadaver organ donor a 95% chance of being transplanted into a compatible recipient.

Because there are two kidneys available for transplantation, the second least incompatible match (donor to recipient 3) would also be accepted because it has been shown that three antigens in common is associated with significantly better survival than if two or fewer antigens are identical. It should be noted that some active transplantation groups feel that histocompatibility matching is unrelated to survival of transplant patients and that only ABO compatibility is required. But the real problem is the fact that present-day histocompatibility matching techniques discriminate only between poor matches and poorer matches by methods that are yet weak and imperfect.

If the clinical situation were less urgent and a relative were available as a donor, then first-degree relatives (preferably) could be screened; in this case not only could the histocompatibility antigen matching be done as in step 3, but a special mixed leukocyte culture test could be performed. As has been stated earlier, the latter test requires 1 week, but if the clinical urgency is not great, it should provide the means for selecting the most compatible relative who is willing to donate one of his kidneys.

Human Leukocyte Antigen Groups

As was stated earlier, the assumption has been made that the leukocytes reflect the antigen content of other tissues, although it is appreciated that there are some organ-specific antigens. There appear to be two series of segregating leukocyte antigen groups. The convention has been followed in Table 17-2 that there are two loci. WHO committee nomenclature has been used.[11] There is agreement as to some of the antigens that belong to each locus and question about others. How many antigen specificities will eventually be determined for these loci probably depends on the amount of effort expended in searching for them.

There is much to be learned about histocompatibility in man. Polymorphism of the HL-A system, especially at locus 2, cross-reaction, and complex specificities make the picture more complicated than this presentation would indicate.[5] The question

TABLE 17-2. Human HL-A Leukocyte Antigen Groups

Locus (Segregant Series) 1	*Locus (Segregant Series) 2*
HL-A1	HL-A5 (Da 5, Te 11)
HL-A2 (Mac, LA2)	HL-A7 (Da 10, 4d)
HL-A3 (LA3)	HL-A8 (Da 8)
HL-A9 (LA3)	HL-A12 (Da 4, Te 9)
HL-A10 (Da 17, Te 12)	W5 } Da 6
HL-A11 (Da 21, Te 13)	W15 } Da 6
W28 (DA15)	Da24 } Da 6

of one or more additional loci or subloci has not been completely resolved. With these deficiencies, however, remarkable advances have been made in serologic typing in just the past 5 years.

ORGAN TRANSPLANTATION

The central problem in transplantation is how to violate a basic biologic law—the recognition and rejection of "nonself"—and get away with it. As we pointed out in the previous section, grafts between identical twins and grafts between other individuals completely compatible in ABO and histocompatibility specificities will not be recognized as "nonself." However, in the clinical setting, the occasion rarely arises for an identical twin to be a donor, and the limitations in present techniques of histocompatibility matching and donor availability do not permit ideal matching.

A brief historical review can mention only that transplantation was described in Greek mythology and early Christian legends. Tagliacozzi, in the sixteenth century, gained a reputation for being able to reconstruct noses (lost in duels and to syphilis). He appreciated (empirically?) that one could not transplant the nose from one person to another and thus devised the operation, used to this day, of utilizing a flap from the patient's own upper arm (autograft). The terminology of transplantation is provided in Table 17-3.

A number of workers have been responsible for the accelerated advancement in knowledge in transplantation in our own century, among them Jensen, Carrel, Murphy, and Medawar. It was the series of classic experiments by Medawar in the 1940's that provided the basis for contemporary transplantation research.[8] Certain principles have emerged from the work of Medawar and other investigators:

1. Allograft immunity is cell-mediated (although humoral mechanisms probably play a role).
2. Grafts between genetically dissimilar individuals may first appear to be accepted, but are then rejected within a period of about 10 days, depending on the strength of the genetic difference (first-set rejection). If another transplant from the same donor (or donor of the same genotype) is attempted, rejection is accelerated (second-set rejection). The process may require only 3–6 days. The recipient has been sensitized (has immunologic memory) and quickly attacks the graft.
3. Tolerance to a graft is an alternative to rejection. The foreign cells may be

TABLE 17-3. Terminology of Tissue Transplantation

New Terminology	*New Adjective*	*Old Terminology*	*Definition*	*Result*
Autograft	Autologous	Autograft	Graft in which donor and recipient are the same individual	Acceptance
Isograft	Isogeneic	Isograft	Graft between individuals with identical histocompatibility antigens (e.g., MZ twins)	Acceptance
Allograft	Allogeneic	Homograft	Graft between genetically dissimilar members of same species (e.g., man to man)	Rejection
Xenograft	Xenogeneic	Heterograft	Graft between species (e.g., ape to man)	Rejection

accepted as "self," especially in the immunologically immature (or deficient) individual, rather than rejected as "nonself." Methods that take advantage of this weakness in the immunologic armor may provide the answer to long-term survival of allografts without resorting to drastic immunosuppression.

Rejection

The mechanism of rejection has been described in the section on development of immunity. Certainly, the cell-mediated immunity of the thymus system plays the major role. Small lymphocytes are transformed to lymphoblasts after detecting the foreign antigen, and they return to the graft as "sensitized" cells capable of participating in the graft rejection. Figure 17-3 illustrates rejection of a skin allograft between genetically dissimilar strains of mice, acceptance of an isograft between mice of the same genetic constitution, and temporary acceptance (overriding of rejection) of an allograft between genetically dissimilar mice under the influence of immunosuppressive medications.

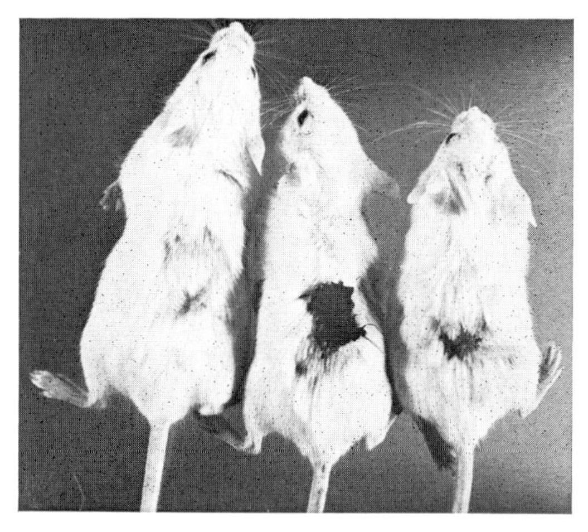

Figure 17-3. From left to right: A/J mouse receiving transplant of A/J skin without rejection; A/J mouse receiving C57 skin without early rejection because of cyclophosphamide immunosuppression; A/J mouse without immunosuppression showing active rejection of C57 skin.

This points up the problem faced by the physician managing a patient with an organ transplant. Because the patient is at risk from rejection, his immunologic mechanisms against "nonself" must be suppressed. But immunosuppression is not yet sufficiently specific, so the patient's immune system is suppressed not only with respect to the transplant, but also with respect to bacterial and viral infections and cancer. He walks a tightrope between rejection and infection.

It has already been pointed out that, if there were complete histocompatibility, as in identical twins, immunosuppressants would be unnecessary. Since this situation rarely occurs in kidney transplants and is not applicable in heart transplants, the need is for techniques that will more accurately define and test the histocompatibility of donor and recipient. Serologic tests are the most popular and are becoming more definitive. Mixed leukocyte cultures add a new dimension to histocompatibility, but they are too time-consuming (1 week) to be applicable to cadaver donors where only hours are available—unless there is an important advance in long-term organ preservation.

What is needed to make organ transplantation an unqualified therapeutic success is an authentic breakthrough in immunology. It is true that 5-year survivors of kidney transplants are becoming more common. However, kidney recipients have advantages over heart recipients in that there is more time to get a good tissue match (including the opportunity to use relatives as donors) and the recipient is initially immunosuppressed by his uremic condition. If heart transplantation is to become a genuine therapeutic alternative, ways to utilize poor histocompatibility matches and even xenografts will be necessary to meet the potentially enormous demands for heart replacement.

It is here that the concept of tolerance

and methods of inducing tolerance might be relevant.[10] The need is for immunologic specificity. The recipient should not have his ability to fight infection and cancer disastrously compromised. The ideal would be to leave the recipient with only one immunologic blindspot, that is, the inability to recognize the transplanted organ as being "nonself."

Graft versus Host

Not only does the recipient recognize the transplant as being "nonself," but the transplant may recognize the recipient as being "nonself" and attempt to attack it. This can occur if the donor tissue is immunologically competent, e.g., lymphoid tissue. In the mouse, a form of this reaction has been called "runt disease." The growth of the host is strikingly impaired, if, for example, spleen cells are injected into the newborn animal. If a transplant of blood-forming tissue is introduced into a subject following total-body irradiation and destruction of host immunity, the transplanted immune-competent cells raise antibodies against the host, leading to wasting and death. A rapidly fatal graft-versus-host reaction may follow a blood transfusion that contains incompatible immune-competent lymphocytes in patients with Swiss-type agammaglobulinemia, who have no immunologic defense.

MATERNAL–FETAL INTERACTION

In the introductory chapter, it was stated that the three major categories of genetically determined diseases were (1) single mutant gene, (2) chromosomal, and (3) multifactorial. To these could be added a fourth category: (4) diseases of maternal–fetal interaction. Some examples are hemolytic disease of the newborn (discussed in Chapter 18) and maternal antithyroid antibodies, a discussion of which follows.

Maternal Antithyroid Antibodies

Familial nongoitrous cretinism has been attributed to maternal antithyroid antibodies crossing the placenta and attacking the thyroid gland of the fetus. Of great interest are the data of Fialkow and others demonstrating the greater frequency of antithyroid antibodies in children with Down syndrome and in their mothers than in their fathers or in control populations. Similar findings have been reported for Turner syndrome. Thyroid disease is also increased in mothers and maternal relatives of children with Down syndrome. This suggests that there is a genetic predisposition to the formation of antithyroid antibodies that is in some way related to the predisposition to nondisjunction. It remains to be seen whether the presence of antibody predisposes to nondisjunction or whether the genetic difference predisposes both the antibody formation and nondisjunction.

AUTOIMMUNE DISEASES

The so-called autoimmune nature of the so-called diseases of connective tissue such as rheumatoid arthritis and systemic lupus erythematosus is well recognized. A variety of other systemic and organ-specific diseases have been attributed to autoimmune mechanisms. A definitive discussion of these diseases exceeds the scope of this presentation. The underlying mechanism of autoimmune disease remains unclear.[6]

Only a single disease, rheumatic fever (the autoimmune basis of which is becoming more apparent), will be briefly discussed. Although there are still some reservations about the immunologic aspects of this disorder, it makes an interesting sequel to our introductory example of the development of immunity to a group A beta hemolytic streptococcal infection. The vast majority of patients who have a streptococcal infection respond as in the example in the first section of this chapter. They develop immunity to the specific type of streptococcus which infected them. Some patients, however, develop rheumatic fever, a systemic disease that is particularly damaging to the heart.

There is evidence that there are circulating antiheart antibodies in patients with active rheumatic fever. It appears that the group A streptococcus shares a common antigen (cross-reactive antigen) with human myocardium. Kaplan first recognized the cross-reactive antigen and by immunofluorescent techniques showed that the sera of patients with active rheumatic fever frequently contain antibodies to this shared antigen. One can thus visualize that in raising antibodies to fight a streptococcal infection, the patient may also raise antibodies capable of attacking the heart. Why this occurs in some patients and not in others appears to be related to a polygenic hereditary predisposition mediated at levels as yet undefined.

SUMMARY

Immunity develops from a basic stem cell through two systems: (1) the bursa system ("B cells"), which is responsible for the production of circulating antibody (immunoglobulins) to combat bacterial and viral infections, toxins, and some particulate antigens; (2) the thymus system ("T cells"), which is responsible for cellular immunity, plays a major role in transplantation as well as representing a factor in the natural defense against bacterial, viral, fungal, and protozoal diseases (and probably cancer).

Except for monozygotic twins, every human being differs genetically from every other human being. However, in the totality of genetic differences between a donor and a recipient in transplantation, only ABO antigens and histocompatibility antigens are of major importance. There is considerable evidence to favor the proposal that there are two closely linked histocompatibility loci. This concept is most applicable in selection of related donors in elective kidney transplantation.

The recognition of "nonself" by a recipient causes rejection of a genetically dissimilar donor organ, but may also cause "rejection" of a recipient if the material transplanted is immunologically competent (graft-versus-host reaction).

Certain disease states that appear to be related to immune reactions that unwittingly attack the "self" are called autoimmune diseases.

REFERENCES

1. Burton, O. C.: Agammaglobulinemia. Pediatrics 9:722, 1952.
2. Burnet, F. M., and Fenner, F.: The Production of Antibodies. London, Macmillan, 1948.
3. Cooper, M. D., Peterson, R. D. A., and Good, R. A.: Delineation of the thymic and bursal lymphoid systems in the chicken. Nature 205: 143, 1965.
4. Edelman, G. M.: Antibody structure and molecular immunology. Science 180:830, 1973.
5. Dausset, J.: Polymorphism of the HL-A system. Transplantation Proc. 3:1139, 1971.
6. Holborow, E. J.: An ABC of Modern Immunology. 2nd ed. London, The Lancet Ltd., 1973.
7. Hume, D. M., Merrill, J. P., Miller, B. F., et al.: Experiences with renal homotransplantation in human: Report of 9 cases. J. Clin. Invest. 34:327, 1955.
8. Medawar, P. B.: Behaviour and fate of skin autografts and skin homografts in rabbits (report to War Wounds Committee of Medical Research Council). J. Anat. 78: 176, 1944.
9. Nora, J. J., Cooley, D. A., Fernbach, D. J., et al.: Rejection of the transplanted human heart: Indexes of recognition and problems in prevention. New Eng. J. Med. 280:1079, 1969.
10. Nossal, G. J. V.: Immunological tolerance in organ transplantation. Fair prospect or fanciful folly? Circulation 39:5, 1969.
11. Van Rood, J. J., and van Leeuwen, A.: Alloantigens of leukocytes and platelets. *In* P. A. Miescher and H. J. Müller (ed.), New York, Grune and Stratton, 1974.

Chapter 18

BLOOD GROUPS AND SERUM PROTEINS

In the discipline of human genetics, few areas of investigation have been more informative than blood groups and, more recently, serum proteins.

The fact that people could be classified into groups by the antigens on their red blood cells was first demonstrated by Landsteiner, who discovered the ABO blood groups in 1900 and was rewarded by a Nobel Prize 30 years later. During the first 25 years after their discovery, the mendelian inheritance of blood groups was intensively studied and firmly established. Now blood groups have broad applications in population and family studies, linkage analysis and chromosome mapping, and forensic medicine.[1] Safe blood transfusion, which represents one of the few indispensable therapeutic options of modern medicine, could not exist without blood-typing, which depends on accurate identification of blood groups.

Serum proteins, as polymorphisms, were found to provide the same sort of genetic information as blood groups when starch gel electrophoresis provided a method for their identification in 1955. Since then, a number of other media have been used in the methodology.

The advantages for genetic investigation offered by blood groups and serum proteins stem from the fact that they are relatively direct expressions of gene action and therefore fall into sharply defined groups with simple inheritance. Codominance is the rule (making heterozygotes as readily detectable as homozygotes) although there are notable exceptions, including the ABO and I systems.

ABO BLOOD GROUPS

Landsteiner found that when the red blood cells of certain individuals were mixed with the serum from certain others, the cells became attracted to one another and formed clumps—the *agglutination reaction.* This was shown to occur when the serum contained antibody specific to an antigen on the surface of the red cell. By this method, people could be separated into four phenotypes, as illustrated in Table 18-1.

It was soon demonstrated that these antigenic differences showed mendelian inheritance, antigens A and B and lack of either antigen (O) being the expression of three alleles, A, B, and O (sometimes referred to as I^A, I^B, and I^O). Note that a per-

TABLE 18-1. ABO Blood Groups and Frequency in a Representative English Population

Genotype	*Phenotype*	*Frequency*	*Red Cell Antigen*	*Serum Antibody*
AA AO	A	.42	A	anti-B
BB BO	B	.09	B	anti-A
AB	AB	.03	AB	none
OO	O	.46	none	anti-A, anti-B

son's serum contains antibodies to whatever antigens are not present on his or her red cells. This is an exception to the rule; antibodies for other blood group systems do not occur unless the person is sensitized by an injection of "foreign" red cells. The exception is a fortunate one—if there were no naturally occurring A and B antibodies, the discovery of blood groups and, indeed, the development of genetics would have been delayed by many years.

The reciprocal relationship between antigen and antibody provides a basis for blood transfusion. If the serum of the recipient contains antibodies against antigens present on the donor cells, the donor cells will be agglutinated and break down, causing a transfusion reaction. Thus group O individuals are called "universal donors"—their red cells will not be agglutinated by either anti-A or anti-B antibody. Conversely, group AB individuals are "universal recipients"; since their serum has no anti-A and anti-B antibody, they can receive cells of any ABO phenotype. Group A people can receive either group A or O blood; group B people, either B or O blood; and group O people, only O blood. The antibodies in the donor serum do not seem to matter, as they become so diluted by the host serum that they do not cause much, if any, agglutination of host cells.

Subtypes

The ABO system can be further divided into subtypes, on the basis of quantitative differences in antigenicity. The most important are A_1 and A_2, increasing the possible phenotypes to 6 (A_1, A_2, B, A_1B, A_2B, and O). B subtypes and other A subtypes also exist.

H Substance

Some group A_1 individuals were found to have an antibody in their serum that agglutinates group O cells, suggesting that the O cells contain an antigen, which was labeled H. Discovery of the Bombay phenotype helped to clarify the situation. This is a very rare phenotype, first discovered in Bombay in 1952, in which the individual's red cells were not agglutinated by anti-A, anti-B, or anti-H, showing that they were not group O cells. A family was then discovered in which a woman of this phenotype could be shown by family studies to be group B, since she had a group AB child by an AO father, though she had no demonstrable antigen B on her red cells. It now appears that the Bombay phenotype results from homozygosity for a mutant allele, h, and that the normal allele, H, is responsible for making H substance, which is a precursor of the A and B antigens.

The Biochemical Basis of the ABO System

The elegant work of Watkins,[4] and others, has demonstrated the biochemical basis of the ABO system. As illustrated in Figure 18-1, there is a glycoprotein precursor, without any demonstrated antigenic activity. The H gene leads to the presence of an enzyme, H transferase, that adds an L-fucose to the precursor, converting it into H substance. In the Bombay phenotype the enzyme is missing, so there is no antigen H (lower arrow) and thus no substrate from which to make A and B antigens.

Once the H substance is produced, it can be modified by the genes at the ABO locus. The A gene provides an enzyme that adds a sugar, N-acetyl-D-galactosamine, convert-

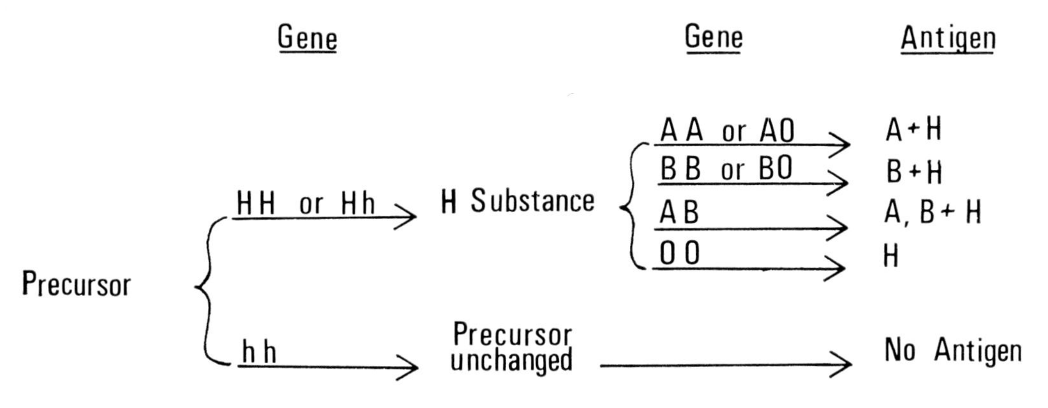

Figure 18-1. Possible pathway for biosynthesis of blood groups.

ing H into antigen A. The B gene provides an enzyme that adds a sugar, D-galactose, which produces antigen B specificity.

The Secretor Locus

In 1930, another polymorphism was discovered that modifies the expression of the ABO system. In about 78% of the population, the ABO blood group substances are water-soluble and occur in body fluids (sweat, tears, saliva, semen) as well as in the red blood cells. In the other 22%, the ABO antigens are limited to the red cells. This difference is determined by the secretor gene Se and its recessive allele se, which when homozygous produces the nonsecretor phenotype. The Se allele appears to be a regulatory gene that allows the H gene to operate in secretory cells; in sese individuals no H substance is produced, so there are no ABO-type antigens in the secretions. The secretor locus is linked to the locus of the Lutheran blood group (see below) with a 15% recombination frequency. This was the first example of autosomal linkage in man.

Frequency of the ABO Groups

The frequency of the ABO blood groups varies widely in different populations. For instance, the frequency of the B allele is high in Mongolia and declines towards the West, being lower in Siberia and still lower in Europe, possibly reflecting the invasion of Europe by the Tartars. A is higher in southern England than in Scotland, perhaps as a result of retreat northward of the aboriginal population that had a high frequency of O, as continental, high-A populations moved in from Europe. Another interesting feature of the blood groups is their association with certain diseases. These do not represent linkages but are probably pleiotropic effects of the genes concerned. For instance, patients with duodenal ulcers are twice as likely to be group O, nonsecretor, as controls from the same population—for instance, their sibs. There is an excess of group A among patients with cancer of the stomach. Although these differences are highly significant, they are not large enough to be of predictive value.

THE RH BLOOD GROUPS

In 1940, Landsteiner and Weiner injected blood from Macaca rhesus monkeys into rabbits and prepared an antiserum, antirhesus, which would agglutinate the red cells of other rhesus monkeys. They were delighted to find that this serum would also agglutinate the cells of about 85% of white New Yorkers, whom they classified as Rh positive. The other 15%, whose cells did not react, were called Rh negative. The Rh-negative quality was soon shown to be recessive to the positive quality. With great perspicacity, Levine associated this difference with erythroblastosis fetalis, more properly termed hemolytic disease of the newborn. This disease causes massive breakdown of the red cells, anemia, jaundice, and other complications often leading to death, deafness, mental retardation, or cerebral palsy. Mothers of such babies were found to have an anti-Rh antibody in their serum, quite similar to the one prepared in rabbits. Thus the disease was shown to result when an Rh-negative mother was sensitized by red blood cells from her Rh-positive baby and developed anti-Rh antibodies. A second Rh-positive child would lead to a great increase in antibody, which would pass across the placenta and react with the cells of the baby to produce the disease. We shall return to this later.

Other antibodies from sensitized mothers were then found that behave genetically as if they were at the same locus, and the situation began to get complicated. Two interpretations of the results were put forward. Fisher and Race postulated three adjacent loci, each with at least two alleles. These were C and c, D and d, and E and e. D corresponds to the original Rh-positive antigen, now called R^o. Antibodies have been found against all these antigenic differences except d.

The other interpretation, passionately defended by Wiener, is that the Rh locus is a complex locus with several antigenic sites, characterized by various combinations of the three kinds of antigenic specificity that Race and Sanger call C, D, and E and their alleles. Table 18-2 sets out the two nomenclatures and the frequencies of the six gene complexes.

TABLE 18-2. The Two Notations for the Rh System, and Frequencies of the Various Gene Complexes in an English Population

Fisher/Race	*Wiener*	*Frequency*
CDe	R^1	0.41
cde	r	0.39
cDE	R^2	0.14
cDe	R^o	0.03
Cde	r′	0.01
cdE	r″	0.01
CdE	r^y	low
CDE	R^z	low

As with the ABO groups, the Rh system has several subgroups, such as D^u, an allele with weak D antigenicity, that may be mistyped as d unless a strong anti-D serum is used. There is also a rare phenotype Rh null, comparable to Bombay, in which there are no Rh antigens. Another rare allele is —D—, in which C and E specificities are missing. According to the Fisher-Race model, this would suggest a deletion involving the C and E regions and imply that C and E were adjacent. For this reason, the CDE locus is sometimes written DCE.

Since each person has two Rh genes, there are 36 combinations. The frequencies of the various combinations can be calculated by multiplying the frequencies of the individual Rh genes. In the case of hemolytic disease of the newborn, it is important to know whether the father is DD or Dd, but unfortunately there is no anti-d antibody, so this cannot be demonstrated directly. Sometimes family studies will demonstrate heterozygosity—for instance, if he has dd children. Otherwise, the probability that he is heterozygous can be calculated if his genotype for the other antigens is known. For example, suppose he types C^-, c^+, D^+, E^-, e^+. Then his genotype must be either cDe/cde or cDe/cDe. The frequency of the first genotype (heterozygous D) is $.03 \times .39 = .117$, and of the second (homozygous D) is $.03 \times .03 = .009$. So the probability of his being heterozygous is $.117/126 = .09$; he is thirteen times as likely to be heterozygous as homozygous D. The blood-typing laboratory would report this as "most probable genotype—cDe/cde."

The Rh genes vary widely in frequency from population to population. The Basques, for instance, have a high frequency of dd individuals—about 30%, while

Orientals and North American Indians have almost none.

Hemolytic Disease of the Newborn

Most hemolytic disease of the newborn (HDN) is caused by maternal–fetal incompatibility for the D antigen. The disease used to affect about 1 in 200 Caucasian babies, but it now can be almost completely prevented. The usual story is that an Rh-negative, or dd, mother has an Rh-positive child. It is not unusual for red blood cells from the fetus to pass across the placenta into the mother's bloodstream, particularly during compression of the placenta at birth. The mother may (though not always) become sensitized and begin to make anti-D antibody. This usually does not reach a high enough concentration to cause trouble in the first pregnancy, unless the mother was sensitized previously by an incompatible blood transfusion. If the mother then has a second D child there is a sharp rise in antibody titer in the mother. The anti-D antibodies may then cross the placenta from mother to baby and coat the baby's red cells, causing their destruction—thus the anemia, jaundice, and other features of the disease.

Formerly, the disease could be treated only by exchange transfusion, in which the baby was transfused with Rh-negative blood, with concurrent removal of its own blood, in an attempt to wash the harmful antibodies out of its system. This was not wholly satisfactory. Now the disease can be prevented by giving the mother RhoGam, an anti-D antibody preparation. If this is given to the mother when the first D-positive baby is born, the anti-D antibody will coat and destroy any D-positive cells in her blood, and thus prevent them from sensitizing her.

Why some incompatible mothers are not sensitized by their babies, while others are, is not fully understood. In part, it results from an interaction with the ABO system. If the D baby's cells are also group A, while the mother is dd O, the baby's cells that get into the mother's circulation will be destroyed by her anti-A antibody and so will not be available to sensitize her against the D antigen.

Hemolytic disease of the newborn can also occur as a result of ABO incompatibility, usually in an O mother and A child. The naturally occurring antibody, an IgM globulin, does not cross the placenta, but antibodies resulting from a previous child or an incompatible transfusion are predominantly IgG and will cross and attack the baby's cells. However, the resulting disease is usually mild and requires no treatment. Occasionally, a mother will develop antibodies against some other blood type antigen, such as K or Fy^a, that may cause HDN in a subsequent child.

THE MNSs BLOOD GROUPS

In 1927, Landsteiner and Levine discovered the MN blood groups after injecting rabbits with human red cells, and used the resulting immune serum to distinguish other human red cell samples. They proposed a two-allele mode of codominant inheritance that is still accepted. The two alleles, M and N, produce three genotypes (MM, MN, and NN) and three phenotypes (M, MN, and N) with frequencies in European populations of 28%, 50%, and 22%, respectively.

About 20 years later, another antibody was found, which was associated with M and N and given the designation S. It was not considered to be an allele of M and N but was thought to be related to M and N as C and E are related to D. Thus, there are MS, Ms, NS and Ns combinations. The genes must be close together, although evidence is mounting that recombination can occur occasionally between the MN and Ss sites.

The MNS antigens do not stimulate antibodies in man, so they are not a problem with respect to blood transfusion or maternal-fetal incompatibility. However, because of its relative frequencies and codominance, it is the most useful blood group system in medicolegal work and other problems of individual identification.

THE P BLOOD GROUPS

Landsteiner and Levine discovered the P blood group system in 1927, using the same type of immunization experiments that identified the MN system. This system contains two phenotypes, P_1 and P_2, with frequencies of 79% and 21%, respectively, in Caucasians. P_1 is dominant to P_2. A very rare allele, p and a p^K allele, perhaps comparable to Bombay, are also known.

THE LUTHERAN BLOOD GROUPS

This blood group system was discovered in 1954 and was named after the person, not the religious sect, in whom the antibody was first found. By this time, the inconsistencies in terminology were great, since many changes had been made as knowledge grew. A new terminology was therefore agreed upon. The phenotypes are designated by the antibodies to which they react. Thus, in this case, there are two antibodies, anti-Lu^a and anti-Lu^b, and three phenotypes, Lu(a+b+), Lu(a+b−), and Lu(a−b+). These result from the segregation of two alleles, Lu^a and Lu^b. Thus the phenotype Lu(a+b−) is the expression of the genotype Lu^aLu^a, since the red cells are agglutinated by anti-Lu^a serum but not by anti-Lu^b serum (Table 18-3). This blood group is distinctive in that it provided the first example in man of autosomal linkage (with secretor) and suggested that crossing-over is more common in the female.

TABLE 18-3. Nomenclature of the Lutheran Blood-Group System and Frequencies in a Caucasian Population

Genotype	*Phenotype*	*Frequency*
Lu^a/Lu^a	Lu(a+b−)	0.001
Lu^a/Lu^b	Lu(a+b+)	0.08
Lu^b/Lu^b	Lu(a−b+)	0.92

THE KELL BLOOD GROUP

The Kell system consists of three pairs of alleles, Kk, Kp^aKp^b, and Js^aJs^b. The k allele is sometimes referred to as Cellano, after the woman in whom the k antibody was first discovered. The Js alleles (Sutter) were originally thought to be at an independent locus. Js^a is very common in Negroes but rare in Caucasians. Hemolytic disease of the newborn is occasionally produced by maternal–fetal interaction in the Kk system.

THE LEWIS BLOOD GROUPS

The Lewis blood group was first described in 1946. The common phenotypes in Caucasians are Le(a+b−) (26%) and Le(a−b+) (69%). Le(a−b−) is quite common in West Africans.

The onset of development of the Lewis antigens usually begins in infancy. Newborn red cells are not agglutinated by either anti-Le^a or anti-Le^b sera, but Lewis antigens are present at birth in the saliva and serum of individuals with the appropriate genotypes. This blood group interacts in a complicated way with the ABO, H, and secretor systems. Individuals with the red cell phenotype Le(a+b−) are nonsecretors of ABH but contain Le^a antigen in their secretions. Individuals with the phenotypes Le(a−b+) and Le(a−b−) are secretors.

One proposal is that the precursor substance is converted in the presence of the Lewis (Le) gene into Le^a substance, which in turn is converted by the H gene in the presence of the secretor (Se) gene into H substance with Le^b activity. In nonsecretors, the conversion of Le^a into H in the secretory tissues does not occur—hence the absence of A or B substances in secretions. The inactive allele le when homozygous leads to an H substance without Le activity. The biochemical basis for this interaction is presented by Morgan and Watkins.[5] Although this sequence is not depicted in Figure 18-1, it is possible to visualize how these steps could be added.

It has been observed that red cells that lacked either Le^a or Le^b antigen would acquire that antigen if suspended in plasma containing it. This was proved *in vivo* when Le(a+b—) donor cells were administered to an Le(a—b+) patient. Following transfusion, the donor cells obtained by differential Rh agglutination tested as Le(a+b+). This information has led to a proposal that the Le antigens may be acquired from plasma antigens. There is an abundance of theories about the Lewis blood groups that cannot be dealt with within the limited scope of this presentation.

DUFFY BLOOD GROUPS

The antibody leading to the discovery of the Duffy group was found in the serum of a patient of that name who had hemophilia and who had received multiple transfusions. The gene was designated Fy^a and its allele Fy^b. Later, a silent allele, Fy, was discovered. The homozygous phenotype Fy(a—b—) was present in 85% of New York Negroes but was very rare in Caucasians. This blood group provides the greatest distinction between blacks and whites. It also has the distinction of being the first locus to be assigned to a particular autosome, chromosome 1, closely linked to the "uncoiler" locus.

THE KIDD BLOOD GROUPS

Jk^a and Jk^b are the alleles in Kidd blood group system. The phenotype Jk(a—b—) may be due to an inhibitory gene or another allele at the Kidd locus. This system is mainly of anthropological interest. Jk(a+) is present in about 95% of West Africans, about 93% of American Negroes, about 77% of Europeans, and about 50% of Chinese. Both anti-Jk^a and anti-Jk^b have been known to cause HDN, and anti-Jk^a has produced transfusion reactions.

THE DIEGO BLOOD GROUP SYSTEM

This blood group system was discovered in Venezuela when it produced hemolytic disease of the newborn in a family possessing some physical characteristics of the native Indians. The antigens are called Di^a and Di^b, and their respective antibodies are anti-Di^a and anti-Di^b. This antigen is reported in Chinese, Japanese, South American Indians, Chippewa Indians, and other phenotypically similar populations, with the notable exception of the Eskimo.

THE I BLOOD GROUPS

This antigen has been studied in patients with acquired hemolytic anemia of the "cold antibody" type—that is, antibodies that are active only at a low temperature (4° C). The antigen I differs in certain respects from other blood group antigens. Almost everyone has some trace of the antigen, and the amount of I antigen on the red cells increases from birth until adult levels are reached at about 18 months of age. The corresponding levels of i decrease as I increases. The i antigen appears to be inherited in a recessive manner, but there seems to be a disturbing excess of i siblings. There are two types of anti-I: auto-anti-I, which occurs in people who have acquired hemolytic anemia with cold antibodies, and natural anti-I, which appears in i phenotype adults. Natural anti-I does not cross the placenta. Examples of anti-i have been found in persons with some types of reticulosis. Transient anti-i is often present in patients with infectious mononucleosis.

THE Xg BLOOD GROUP SYSTEM

The Xg blood group differs from previous groups in that it is X-linked and dominantly inherited. The discovery of this X-linked blood group offers more hope for mapping of the X-chromosome than has yet been realized. A large number of X-linked conditions have been studied and found not to be measurably linked to the Xg locus. X-linked ichthyosis and ocular albinism have thus far been reasonably well established as linked with the Xg locus, but the precise location of the Xg locus on the

X chromosome is still a matter of conjecture.

An attractive theory, based on modest data, is that it may be located at the distal end of the short arm. Although there is conflicting evidence on Lyonization of the Xg blood group, evidence against Lyonization could support the concept of lack of inactivation of the Xg locus and adjacent segments of the short arm of the X chromosome, which would account nicely for the phenotypic abnormality in XO Turner syndrome.

Xg(a+) hemizygous males react as strongly as homozygous females. The heterozygote female may or may not produce a weaker reaction. How this evidence can be made to agree with the speculation in the preceding paragraph remains an open question.

OTHER BLOOD GROUPS, PUBLIC AND PRIVATE

The Y+ blood groups, discovered in 1956, the Auberger, in 1961, and the Dombrock, in 1965, have been the subject of considerable investigation. A number of other "public" antigens have been described, including August, Colton, Gerbich, Gregory, Lan, and Vel. The term "public" antigen is used to describe antigens that are encountered frequently, as opposed to a "private" antigen, which is limited to a single kindred.

SERUM PROTEINS

A number of genetically informative serum protein polymorphisms have been determined by electrophoresis, including haptoglobins, immunoglobins, complement, transferrins, and the X-linked Xm system. Figure 18-2 is an illustration of a simple electrophoretic separation of serum proteins into albumin and globulin fractions. Further separation of these fractions is achieved by special methods. For example, gamma globulin is separated into IgA, IgD, IgE, IgG, and IgM. These *immunoglobulins* are discussed in Chapter 17. *Transferrins* are

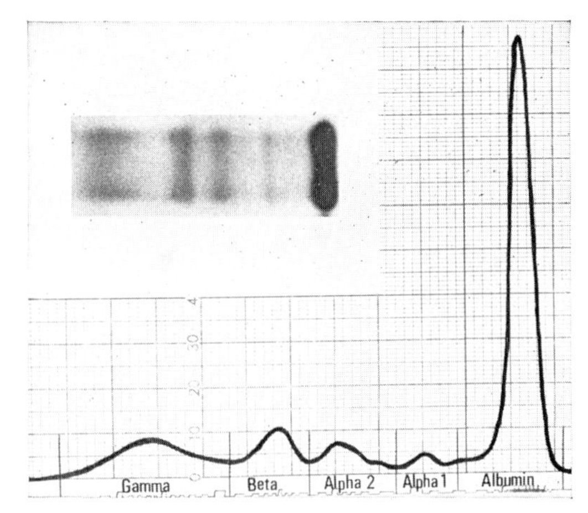

Figure 18-2. Serum proteins separated by electrophoresis on agar gel with densitometric display.

beta globulins that bind iron. *Xm serum system* has the potential to be another useful marker in mapping the X chromosome. Recent findings suggest linkage between the Hunter and Xm loci.

The haptoglobins are the earliest and most extensively studied of serum protein polymorphisms.[3] These proteins are alpha-2-globulins and have the property of binding hemoglobin. Two allelic genes Hp^1 and Hp^2 determine three main phenotypes: Hp1—1 (genotype Hp^1Hp^1); Hp2—1 (genotype Hp^1Hp^2); and Hp2—2 (genotype Hp^2Hp^2). In addition to these three common types of haptoglobin, that form characteristic patterns through electrophoresis on starch gel and certain other media, a number of other variants have been described. Two loci appear to be involved in haptoglobin synthesis, one for alpha and one for beta chains; each chain is susceptible to point mutations, as are the hemoglobin chains. There is evidence that some of the alleles at the haptoglobin locus have arisen through duplication by nonhomologous crossing-over. Haptoglobins may prove to be useful in linkage studies. The alpha locus has been proposed by one group of investigators to be on chromosome 13 and by another group to be on chromosome 16. The common evolutionary origin for the alpha chain of haptoglobin and the light chain of gamma globulin has been postulated; if this is so, it would represent a step toward formulation of a unifying concept in the development of serum proteins.

The existence of *complement* has been recognized for almost a century and its participation in antigen–antibody reactions for decades, but it has been only within the last 5 years that polymorphism in the various components of complement has been investigated.[2] Currently, there are nine recognized components of complement, designated C1 through C9, in conformity with a standardized nomenclature for variants of complement recommended by the World Health Organization. Deficiencies of identifiable components of complement have been identified in kindreds and are associated with immunologic abnormalities.[2]

An interesting disease, hereditary angioneurotic edema (autosomal dominant) has recently been shown to result from a deficiency of the normally occurring inhibitor of the first component. Affected individuals have intermittent episodes of localized edema (swelling of tissues due to the accumulation of extracellular water), which can cause a variety of problems, depending on where the swelling occurs, including possible death if the larynx is involved.[2] Treatment of attacks with fresh frozen plasma, thus restoring the inhibitor, is fairly successful.

SUMMARY

Blood groups and serum proteins have proved to be highly informative polymorphisms for genetic investigation, family and population studies, linkage analysis and chromosome mapping, and forensic medicine. The ability to identify blood types accurately has made safe blood transfusion possible.

The ABO blood groups were first discovered in 1900. The antigenic specificities A and B are co-dominant, and O (lack of antigen) is recessive. A reciprocal relation exists, in that an individual possessing an antigen (A, for example) has no antibodies against A, but does have antibodies against B—and vice versa. A type AB individual has no antibodies against A or B, but a type O individual has antibodies against both A and B.

The Rh blood groups are next in importance for typing, from the point of view of transfusion. Hemolytic disease of the newborn (HDN) is most often associated with this blood group. A recent advance, the use of RhoGam, now makes HDN a preventable disease, for those mothers who have not yet been sensitized.

A great deal of useful genetic information can be derived from the study of the many other public and private blood groups. Recently, serum proteins such as the haptoglobins, the Xm system, and complement polymorphisms have been the subject of genetic investigation.

REFERENCES

1. Race, R. R., and Sanger, R.: Blood Groups in Man. 5th ed. Philadelphia, F. A. Davis, 1968.
2. Ruddy, S., and Austen, K. F.: Inherited abnormalities of the complement system in man, p. 69. *In* A. G. Steinberg, and A. G. Bearn, (ed.), Progress in Medical Genetics. Vol. 7. New York, Grune & Stratton, 1970.
3. Sutton, H. E.: The haptoglobins, p. 163. *In* A. G. Steinberg, and A. G. Bearn (ed.), Progress in Medical Genetics. Vol. 7. New York, Grune & Stratton, 1970.
4. Watkins, W. M.: The possible enzymic basis of the biosynthesis of the blood group substances, p. 171. *In* J. F. Crow and J. V. Neel (ed.), Proceedings of the Third International Congress on Human Genetics. Baltimore, Johns Hopkins Press, 1967.
5. Morgan, W. T., and Watkins, W. M.: Genetic and biochemical aspects of human blood group A-, B-, H-, Le^a- and Le^b-specificity. Br. Med. Bull. 25:30, 1969.

Chapter 19

GENETICS AND CANCER

Normal cells show an extraordinarily precise regulation of their growth. During development, organs grow to their appropriate size and then miraculously stop. The skin, intestinal lining, and other epithelia keep themselves in a dynamic equilibrium by replacing the cells that flake off at the surface by division of cells in the basal layer. If the skin is cut, cell division increases until the gap is closed. If a kidney is removed, the other kidney enlarges to the point where it can compensate for the loss. The mechanisms by which this remarkably sensitive control is maintained are not well understood. But it is clear that cells may sometimes escape from them and grow in an unregulated manner. Such uncontrolled growth is called *neoplasia,* and the resulting growth is called a *neoplasm.* Loosely speaking, a neoplasm is referred to as a "tumor," which simply means a swelling.

Neoplasms may be benign or malignant. *Malignant* neoplasms can spread both through the adjoining tissues and by *metastasis,* when a few neoplastic cells may enter the blood or lymph stream and float to some other part of the body where they may establish a new focus of malignant growth. They are divided into *carcinomas,* neoplasms of epithelial tissues, and *sarcomas,* neoplasms of connective tissues. *Benign* neoplasms do not spread into adjacent tissues, although they may cause trouble by mechanical pressure. Neoplasms are named by the tissue of origin. For instance, adenomas and myomas are benign tumors of glands and muscle, respectively, and a bronchogenic carcinoma is a malignant neoplasm of the bronchial epithelium. "Cancer" is the Latin word for "crab," and strictly speaking, the term refers to carcinomas, but it is often used for other neoplasms too, as we will do in this chapter. The initiation of a neoplasm may also be referred to as *carcinogenesis,* or *oncogenesis.*

Although about one in every five deaths is caused by cancer, little is known of the factors involved. Many environmental agents are known to predispose to cancers, including certain viruses, radiation, chronic irritation, and several groups of chemicals, called carcinogens. There are also a number of mutant genes that cause specific types of cancer. Furthermore, there are a number of "cancer-prone" families with extraordinarily high concentrations of cancers of various types. These are probably too unusual to be accounted for by chance accumulations of cancer cases, even though the population frequency is so high. Neither do they fit the Mendelian laws very well; one is tempted to invoke major mutant genes with irregular expression, or an unusual concentration of polygenes, but the possibility of a virus, perhaps interacting with the genome, should not be neglected. Finally, there is an association between certain cancers, particularly the leukemias (neoplasias of the white-blood-cell-forming tissues), and chromosomal aberrations. Thus it is not unreasonable to approach the genetics of cancer as one would approach the genetics of other common diseases: We may postulate that some may be caused by mutant genes, some by chromosomal aberrations, and some by major environmental agents, but that the majority have a multifactorial basis involving gene–environment interactions. However, the borders between the groups are less clearly defined than they are for some disease categories, as are our ideas about etiology, for that matter.

Perhaps the most useful concept of the nature of cancer, first formulated by Tyzzer in 1916, is that neoplasms arise by *somatic mutation,* occurring in a single cell that, released from its regulatory control, multiplies rapidly and forms a tumor in which all the cells are descendants of the original mutant cell. Evidence comes from several sources. The karyotypes of cancers often show chromosomal changes that may differ from tumor to tumor but that, within one tumor, appear to arise by a series of changes from a single cell. Another approach makes use of Lyonization of the X chromosome by showing that certain tumors occurring in females heterozygous at the X-linked G6PD locus have the phenotype of either one or the other allele but not both and must therefore arise from a single cell (p. 218). Further ramifications of the somatic mutation hypothesis are discussed in relation to retinoblastoma (p. 208).

The somatic mutation theory does not deny the importance of nongenetic factors in neoplasia. For example, there is evidence for an important viral role in certain human malignancies, such as leukemia, and in many kinds of neoplasms in lower animals—e.g., the Rous sarcoma virus in chickens and the Shope papilloma virus in rabbits. This should not be taken to mean that cancers are infectious, however. There are well-documented families with cases of leukemia in successive generations with no direct personal contact between affected individuals. To the traditional concept of horizontal transmission must be added the concept of vertical transmission of viruses. Viruses and other cancer-producing agents may eventually be shown to act as environmental triggers in individuals with a hereditary predisposition to cancer. The hereditary predisposition could assume many forms, from immunologic abnormalities to chromosomal instability, to defects in cellular regulation, to the harboring of temperate viruses in the host genome through successive generations. Given these sorts of hereditary predisposition, a superimposed infection, chemicals, or radiation may trigger the neoplastic growth.

Case History

This 6-year-old boy presented to his primary physician with a history of fever, bruising, and listlessness for a period of about a week. The physician obtained a peripheral blood count which revealed a low white blood cell count and immature forms. The patient was referred to the university medical center, where the mother confided to her consultant physician that she had suspected the diagnosis of leukemia before taking the child to the doctor, because this was the way her daughter by a previous marriage had behaved before she had been diagnosed as having leukemia.

A bone marrow study confirmed the diagnosis of acute stem cell leukemia of childhood, and standard therapy was begun. Several months later, the paternal grandmother was diagnosed as having acute myelogenous leukemia. The grand-

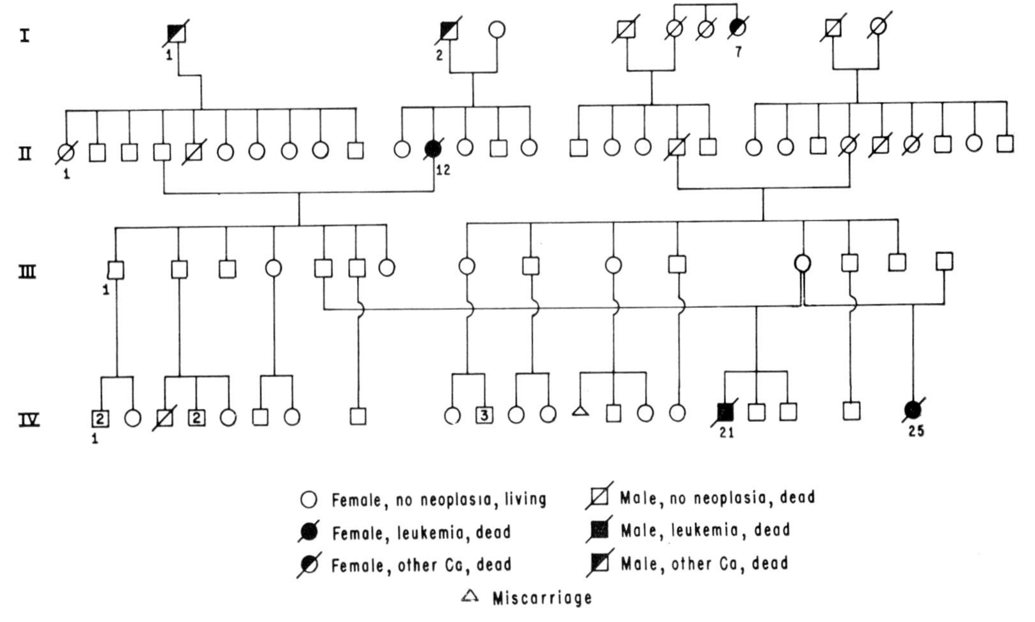

Figure 19-1. Pedigree of the family with leukemia described in the case history.

mother's clinical course was rapidly downhill, and she died within 6 weeks. The half-sister of the proband had died before he was born with a rather fulminant course, lasting 6 months (see Fig. 19-1).

This mother of four children by two fathers was understandably deeply distressed by having a second child with leukemia and wanted to know what the chances were of her remaining children developing this disease. She had not been contemplating future pregnancies. She was counseled that the risk was still very small, but that there were no data regarding the risk to a third sib after two affected sibs or half-sibs. The risk after one affected sib is increased fourfold (to 1/720) over the population risk of 1/2880 for leukemia in childhood.[6]

Her son survived with leukemia for almost 2 years. Shortly after his death, she returned to the center with another son who had a low-grade fever and was acting listless. An evaluation for leukemia was undertaken. This time, the family was happily assured that their child did not have leukemia.

CANCERS ASSOCIATED WITH MUTANT GENES

There are a number of diseases in which a mutant gene causes its carrier to develop a specific type of malignancy.[4] Perhaps the best known of these is *multiple polyposis of the colon,* in which mushroom-like growths, or polyps, cover the internal lining of the colon. Sooner or later, one or more of these becomes neoplastic and forms a carcinoma that may be fatal if not detected in time. Here is an example where family follow-up of affected individuals is mandatory. Regular examination of the colon in high-risk individuals and surgical removal of the bowel if polyps appear will save many a life. In some families, the multiple polyposis is associated with tumors of bone (osteomas) and connective tissue (fibromas) and sebaceous cysts. This type, Gardner's syndrome, also shows autosomal dominant inheritance.

In *adenocarcinomatosis,* a dominant gene causes tumors of a variety of endocrine glands. One form of *tylosis* (thickening of the horny layer of the skin on hands and feet) is associated with cancer of the esophagus. In these and other dominant mutations leading to cancer of specific sites, there must be an abnormal protein that should tell us something about the nature of neoplasia, but so far none has been identified.

Retinoblastoma is a very malignant neoplasm of the retina, appearing in early childhood and leading to death if not treated early. Treatment, by removal of the eye, or irradiation if diagnosed early, can be successful, with luck. The inheritance is autosomal dominant, but most cases are sporadic (90% or more), and of these about 60% are unilateral. The unilateral cases are likely to be somatic mutations, not involving the gonads, but, unfortunately, about 15% are hereditary, so the counseling has to be guarded. The risk for the child of a sporadic unilateral case is around 5%. For the bilateral cases, the counseling is complicated by reduced penetrance, estimated to be 60 to 90% in various series. The risk for sibs of a sporadic bilateral case may be up to 10%, and for the offspring about 40%.

The characteristics of age of onset, laterality, and familial transmission of retinoblastoma led Knudson[4] to propose that the malignant change required two mutational events. Persons inheriting the gene for retinoblastoma carry the first mutation in all their cells, and a single additional mutation in any one of the many thousands of cells in the retina will initiate a tumor. In persons not carrying the mutant gene, two mutations are required for malignant change, which will therefore be a much rarer event. This would explain why the inherited type has an earlier age of onset and is usually bilateral and why the sporadic case has a later onset and is almost always unilateral.

Knudson and colleagues[4] present an impressive amount of evidence that this model applies to many other human cancers. They point out that virtually every cancer that occurs in man occurs in a genetic (autosomal dominant) form as well as a nongenetic (sporadic) form, and they hypothesize that in all forms of cancer two or more changes are necessary. The first change is mutational and is specific for one or more tissues, and the second change may be mutational or involve some other kind of change. In the genetic forms, at least one additional change must occur before a tumor develops. In the nongenetic forms, two changes are necessary, both occurring in a somatic cell. Thus the genetic forms of tumors occur earlier than the nongenetic forms and are frequently multiple. Here we have strong support for the somatic mutation theory of cancer. Evidence from the induction of cancers by chemical carcinogens and radiation also favors a two-hit model.

Xeroderma pigmentosa is a recessively inherited condition in which there are many freckles, horny lumps on the skin, and areas of atrophy. There is marked sensitivity of the skin to sunlight, and, eventually, malignant change occurs in the skin. This disease has recently created great interest, as the deficient enzyme has been identified as an endonuclease that repairs DNA broken by ultraviolet light. This also should tell us something about neoplastic transformation.

Finally, there is an interesting group of three diseases caused by autosomal recessive inheritance in which the gene seems to cause chromosomal instability, as manifested by an increased number of breaks and rearrangements of chromosomes. There is also an increased susceptibility to cancer, particularly leukemia.

These diseases are *Bloom syndrome*—growth retardation, spiderlike capillary dilatations in the skin of the cheeks (telangiectases), and reddening in the sun; the *aplastic anemia of Fanconi*—malformations of the limbs (typically a variable reduction or absence of radius to thumb) and of the kidneys, growth retardation, microcephaly, skin pigmentation, and progressive reduction of blood cell formation in the marrow; and *ataxia-telangiectasia*—a progressive ataxia (loss of balance), capillary dilatations of skin and the conjunctiva of the eye, and immunologic deficiency involving IgA and IgE. The defective enzymes have not been identified, but the obvious relation of gene-induced chromosomal instability to cancer susceptibility is an intriguing one. Environmental chromosome-breaking agents (radiation, benzene) also increase the risk of cancer. Does the essentially random chromosome breakage eventually involve a particular chromosome locus that determines malignant transformation? Do the genes causing chromosome breakage do it by direct interference with a DNA repair enzyme, or do they make the cell susceptible to an environmental agent, such as a virus? In Bloom's syndrome, there is preliminary evidence for an impaired DNA polymerase. On the other hand, the SV_{40} virus, which will cause a malignant transformation in cells in tissue culture, transforms cells from Fanconi's anemia patients much more effectively than normal cells.[7] The heterozygotes are intermediate. The cells of patients with Down syndrome and of certain members of "high-cancer" families are also sensitive to the transforming virus. A lot of intriguing facts are accumulating that may eventually fall into place and elucidate the etiology of carcinogenesis.

To add to the mystery, there are a number of genetically determined diseases involving immunologic defects (Bruton's X-linked agammaglobulinemia, Wiskott-Aldrich X-linked eczema and thrombocytopenia) in which there is an increased susceptibility to malignancies, although no chromosomal breakages have been reported. This supports the idea that immunologic mechanisms play a part in our defenses against neoplasia.

CANCERS ASSOCIATED WITH CHROMOSOMAL ABERRATIONS

In this category we might have included the three recessively inherited diseases associated with chromosome breaks, if we had not already put them in the mutant gene category.

There appears to be a relation between Down syndrome and leukemia, with Down patients having a thirtyfold increase in acute leukemia over the population rate of 1 in 2880. It is also suggested that leukemia may occur more often in the families of Down syndrome patients than in controls. 13-trisomy and Klinefelter syndrome may also predispose to leukemia.

The Philadelphia chromosome (mentioned in Chapter 3) is found in the blood-forming cells of most patients with chronic myelogenous leukemia (a cancer of the myeloid white blood cells, precursors of the polymorphonuclear leukocytes). It is a number 22 chromosome with a deletion of part of the long arm (the missing piece may be translocated to chromosome q) that appears in most patients when the leukemia appears and disappears during remission of the disease. Thus it seems to be a concomitant rather than a precursor of the disease.

CANCERS OF PROBABLE MULTIFACTORIAL ETIOLOGY

This group includes a number of relatively common cancers that show a familial predisposition, but no clear-cut Mendelian pattern.[4]

Breast Cancer

Carcinoma of the breast has been the subject of many family studies,[11] almost all of which have demonstrated a twofold to threefold increase in frequency in the near relatives over that in controls. The increase is greater if the proband has a premenopausal onset rather than postmenopausal and if the proband's disease is bilateral rather than unilateral (about 3% of breast cancers are bilateral). For instance, in one study,[2] the frequency in sisters and daughters of probands with premenopausal onset was 6.7% versus 2.3% in controls; if the proband had a postmenopausal onset, there was no significant increase. If the proband's disease was bilateral, the frequency in the relatives was 13%, and if the proband's disease was both premenopausal and bilateral the risk for relatives was 17%. Familial cases had an earlier age of onset.

In addition, there are a small number of families on record in which the pattern was consistent with an autosomal dominant gene producing one or more of the following conditions: early, multiple breast cancer, leukemia, brain tumor, and sarcoma and/or carcinoma of the lung, pancreas, skin, and, possibly, ovary.

Twin studies have been rather unrewarding, because of the difficulty of ascertaining twins with cancer in an unbiased manner and the long period of observation necessary to determine whether the co-twin is concordant or discordant. For instance, a large Danish study of 4368 like-sexed twins ascertained 70 pairs in which at least one had breast cancer; 4 of 23 monozygotic pairs and 6 of 47 dizygotic pairs were concordant. Pooled data from several studies provide concordance rates of 0.28 for MZ and 0.12 for DZ pairs.

Lung Cancer

Bronchogenic carcinoma was a relatively rare disease at the beginning of this century, but it is now the greatest cause of death from cancer in the United States. Smoking is the major culprit, but predisposing genetic factors may also exist. If you have a near relative with lung cancer and you smoke cigarettes, your risk of lung cancer is increased 14 times over that of the general population, to say nothing of the fact that smoking also increases your risk of heart disease.

An exciting new development is the discovery of a genetic polymorphism associated with susceptibility to lung cancer. Aryl hydrocarbon hydroxylase is an inducible microsomal enzyme involved in the metabolism of polycyclic hydrocarbons, among other things. The enzyme converts certain polycyclic hydrocarbons into the carcinogenic epoxide form. The extent of induction shows genetic variation, 45% of the population showing low, 46% intermediate, and 9% high inducibility. Patients with bronchogenic carcinoma showed almost none of the "low" and 30% of the "high" phenotype. This suggests that individuals with easily inducible enzymes will more readily convert non-carcinogenic hydrocarbons from cigarette smoke and other sources into the carcinogenic forms. Thus this polymorphism is an important genetic determinant of susceptibility to lung cancer.[3a]

Leukemia

Videbaek, in 1947,[12] published a study of 209 families ascertained through a proband having leukemia and concluded that there was a hereditary basis in this disease. His study was criticized because his control group contained fewer cases of leukemia than did the general population. The major problem with his study from the point of view of contemporary genetic analysis would be with his efforts to interpret the data in Mendelian terms. Subsequent studies by a number of investigative teams have led to the general conclusion that there are significant familial aggregates in leukemia, although at least one fairly recent study completely failed to establish a familial tendency role for acute childhood leukemia.[8]

As mentioned earlier, the risk of leukemia is increased in at least one chromosomal anomaly, Down syndrome. Suggestions that an increased frequency of leukemia may exist in other chromosomal syndromes (Klinefelter and 13 trisomy) have been very tentative. The risk of leukemia in Bloom syndrome may be as high as 1 in 8, but the majority of leukemia cases are non-chromosomal and non-Mendelian. The risk of recurrence of leukemia in an identical twin is 1 in 5 and in a sib is 1 in 720.

RECENT DEVELOPMENTS

A number of recent developments have greatly increased the tempo of cancer research. Some of these developments are clearly related to one another, and in other cases the relationship is not clear, but perhaps it will not be long before they all fall neatly into place. We can only refer briefly to them here.

It had been known for some years that certain RNA viruses will cause tumors in experimental animals. The Rous chicken sarcoma virus is the prototype. Now there are about 100 RNA viruses and 50 DNA viruses that are known to cause a variety of tumors in a variety of species. When Rous sarcoma virus is added to cultures of rat fibroblasts, some of the cells become *transformed*—that is, their colony morphology changes and they become malignant. A series of elegant experiments[9] showed that the virus contains an enzyme, RNA-directed DNA polymerase, or reverse transcriptase, that (contrary to the Watson-Crick dogma) will synthesize DNA from an RNA template. The DNA remains in the transformed cell as a "provirus," perhaps becoming integrated into the host cell's DNA, and, under certain conditions, it can be made to synthesize new viral RNA, leading to the formation of new virus particles. This would account for the phenomenon of viral latency, in which a virus disappears after infecting an organism, reappearing months or years later. This work removed the dichotomy between the viral and genetic theories of cancer. It also provided a powerful new biological tool, reverse transcriptase.

Another landmark was the discovery, in 1964, by Gold and Freedman, of McGill University, of a "carcinoembryonic antigen," a glycoprotein present in the digestive tract of human fetuses between 2 and 6 months of age, in adenocarcinomas of human colon, and in the serum of most patients with colonic tumors! Besides its diagnostic importance, it drew attention to many similarities between neoplastic and embryonic tissues and suggested that malignant transformation may involve the desuppression of genes normally active only in the embryo.[1]

Some years later, the viral oncogene hypothesis was proposed by Huebner and Todaro. Briefly, there is evidence to suggest that the genomes of RNA tumor viruses are present in the cells of most vertebrate species, are vertically transmitted from parent to offspring, and, depending on the host genotype and various environmental modifiers (radiation, carcinogens, other viruses) cause malignant transformation of the host cell or production of virus.[10]

Finally, there is the exciting work of Spiegelman's group, relating RNA tumor viruses to human neoplasms.[3] For instance, it was possible to isolate from a human breast cancer particles that had the characteristics of mouse mammary tumor virus. These particles contained 70S RNA and an RNA-directed DNA polymerase. This RNA template and enzyme were used to synthesize DNA, heavily labeled with tritium, with a base sequence complementary to that of the viral RNA. This DNA hybridized with mouse mammary tumor virus RNA, showing that the mouse virus had RNA sequences homologous to the RNA of the human breast cancer particles! Similar homologies between human and mouse leukemias, sarcomas, and lymphomas have been shown. Furthermore, by similar methods, it was shown that the DNA of leukemic cells has sequences not found in normal cells, suggesting that the viral genome can be incorporated into the host genome. The DNA of the normal co-twin of a leukemic patient did not show these sequences, suggesting that the virus was incorporated into the leukemic twin's cells after splitting of the zygote—contrary to the oncogene hypothesis.

There are thus many consistencies between the evidence for the somatic mutation theory of neoplasia, carcinoembryonic antigens, the oncogene hypothesis, and the homologies between animal tumor viruses and human neoplasms. But there are still puzzling inconsistencies. The next few years should be interesting ones for oncologists!

SUMMARY

Cancer appears to result from a failure of the precisely regulated growth of cells.

Uncontrolled growth results in neoplasia.

The genetics of cancer may be approached as are the genetics of other common diseases: Some cancers may be caused by mutant genes, some by chromosomal aberrations, and some by major environmental agents, but the majority have a multifactorial basis involving gene-environment interactions. In cancer, while there are examples from each of these groups, the boundaries are less clearly defined than in some other disease categories.

The somatic mutation theory, the viral oncogene hypothesis, carcinoembryonic antigens, and homologies between animal tumor viruses and human neoplasm are areas of investigation that hold considerable promise.

REFERENCES

1. Alexander, P.: Foetal "antigens" in cancer. Nature 235:137, 1972.
2. Anderson, D. E.: A genetic study of human breast cancer. J. Nat. Cancer Inst. 48:1029, 1972.
3. Baxt, W., Yates, J. W., Wallace, H. J., Holland, J. F., and Spiegelman, S.: Leukemia—specific DNA sequences in leukocytes of the leukemic member of identical twins. Proc. Natl. Acad. Sci. U.S. 70:2629, 1973.

3*a*. Editorial. Lancet i:910, 1974.

4. Knudson, A. G., Strong, L. C., and Anderson, D. E.: Heredity and cancer in man, p. 113. *In* A. G. Steinberg, and A. G. Bearn (ed.), Progress in Medical Genetics. Vol. 9. New York, Grune & Stratton, 1973.
5. Miller, R. W.: Relation between cancer and congenital defects in man. New Eng. J. Med. 275:87, 1966.
6. Miller, R. W.: Persons with exceptionally high risk of leukemia. Cancer Res. 27:2420, 1967.
7. Miller, R. W., and Todaro, G. I.: Viral transformation of cells from persons at high risk of cancer. Lancet i:81, 1969.
8. Steinberg, A. G.: The genetics of acute leukemia in children. Cancer 13:985, 1960.
9. Temin, H. M.: RNA-directed DNA synthesis. Sci. Am. 226:25, 1972.
10. Todaro, G. J., and Huebner, R. J.: The viral oncogene hypothesis: New evidence. Proc. Natl. Acad. Sci. U.S. 69:1009, 1972.
11. Vakil, D. V., and Morgan, R. W.: Etiology of breast cancer. I. Genetic aspects. Can. Med. Assoc. J. 109: 29, 1973.
12. Videbaek, A.: Heredity in Human Leukemia. Copenhagen, Nyt Nordisk Forlag, 1947.

Chapter 20

GENETICS OF BEHAVIOR

Man's behavior is probably his most important phenotypic feature, but little is known of its genetic basis.[2,6] The literature on human behavioral genetics deals mostly with the psychoses and the behavioral changes resulting from mutational or chromosomal causes of mental retardation. Yet the psychological effects of phenylketonuria or trisomy 21 will teach us no more about the genetics of normal behavior than tone-deafness will teach us about the genetics of musical ability or throwing a monkey wrench into a moving engine will teach us about the engine's function. The same applies to major mutant genes altering structure in experimental animals such as the waltzing mouse or vestigial-wing Drosophila.

One would expect that genes affecting "normal" behavior would be subtle in their effects and that the primary biochemical effect of the gene at the polypeptide level might be far removed from the behavioral effect. Thus there is a gene controlling the ability to taste phenylthiocarbamide (PTC), the allele for inability to taste being recessive. Presumably, the gene affects some enzyme, but it would be hard to deduce the existence of such an enzymatic difference from a genetic study of preference for cabbage, which contains a PTC-like chemical.

Nevertheless, some functional defects resulting in behavioral differences do have a fairly simple genetic basis, such as the specific dyslexias[5] (reading disabilities), and some have contributed significantly to the elucidation of normal function—for instance, the defects of color vision.

Furthermore, a number of *pathologic genes* have more or less specific effects on behavior. The phenylketonuric child (untreated) is hyperactive and irritable and has an uncontrollable temper, abnormal postural attitudes, and agitated behavior. About 10% show psychotic behavior. Multiple-discriminant analysis of a number of test scores permits discrimination of PKU children from those with other types of mental retardation, so the biochemical defect must have certain specific effects on behavior, from which we should be able to learn something. At least the monkey wrench is always thrown into the same part of the engine, so the resulting damage may tell us something of the mechanism. Similarly, characteristic behavioral changes often precede the choreic movements in Huntington's chorea. Congenital cretinism, some types of which are recessively inherited, produces its familiar effects on personality. Perhaps the most striking example of a gene-induced behavioral defect is the bizarre tendency to self-mutilation in the Lesch-Nyhan syndrome. We are still far removed from a complete understanding of the relation between the gene-determined biochemical change and the behavioral response, but the rapid advances being made in neurobiochemistry may make this approach very rewarding.

Finally, information from animal experiments tells us that mutant genes known by their prominent effects on the physical phenotype, such as albinism, may have much more indirect and subtle effects on behavior. Thus a particular behavioral parameter, such as aggression, may be influenced by the indirect effects of a large number of genes with major effects on other parameters. In this sense, the genetic basis of the behavioral parameter is polygenic.

Chromosomal aberrations also have effects on behavior.[1] A child with Down syndrome tends to be happier and more responsive to his environment than other children of comparable IQ, and he is often musical. Girls with Turner syndrome rate high on verbal IQ tests but low on motor performance and seem to have a deficit in perceptual organization. Males with an XYY chromosome complement seem to be predisposed to behavioral disturbances involving crimes against property, such as arson, rather than personal aggression, such as murder, though the picture is far from clear. Again the series of causal links between the excess or deficiency of chromosomal material and the behavioral phenotype is entirely obscure.

There is such a wealth of literature on the genetics of *intelligence* that an adequate review would be longer than this book.[2] Suffice it to say that the evidence is strongly in favor of a substantial genetic component. The degree of phenotypic correlation is at least as high as it is for physical traits such as height and weight. The picture is complicated by a high degree of assortative mating and the interaction of genotype and environment (see Chapter 16).

Much of the early work on the genetics of intelligence has considered it as an entity, measured more or less accurately by a variety of performance tests, more or less "culture free." More recently, the trend has been toward identification and description of its various components, and this provides an opportunity for defining more specifically the genetic basis of these components. Vandenberg[8] has combined the results of several studies and concluded that heredity was most important for verbal ability and less so, in descending order of importance, for word fluency, perceptual speed, spatial perception memory, number ability, and reasoning (as defined specifically for the purposes of the test). He suggests that the explanation for the rather low degree of heritability for reasoning may be that, for several tests, the right answer may be reached by several routes, which require different amounts of time. Identical twins often select different, equally correct routes, which tends to lower the concordance rate. We point this out mainly to illustrate the difficulty in interpreting this kind of data.

Specific *dyslexia,* or congenital word-blindness, is one of the few common behavioral disorders that shows a simple mode of inheritance, as first shown in the thorough study by Hallgren.[5] It is undoubtedly heterogeneous, and there may be a clinically indistinguishable nongenetic type. Affected children show the normal distribution of intelligence but have specific reading and writing disabilities, such as confusion of letters (p/q or b/d, for instance), and may have "nervous" symptoms or turn out to be "problem children." There is an increased frequency of speech defects, such as stuttering, lisping, or retarded speech development, but probably not of left-handedness. The familial cases (which are the great majority) fit the expectation for autosomal dominant inheritance.

The *psychoneuroses* are so common that they might almost be considered normal; their genetics is correspondingly complex. There is considerable confusion even as to their definition. The few genetic studies available agree that the psychoneuroses are familial, with some degree of specificity for subtypes—that is, if the proband has an anxiety state, most of the affected relatives have an anxiety state, and there is a similar correspondence for hysteria and obsessional neurosis. This finding is corroborated by twin studies, but the importance of environmental factors is also demonstrated. Rosenthal[7] summarizes the situation in the following terms:

> By and large we are limited in our conclusions about the heredity issue in neurosis because of the sparseness of studies, their relative lack of variety, their failure to take various diagnostic precautions, and the difficulty involved in assessing the role of environmental factors. However the overall evidence points to the likelihood that heredity plays a role in the development of psychoneurotic symptoms, but we can say very little about the genetics involved, except that various polygenic systems may be involved in a more or less low-keyed way. . . . Further studies with increased methodological sophistication may help us to understand more clearly the diathesis-stress interactions and their relation to subtype syndromes.

Diagnostic difficulties also confuse the literature on the psychoses, but there is evidence of a major hereditary component in both schizophrenia and manic-depressive psychosis.[3] (See Chapter 12.) We have not attempted to give recurrence risk figures, since the data vary so widely from study to study that any attempt to average them would be bound to be misleading.

The situation for *homosexuality* is even less clear. Twin studies suggest that heredity plays an important role with respect to whether males become homosexuals, but family studies show a wide range of psychopathology in the families of homosexual probands, which raises the question of whether the homosexual behavior is secondary to such psychopathology or vice versa.[7]

For *alcoholism,* the situation is also complicated. Several family studies show a familial tendency; twin studies show that monozygotic pairs have a higher concordance than dizygotic pairs, but dizygotic pairs also have a fairly high rate. One study, using children of alcoholics who had been adopted by nonalcoholics, found a low rate of alcoholism in this group and concluded that the contribution of environmental factors is of overriding importance with respect to whether addiction occurs. It may be that some aspects of drinking behavior are heritable and others not.[7]

Family studies of *psychopathy and criminality* show that most children with antisocial behavior come from broken homes; thus it is impossible to tell from such studies how much of the antisocial behavior in these children is biologically transmitted and how much culturally acquired. Twin studies seem to have been devoted more to criminality than to the broader category of psychopathy. Monozygotic pairs show a higher concordance rate than dizygotic pairs, but it is clear that criminality must be a highly heterogeneous category. There are undoubtedly predisposing factors, such as EEG abnormalities, low IQ, chromosomal anomalies, and the so-called "constitutional psychopathic state"—criminals tend to be predominantly mesomorphic. Each of these factors is under some degree of genetic control. This would account for at least part of the estimated heritability. However, the largest group contributing to criminality are those classified as having a psychopathic or sociopathic personality, and the genetics of this condition is still almost completely obscure. In any case, the role of the environment is clearly of major importance in criminality.[7]

Finally, there is the question of *personality* and whether it has any genetic basis. The question is of some eugenic interest in this time of population crisis. For instance, if personality traits such as aggression and altruism were genetically determined, one would expect the former to be selected for and the latter selected against, since the altruistic would be more likely to limit their family size than the aggressive. Conversely, individuals with more "altruism genes" might be more efficient at living in groups, which would have had a positive selective value in the days of primitive man.

According to Eysenck,[4] personality may be classified along two relatively independent dimensions: various grades of neuroticism or instability, on the one hand, and extroversion–introversion on the other. Unstable extroverts are more likely to become delinquent; unstable introverts are more likely to become neurotic. Several twin studies have shown quite high heritability estimates for these dimensions, both by questionnaire and laboratory measurements. Family studies also show significant correlations between near relatives, and there seems little doubt that heredity is important in determining individual differences in personality. Just how important, and by what mechanisms, remains to be seen.

SUMMARY

The disorders caused by certain mutant genes and chromosomal aberrations are accompanied by characteristic behavioral features, but such findings have revealed little about the role of genes in determining normal behavior or the common kinds of psychopathology. Data from family, twin, and adopted or foster child studies suggest a substantial degree of heritability for personality characteristics and for various categories of abnormal behavior, but conclusions must be guarded, in view of the limitations of both the data and the methodology.

REFERENCES

1. Court Brown, W. M., Jacobs, P., and Price, W. H.: Sex chromosome aneuploidy and criminal behaviour, p. 180. *In* J. M. Thoday and A. S. Parkes (ed.), Genetic and Environmental Influences on Behaviour. Edinburgh, Oliver and Boyd, 1968.
2. Ehrman, L., Omenn, G. S., and Caspari, E.: Genetics, Environment, and Behavior. New York, Academic Press, 1972.
3. Erlenmeyer-Kimling, L. (ed.): Genetics and mental disorders. Int. J. Mental Health 1:5, 1972.
4. Eysenck, H. J.: Genetics and personality, p. 163. *In* J. M. Thoday and A. S. Parkes (ed.), Genetics and Environmental Influences on Behaviour. Edinburgh, Oliver and Boyd, 1968.
5. Hallgren, B.: Specific dyslexia (congenital word-blindness). A clinical and genetic study. Acta Psychiatr. Neurol. Suppl. 65:1, 1950.
6. McLearn, G. E.: Behavioural genetics. Ann. Rev. Genet. 4:437, 1970.
7. Rosenthal, D.: Genetic Theory and Abnormal Behavior. New York, McGraw-Hill, 1970.
8. Vandenberg, S. G.: Primary mental abilities or general intelligence? Evidence from twin studies, p. 146. *In* J. M. Thoday and A. S. Parkes (ed.), Genetic and Environmental Influences on Behaviour. Edinburgh, Oliver and Boyd, 1968.

Chapter 21

SOMATIC CELL GENETICS

This is another chapter that will be out of date before this book appears. Techniques for the study of mammalian cells in tissue culture have been available for many years, but recent developments have opened up exciting possibilities for their application in human genetics.

Human fibroblasts provided the material from which the correct human chromosome number was established, as well as the first demonstration of human chromosomal disease. Later it was discovered that lymphocytes could be cultured from peripheral blood, following stimulation by phytohemagglutinin, and this is now the technique most widely used for diagnostic cytogenetics, supplemented where necessary by preparations from fibroblasts or bone marrow (Chapter 2).

The great advantages of somatic cell lines grown in culture are that one can perform therapeutic trials and other procedures on cell cultures that are inappropriate on patients; that the cell environment can be much more accurately defined and controlled than that of the cell *in vivo;* and that cell lines from patients with rare genotypes can be maintained for study long after the patient becomes unavailable. For these reasons, fibroblasts and blood cell cultures have become useful in the antenatal diagnosis (Chapter 22) and investigation of inborn errors of metabolism.[5]

Most cultures of normal human fibroblasts will become senescent after a certain number of generations—usually about 50—whereas lines developed from malignant tissues may have an indefinite life span. However, the latter lines are aneuploid and therefore not representative of the normal cell. It has recently been discovered that leukocyte cultures may be induced to become permanent, euploid lines, which will be a boon to further studies.

Differentiated cells usually do not grow in culture, or, if they do, they usually lose their differentiated properties, though in some cases (retinal pigment cells, cartilage cells), they can be coaxed to redifferentiate under special culture conditions. Furthermore, cultures from some endocrine gland tumors (adrenal, pituitary, thyroid) will continue to produce hormones, offering the possibility of commercial hormone production from such sources.

The ability to "clone" cells—that is, to grow colonies all descended from a single progenitor—provided critical proof of the Lyon hypothesis (Chapter 4). It has also been used to demonstrate the somatic mutation theory of neoplasia.[1] Leiomyomas of the uterus (a kind of muscle cell neoplasm) were taken from women who were heterozygous for G6PD types A and B and who would therefore be a mosaic of cells with either the A or B allele active. Any particular leiomyoma typed either A or B, showing that all the cells from one tumor arose from one progenitor cell. The same is true for many other neoplasms. On the other hand, the same experiment with trichoepitheliomas and neurofibromas (dominantly inherited skin tumors) showed that the tumors had both G6PD types, suggesting that they arose from multiple foci.

Cultured human cells have also been extensively utilized in the study of those inborn errors of metabolism in which the error is manifest in culture. Intensive biochemical studies on such diseases as the Lesch-Nyhan syndrome and orotic aciduria are beyond the scope of this book, but they are throwing new light on the important question of gene regulation.[5]

Another use of somatic cell genetics has been in the detection of heterogeneity through complementation. In Chapter 6 we mentioned the demonstration that cells from patients with the Hurler's type of mucopolysaccharidosis would prevent storage of mucopolysaccharides in cells from patients with the Hunter type and vice versa, thus showing that each provided something the other lacked and that they were therefore nonallelic. On the other hand, cells from the Hurler-type patient did not complement cells from patients with the Scheie type, suggesting that they are allelic.

The most dramatic recent development in somatic cell genetics has been the discovery that fibroblasts may fuse with one another to form a heterokaryon (with two individual nuclei), which may be followed by nuclear fusion.[3] The cell fusion may occur spontaneously, at a low frequency, and the rate is markedly increased by the use of inactivated Sendai virus. Fusion may occur between cells of quite different species (for instance, man and mosquito!). Following nuclear fusion, the chromosomes of one parental type tend to get lost, a fact that has been ingeniously made use of for chromosome mappings.

For example, a line of mouse cells was developed by selection that lacked the enzyme thymidine kinase (TK). These were mixed with cells from a human line lacking the enzyme hypoxanthine-guanine-phosphoribosyl transferase (HGPRT). Both of these lines will grow on normal medium since the enzymes are not involved in the major biosynthetic pathways for thymidylic acid and purines, respectively. However,

TABLE 21-1. Autosomal Linkages in Man[a]

A. Linkage groups assigned to specific chromosomes[b]	
Chromosome 1	6-phosphogluconate dehydrogenase (F)
	phosphopyruvate hydratase (S)
	Rh blood group (F)
	elliptocytosis-1 (F)
	phosphoglucomutase$_1$ (F)
	amylase, pancreatic (F)
	adenylate kinase$_2$ (S)
	cataract, zonular pulverent (F)
	Duffy blood group (F)
	auriculo-osteodysplasia (F)*
	centromere
	"uncoiler" region (C)
	fumarate hydratase (S)*
	guanylate kinase (S)*
	peptidase C (S)
	uridyl diphosphate glucose pyrophosphorylase (S)*

TABLE 21-1. Autosomal Linkages in Man *(Continued)*

Chromosome	Locus
Chromosome 2	acid phosphatase$_1$ (FC)*
	Dombrock blood group (F)
	interferon$_1$ (S)
	MNS blood group (FC)*
	sclerotylosis (F)
Chromosome 5	hexosaminidase B (S)*
	interferon$_2$ (S)*
Chromosome 6	HL-A histocompatibility region (S)
	phosphoglucomutase$_3$ (S)*
	malic enzyme (S)
	indophenoloxidase-B (S)
Chromosome 7	mannosephosphate isomerase (S)*
	pyruvate kinase$_3$ (S)*
	hexosaminidase A (S)*
Chromosome 10	glutamate oxalacetic transaminase (S)*
	hexose kinase (S)*
Chromosome 11	lactate dehydrogenase A (S)
	esterase-A$_4$ (S)
	acid phosphatase, lysosomal (S)*
	Killer antigen (S)*
Chromosome 12	lactate dehydrogenase B (S)
	peptidase B (S)
	citrate synthase, mitochondrial (S)
	serine hydroxymethylase (S)*
	triosephosphate isomerase (S)*
Chromosome 14	nucleoside phosphorylase (S)
Chromosome 16	lethicin cholesterol acetyltransferase
	haptoglobin alpha (FC)
	adenine phosphoribosyltransferase (S)
Chromosome 17	thymidine kinase (S)
	galactokinase (S)*
Chromosome 18	peptidase A (S, FC)
	poliomyelitis virus sensitivity (S)
Chromosome 19	phosphohexose isomerase (S)
Chromosome 21	anti-viral protein (S)
	indophenoloxidase-A (S)
	lipoprotein-Ag (FC)

B. Unassigned linkage groups (numbers represent estimated centimorgans)

Lutheran blood group (13) secretor (4) myotonic dystrophy
ABO blood group (13) nail-patella locus (0) adenylate kinase
beta (0) delta (0) gamma hemoglobin loci
AM$_2$ immunoglobulin (0) Gm immunoglobulin (28) alpha 1-antitrypsin
transferrin (15) pseudocholinesterase
albumin (3) group-specific component
PTC taster locus, Kell blood group
glutamate pyruvate transaminase, epidermolysis bullosa
Lewis blood group, complement component$_3$

[a] Data from McKusick.[2]

[b] F indicates linkages established by family studies, S those by somatic cell hybridization, and C those by cytogenetic means. An asterisk denotes "provisional" assignments, that is, those reported by a single laboratory or from a single family.

the mutants do not utilize exogenous thymidine and hypoxanthine, respectively.

Neither mutant will grow in the presence of aminopterin, which blocks endogenous purine synthesis. Cells derived from fusion of these two lines can grow in the presence of aminopterin if thymidine and hypoxanthine are provided in the medium. To obtain hybrid cells, the mixture of cells from the two lines is therefore grown in the "HAT" medium, containing hypoxanthine (H), aminopterin (A), and thymidine (T). Only hybrid cells between the two lines would be able to grow on the medium, since only they would have both TK and HGPRT, and thus we have a rapid and efficient way of selecting hybrid cell lines. Sublines are then maintained, in which progressive loss of human chromosomes occurs. Only lines that retain the human chromosome with the gene for TK will grow in the HAT medium, since the mouse chromosome carries the inactive mutant. It was soon apparent that only lines retaining chromosome 17 grew on the HAT medium, and it was thus demonstrated that the gene for thymidine kinase is on chromosome 17.[3]

A recent paper by Ruddle and associates[4] shows how rapidly mapping is progressing. They report evidence from cell hybridization experiments that the locus for peptidase C is on chromosome 1. Similar studies have shown that the gene for phosphoglucomutase$_1$ (PGM_1) and for 6-phosphoglucomutase (6PDG) are also on chromosome 1. Classical linkage studies have shown that the PGM_1 locus is linked to the Duffy (Fy) blood group locus, which in turn is linked to the salivary and pancreatic amylase loci Amy-1 and Amy-2 and to the locus for nuclear cataract Cae, and further that the 6PDG locus is linked to the Rh locus, which in turn is linked to the elliptocytosis locus. Thus chromosome 1 is the best-mapped autosome, and at least 16 genes have now been assigned to it. Progress in human chromosome mapping should be rapid in the next few years. Table 21-1 illustrates the rapid progress being made in the field of chromosome mapping by the cell hybridization technique.

Cell hybridization is also throwing new light on the question of differentiation and gene regulation.[5] For instance, when cells from a heteroploid mouse strain, which are rapidly synthesizing DNA and RNA, form heterokaryons with nucleated erythrocytes of the chicken, which are synthesizing virtually no nucleic acids, the hen nuclei immediately begin synthesizing nucleic acids, and after about a week they develop nucleoli, at which time hen-specific proteins begin to be synthesized. This suggests that the cytoplasm regulates nuclear activity and that the nucleolus plays an important role in transferring genetic information from nucleus to cytoplasm.

A final example of the use of cultured cells in somatic cell genetics relates to the question of genetic engineering, or directed genetic change. This has been demonstrated in the case of galactosemia, where a virus was used to transfer DNA from a normal cell line to a strain from a galactosemic patient; cells from the treated galactosemic line were then shown to be producing the missing enzyme (transduction). This approach opens exciting therapeutic possibilities and also presents some difficult questions for those concerned about the possible dangers of tampering with the human germ plasm.

SUMMARY

Somatic cell genetics is a rapidly developing area in human genetics that is providing new opportunities for antenatal diagnosis, investigation of basic problems in cytogenetics, gene regulation and inborn errors of metabolism, and procedures and therapeutic trials that are inappropriate to perform on patients.

REFERENCES

1. Fialkow, P. J.: The origin and development of human tumors studied with cell markers. N. Eng. J. Med. 291:26, 1974.

2. McKusick, V.: Personal communication. Professor of Medicine, Johns Hopkins University.
3. Migeon, B. R., and Childs, B.: Hybridization of mammalian somatic cells, p. 1. *In* A. G. Steinberg, and A. G. Bearn (ed.), Progress in Medical Genetics. Vol. 7. New York, Grune & Stratton, 1970.
4. Ruddle, F., Ricciuti, F., McMorris, F. A., Tischfield, J., Creagan, R., Darlington, G., and Chen, T.: Somatic cell genetic assignment of peptidase C and Rh linkage group to chromosome A-1 in man. Science 176:1429, 1972.
5. Ruddle, F., and R. S. Kucherlapati.: Hybrid cells and human genes. Sci. Amer. 231:36, 1974.

Chapter 22

ANTENATAL DIAGNOSIS OF GENETIC DISEASE

One of the most exciting chapters in the history of genetics as applied to medicine is the development of techniques for the diagnosis of congenital disease before birth.[4] Until recently, it was possible to provide genetic counseling only in terms of probabilities of recurrence—based either on the Mendelian laws or on empirical estimates. Now, for some few diseases, it is possible to provide a definite answer—yes or no; the child is unaffected or it is affected. If the child is affected, the pregnancy can be terminated, and the parents can begin again. Thus some couples who could otherwise enjoy the privilege of children only by facing an increased risk of having an abnormal child can now raise a family without fear of the disease in question.

Prenatal diagnostic techniques are being developed along several lines: examination of cells from the amniotic fluid for chromosomal or biochemical abnormality; examination of the amniotic fluid itself for tell-tale biochemical aberrations; and visualization of the fetus either directly, with a fetoscope, or indirectly, by special radiographic techniques.[2,5]

AMNIOCENTESIS

Amniocentesis is the removal of fluid from the amniotic sac. It has been used diagnostically since the mid-1930's for detecting fetal distress and, more recently, for following the progress of Rh hemolytic disease. In Israel, in the mid-1950's, prenatal sex determination was done by examining cells from the amniotic fluid for sex chromatin, and some male fetuses were aborted in cases where the mother was heterozygous for a sex-linked recessive deleterious gene. In the past 10 years, techniques have been developed for culturing and karyotyping cells from the amniotic fluid and for detecting a number of inborn errors of metabolism.

The fluid is obtained by sticking a needle through the anterior abdominal wall into the amnion. In skilled hands, and with the help of ultrasonography (a kind of sonic radar), this can be done with virtually no danger to the fetus or mother. The optimal time is around 14–15 weeks after the beginning of the last menstrual period; before this, the amount of fluid is quite small (there is about 125 ml at 15 weeks) and the cells do not grow well. The longer one

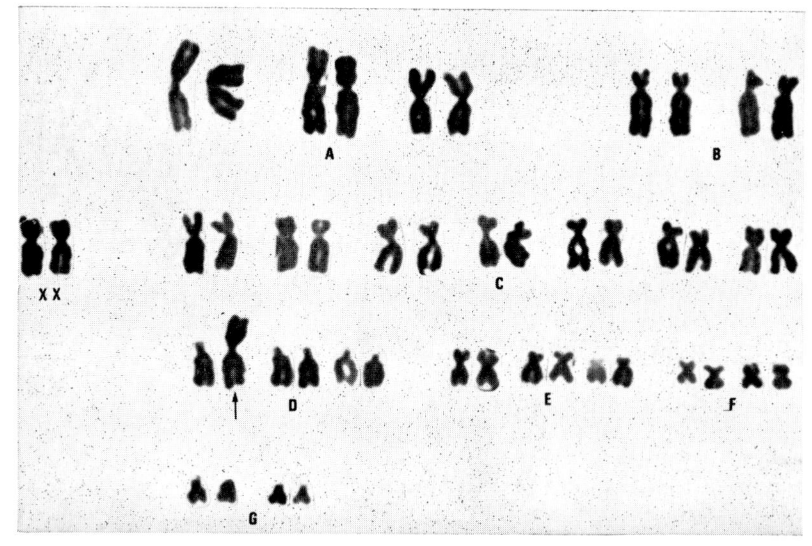

Figure 22-1. Karyotype of proband in clinical example with unbalanced D/D translocation. Note that there are 46 chromosomes and 5 chromosomes in the D group, but that one chromosome has essentially all the genetic information possessed by the long arms of two D chromosomes (arrow). The proband has a D chromosome in triplicate and the stigmata of trisomy 13.

waits after this, the less time there is to grow the cells, make the appropriate tests, and obtain a diagnosis while there is still time to perform an abortion (before 21 weeks). Since some biochemical tests take 4–6 weeks, the schedule may be quite tight.

Clinical Example

A 23-year-old mother delivered a female infant who had the classic stigmata of 13 trisomy. A karyotype of the infant, who survived only one month, revealed an unbalanced D/D translocation, 46,XX,—D,+t (DqDq) (Fig. 22-1). The parents and their one living child were studied. The mother was found to be a balanced D/D translocation carrier, 45,XX,—D,+t(DqDq) (Fig. 22-2), and the father and sister were normal. In the course of investigating this family (Fig. 12-3), a number of balanced translocation carriers were identified through four generations.

Approximately half the living members investigated so far are balanced D/D translocation carriers. In generation III, there are an equal number of living members and spontaneous abortions. The proband is the only confirmed case of unbalanced D/D translocation, but one other infant is known to have died at 2 days of age with multiple anomalies consistent with 13 trisomy.

The recurrence risk for unbalanced D/D translocation within this family appears to be of the order of 10%, and the risk of balanced D/D translocation approximates 50%. As was mentioned in Chapter 3, D/D translocation is one of the most common autosomal rearrangements, and the vast majority are balanced. The risk of 1% of unbalanced translocation to the offspring of a balanced D/D translocation carrier is mentioned in the literature. This family and others we have studied lead us to believe that the risk is higher and that amniocentesis is certainly justified.

The parents were counseled about the prospects for future children and about the possibility of monitoring any future pregnancies by amniocentesis. While the karyotyping of the various family members was still in progress, the mother informed her

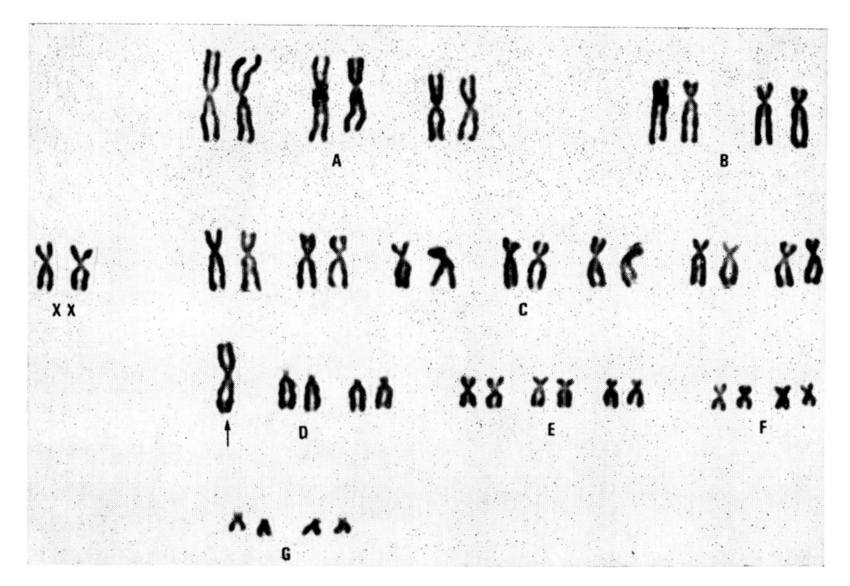

Figure 22-2. Person with balanced D/D translocation from family in clinical example. Note that, although there are only 45 chromosomes, one chromosome has the equivalent genetic material of two D chromosomes and the person is normal.

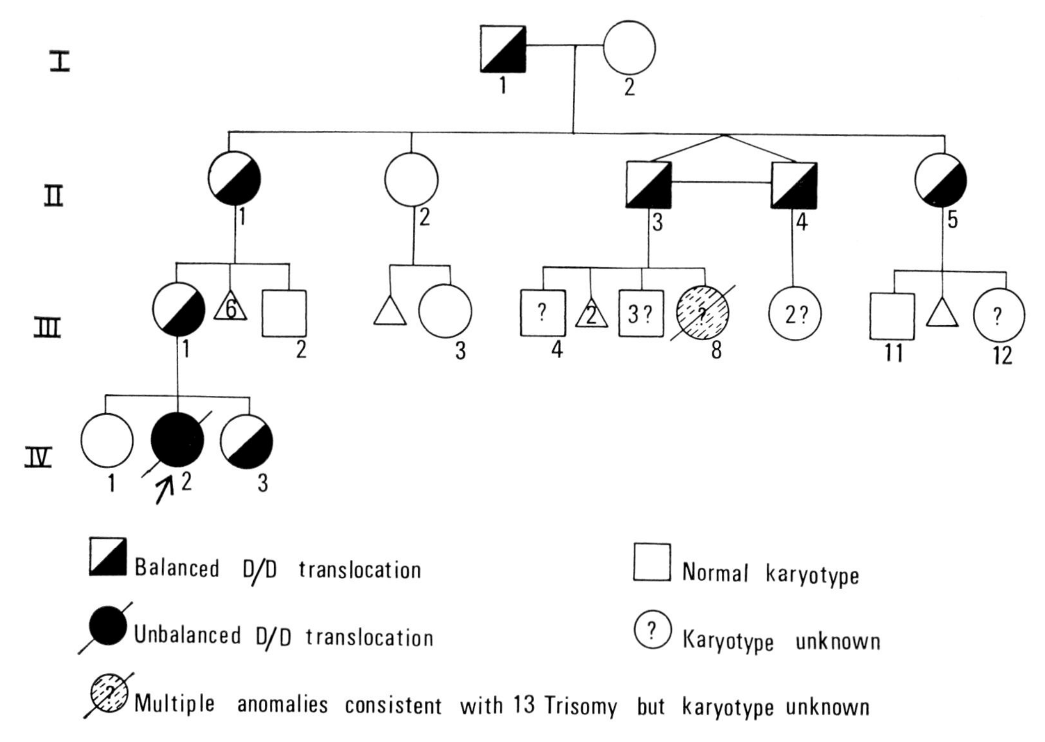

Figure 22-3. Family with D/D translocation transmitted through four generations.

physician at the genetics clinic that she was 8 weeks pregnant. An attempt at amniocentesis was made at 12 weeks, but the cells did not grow. Another attempt at 14 weeks was also unsuccessful. The amniocentesis performed at 16 weeks showed that the fetus was a female carrier of a balanced D/D translocation. The mother elected to continue the pregnancy. A healthy female infant was delivered, and confirmation of the balanced D/D translocation was made from peripheral blood cells.

Prenatal diagnosis by amniocentesis is so new a weapon in the medical armamentarium that there is still much to learn about the long-term effects of the procedure on the baby, the mother, the family, and society.

CHROMOSOMAL ABERRATIONS

Except for Rh disease, chromosomal problems are the commonest reason for performing diagnostic amniocentesis. Though there are some pitfalls, prenatal screening for cytogenetic abnormalities is in most cases highly reliable. The pitfalls include undetected mosaicism, twins, normal variants that may be misinterpreted as abnormal, contamination by other cells, and polyploidy.[4]

There is still some uncertainty as to how high the fetal risk should be before one undertakes an amniocentesis. The answer depends partly, of course, on the available resources and needs. In the United States, it is generally agreed that women in the following categories deserve prenatal diagnosis:

1. Carriers, or wives of carriers, of balanced translocations, as in our clinical example.
2. Women who have had a previous aneuploid child, where the risk is estimated as around 2%. These two categories constitute a relatively small proportion of the high-risk population.
3. Women in the older age groups. The exact cutoff point differs from center to center, but in many units any woman of age 35 or over is considered eligible.

Since, in the United States, about one third of all cases of Down syndrome are born to women of age 35 or over (who represent only 7% of childbearing women), monitoring of all such women, and termination of affected pregnancies would lead to a substantial reduction in the number of affected children. It has been seriously proposed that aneuploidies are so common that it would be economically justified to screen all pregnancies for chromosomal anomalies.[8] However, this view has not yet received much support, particularly from obstetricians, and certainly no such widespread program should be undertaken before the long-term effects of amniocentesis on the surviving child have been thoroughly evaluated.

SEX-LINKED DISEASES

Another category of women eligible for prenatal chromosomal examination are carriers of X-linked genes causing serious diseases. Most X-linked diseases cannot be diagnosed in cell culture, but the chromosomal sex can be determined by karyotyping. (This is preferable to sex chromatin determination on uncultured cells, which can be misleading.) Thus a carrier female has the option of abortion if the fetus is a male who has a 50:50 chance of being affected—and also a 50 : 50 chance of being normal. This is a much more difficult problem for the parents than the prospect of aborting a fetus known to be affected, but some couples do find this an acceptable option, to ensure that they will not have affected sons.

INBORN ERRORS OF METABOLISM

Diagnosis of inborn errors of metabolism *in utero* usually depends on demonstration of the biochemical defect in fetal cells obtained from the amniotic fluid. Usually this is done on cultured cells; in some cases, it may be possible to do it by histochemical procedures directly on the amniotic fluid cells.

Thus any genetic disease in which the biochemical defect is demonstrable in cultured fibroblasts is *eligible* for prenatal diagnosis, and there are now as many as 50 such diseases. Detection of the enzyme defect in cultured fibroblasts is only the first criterion, however. It must also be shown that the enzyme is normally present in fetal cells at the time the diagnosis is to be done and that the deficiency in mutant cells is great enough to permit discrimination from the heterozygote as well as the normal individual. Furthermore, the amniotic fluid contains epithelial cells as well as fibroblasts, and these may differ in their biochemical characteristics.

All of the eligible diseases are rare, and so far only about a dozen have actually been diagnosed *in utero.* In most cases, the parents identify themselves as heterozygotes by having an affected child, and only subsequent pregnancies can be monitored, so that only a minority of affected cases can be prevented. An exception is Tay-Sachs disease, a recessively inherited error of brain lipid metabolism leading to progressive degeneration, blindness, and death within the first 2 or 3 years. In this case, the gene is sufficiently frequent in some populations (Ashkenazi Jews) that it is practical to screen couples before reproduction, detect those in which both members are heterozygous, and screen their pregnancies. Several such screening programs are under way. Although much remains to be learned about the psychological and social consequences of discovering that one is a carrier of a harmful (in the homozygote) mutant gene, there is no doubt that the ability to have a normal family, without facing a 1-in-4 risk for each child of developing this dreadful disease, is considered a boon by most heterozygous couples. If all heterozygous couples chose selective abortion of affected children and had the same number of liveborn children as they would have had if they were not heterozygous (reproductive compensation), the mutant gene frequency would increase, since 2 out of 3 children would be heterozygous. However, this dysgenic effect would be so small as to be negligible.[6]

Cystic fibrosis of the pancreas, the most common recessive disease in Caucasians (1/1500 in some populations) cannot be diagnosed with confidence in fibroblast cultures, and heterozygote detection is still not practical. When these difficulties have been overcome, this disease will certainly be a candidate for population screening and selective abortion, and the disease frequency can then be drastically reduced.

Sickle-cell disease and thalassemia can now be diagnosed from very small amounts of fetal blood. It will probably not be too long before it is possible to obtain samples of blood from the fetus without harming it, and it will then be possible to reduce dramatically the frequency of these lethal anemias (which may affect as many as 1 in 400 Negroes or Greeks, respectively)—once the psychological and social problems of population-screening programs have been resolved. Presently existing screening programs for detecting heterozygotes can only offer heterozygous couples the options of taking a 1-in-4 risk, or not having children, at least with each other. Prenatal diagnosis would be a great advantage.

AMNIOTIC FLUID EXAMINATION

Until recently, examination of the amniotic fluid has not been very useful in prenatal diagnosis, except in the case of Rh disease. It now seems likely that anencephaly and spina bifida may be diagnosed by this means. In the second trimester, the fetal liver synthesizes large amounts of a protein, alpha-fetoprotein, which circulates in the blood and, presumably, in the cerebrospinal fluid. When the neural tube is open, as in anencephaly and spina bifida aperta, this protein may be found in excessive amounts in the amniotic fluid. Mothers who have had an affected child may then have subsequent pregnancies monitored; a number of affected fetuses have already been diagnosed and aborted in such cases.[1] It remains to be seen how sensitive and reliable the technique is; if it is successful, it will be a boon to many parents who have had a child with such a defect and face about a 5% risk of recurrence. Recent evidence suggests that the protein may also be elevated in the mother's serum, raising the possibility of routine screening for this defect.[1]

FETOSCOPY

A further advance in prenatal diagnosis will be the development of means for direct examination of the fetus. Already, some of the grosser malformations can be detected by x-ray examination, and ultrasonography can demonstrate a surprising number of malformations, such as anencephaly, hydrocephaly, and even certain kinds of heart malformation. Screening by this means may become a routine procedure before too long.

Progress is already being made in the development of a "fetoscope," an instrument that can be put directly into the amnion with an optical system that allows direct visualization and photography.[7,9] This may be of more value for obtaining fetal blood than for screening for malformations.

ETHICAL PROBLEMS

The advent of prenatal diagnosis has generated considerable concern about the ethical implications, which has led to much soul-searching and lengthy discussion.[2] Will the increase in "therapeutic" abortion lead to further changes in attitude that some will interpret as diminished respect for human life? If one is willing to kill a human being at 4½ months of gestation because it will be diseased, why not kill a diseased child at birth? Or at age 2. . . ? (The "camel's nose" argument—if you are willing to let the camel's nose inside the tent door, you'd better be prepared to have the whole camel in your tent.) How severe must a disease be to justify abortion—for instance, suppose you unexpectedly detected an XYY karyotype while looking for something else?

These questions are too complex to be settled here. We say only two things: First, we do not accept the "camel's nose" argument, which could have been (and probably was) used against almost any scientific advance. One should not deny mankind the benefits of new knowledge because there might be some bad side effects. The important thing is to be aware of the possible harmful effects and to take steps to prevent *them*, not the initial advance. And second, public awareness is the best protection against the possible abuses of new knowledge.

SUMMARY

A new approach to antenatal diagnosis is through amniocentesis utilizing desquamated fetal cells from the amniotic fluid for obtaining cultures on which to perform karyotypes or biochemical tests. Diagnostic tests may also be done directly on the amniotic fluid (as in Rh disease) and actual visualization of the fetus may be performed through fetoscopy.

Chromosomal problems are (after Rh disease) the most common reason for performing amniocentesis. Carriers or wives of carriers of balanced translocations and carriers of serious X-linked disorders, women who have had a previous aneuploid child, and mothers over 35 years of age are candidates for antenatal karyotyping of fetal cells. Diagnosis should also be feasible in any genetic disease in which the biochemical defect is demonstrable in cultured fibroblasts, and currently over 50 such diseases can be so diagnosed.

REFERENCES

1. Brock, D. J. H., Bolton, A. E., and Scrimgeour, J. B.: Prenatal diagnosis of spina bifida and anencephaly through maternal plasma-alpha-fetoprotein measurement. Lancet i:767, 1974.

2. Emery, A. E. H. (Ed.): Antenatal diagnosis of genetic disease. Edinburgh, Churchill and Livingstone, 1973.
3. Hilton, B., Callahan, D., Harris, M., Condliffe, P., and Berkely, B. (ed.): Ethical Issues in Human Genetics. New York, Plenum Press, 1973.
4. Hsu, L. Y., Dubin, E. C., Kerenyi, T., and Hirschorn, K.: Results and pitfalls in prenatal cytogenetic diagnosis. J. Med. Genet. 10:112, 1973.
5. Milunsky, A., and Littlefield, J. W.: The prenatal diagnosis of hereditary disorders. Springfield, Ill., Charles C Thomas, 1973.
6. Motulsky, A. G., Fraser, G. R., and Felsenstein, J.: Public health and long-term genetic implications of intra-uterine diagnosis and selective abortion. Birth Defects VII:22, 1971.
7. Patrick, J. E., Perry, T. B., and Kinch, R. A: Fetoscopy and fetal blood sampling: a percutaneous approach. Am. J. Obstet. Gynecol. 119:593, 1974.
8. Stein, Z., Susser, M., and Guterman, A.: Screening programme for prevention of Down's syndrome. Lancet i:305, 1973.
9. Valenti, C.: Endoamnioscopy and fetal biopsy: A new technique. Am. J. Obstet. Gynecol. 114:561, 1972.

Chapter 23

GENETIC COUNSELING

THE NEED FOR GENETIC COUNSELING

Genetic counseling usually begins with someone wanting to know whether a disease suspected of being genetic will recur in the near relatives of a patient with the disease. The role of the counselor is, first, to estimate *P*, the probability of recurrence. Some authorities believe that the task ends here, but we believe that the counselor, when asked, should be ready to assist the person concerned in deciding what to do and in taking the appropriate action. However, the final decision must be left to the family.

Most counseling involves the occurrence of a particular disease in a child and the question of whether future children might be similarly affected. The parents may be referred because they have expressed their concern to a physician, or sometimes because the physician thinks the disease has an appreciable recurrence risk. Sometimes they approach the counselor directly, about whether they should have another baby. Parents may also want to know about the risk for affected children's children, or for the children of the unaffected brothers and sisters. A person contemplating marriage may be concerned about a specific disease or history of racial admixture in the fiancé's family. Cousins contemplating marriage may be worried about the possible genetic hazards of consanguinity. Occasionally, the question may involve a child being considered for adoption and the presence of a disease or racial admixture in the family. And sometimes the counselor may be concerned with the near relatives of people he had counseled, who are at risk for developing, or having children with, a particular disease.

In certain populations in which a severe genetic disorder is unusually high, screening programs are now being initiated to detect heterozygotes and to encourage them not to marry other heterozygotes for the same gene. This involves "prospective" counseling rather than retrospective (when the affected child already has been born); it will require a quite different approach and raise quite different problems.

THE GENETIC COUNSELOR

In many cases, the family doctor is the most appropriate person to do the counseling because he knows the family, its attitudes, and socioeconomic background better than a consultant. However, he may have neither the genetic knowledge nor the time for several interviews, and in many cases the family does not have a family doctor. Finally, some cases may be so complex or may require sufficiently specialized tests that the services of a professional genetic counselor are desirable.

At present, a genetic counselor requires no formal qualifications. Some counselors have a higher degree, such as a Ph.D. in genetics, some have a medical degree, and some have both. A good counselor needs a sound grasp of genetic principles, a wide knowledge of the scientific literature on diseases of possible genetic origin, and much sympathy, tact, and good sense. He should be associated with a university hospital, so that he can take advantage of its diagnostic resources and the expertise of its staff and so that he is available to his colleagues and to patient's families. Those in need of genetic counseling are more likely to get it if the counselor is on the hospital staff than if they have to find one elsewhere. Often the family may not know that such a service is available unless they are referred to one by an interested physician.

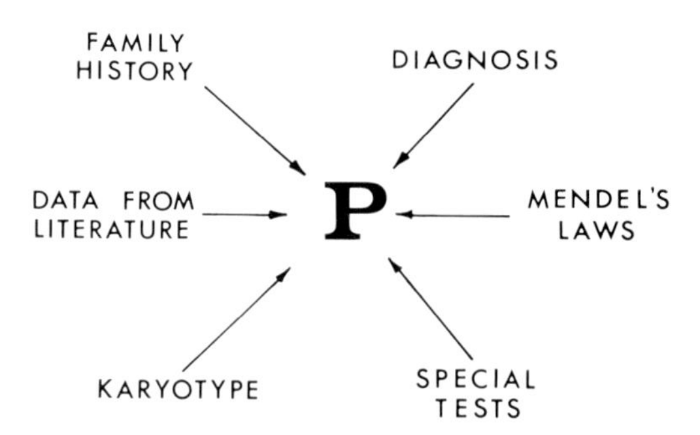

Figure 23-1. The sources from which the counselor estimates *P,* the probability that a given person will develop the disease in question.

THE COUNSELING INTERVIEW

The first interview usually takes the better part of an hour or more; in addition to the collection of information, it allows the counselor to begin to get to know the parents and vice versa. For this reason, it is best that the counselor, rather than an assistant, take the family history, time-consuming though it may be. It is also preferable to interview both parents; together they may present a more accurate family history than separately, and the counselor has a chance to get some impression of how they interact, which may be helpful later in the counseling process. This also guards against misinterpretation of information given by the counselor when it is communicated by the interviewed parent to the other parent. The facts given by the counselor should be written down and placed on record in the patient's hospital chart and in a letter to the family doctor, or in a letter to the parents, or both.

The first requirement, of course, is that the disease causing the concern be accurately diagnosed. This may already have been done, in which case the counselor need only be able to evaluate the reliability of the diagnosis and know when to ask for confirmatory tests or opinions. In other cases, the counselor may himself aid in making the diagnosis, either by performing special tests, such as examination of the chromosomes, or through his familiarity with diseases so rare that the practicing physician may not know of them. For instance, he must take into account the problem of genetic heterogeneity—the fact that diseases that present very similar features may have different causes and therefore different risks of recurrence. Further special tests and interpretation of the family history may permit a distinction to be made. Thus taking the family history may help to establish a diagnosis, and it is usually essential to the estimation of *P,* the probability of recurrence (Fig. 23-1) .

Taking the Family History

This involves construction of a pedigree and listing the patient's near relatives by sex, age, and state of health, particularly with reference to the occurrence of relevant diseases in the family. A special form is used to ensure systematic recording of the data. It is often necessary to correspond with doctors and hospitals or to examine medical records directly to confirm diagnoses of possible relevant disease occurring in family members. In most cases, carrying the family study beyond first cousins and grandparents is not useful, both because the information gets progressively less reliable with increasing distance of relationship and because diseases only in relatives more distantly removed than this are not likely to be relevant to the patient. Depending on the nature of the disease involved, the counselor may want to amplify the family history by doing special examinations or tests on particular family members (e.g., to detect whether certain individuals are carrying a mutant gene or a chromosomal rearrangement).

The counselor may be able to estimate P reliably by the end of the first interview, and if the estimate is reassuringly low and the parents are happy about it, there may be no need for further interviews. However, if the risk is not reassuring or if the parents have any doubts about it or show any signs of uneasiness, a second interview is indicated. This may be necessary anyway if records have to be checked or special tests done. The interval provides an opportunity for the parents to absorb the information given, clarify their thoughts, and define the questions they want to ask.

Establishing the Recurrence Risk

The process by which the counselor establishes the recurrence risk in question has been reviewed in Chapter 5 and elsewhere.[2,3,5] It involves placing the disease in one of four etiologic categories: diseases due to major mutant genes, chromosomal aberrations, major environmental agents, and multifactorial causes. The recurrence risk can then be calculated from either Mendelian laws or selection of the appropriate empiric estimate as outlined in previous chapters.

Interpreting the Recurrence Risk

The next step in the counseling process is to make certain that the parents know what the probability figure means in their situation. Some have trouble understanding the very concept of probability, and some, even though they have an affected child, may not grasp the full implications of having another one.

Once they clearly understand the meaning of the probability they have been given, the parents must convert the probability into a decision—whether to have another baby, seek sterilization, marry, adopt, or whatever their particular problem indicates (Fig. 23-2). The counselor may be able to help the family reach a wise decision, but he should avoid making it for them. For instance, he can point out the various factors to be considered: the severity of the disease in relation to the risk of recurrence, the impact of the disease on the rest of the family, the social and moral pressures they may feel, the possibility of adoption or artificial insemination as an alternative to having their own children, the pros and cons of sterilization, and the possibility of monitoring the next pregnancy by amniocentesis.

The counselor should not try to impose his view of the appropriate decision on the parents. However, it is not enough simply to present the required probability and leave it at that. In the ensuing discus-

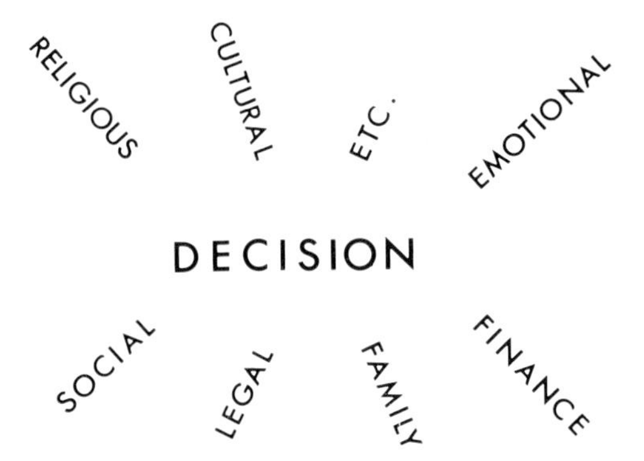

Figure 23-2. Some of the factors entering into the decision that will be made in the light of P.

sion the parents may eventually ask the counselor what he would do in the same situation, in which circumstance we feel that he should say that no one really knows what he would do without actually being in the same situation and then he may be justified in saying what he thinks he *might* do. More than one of our counselees has expressed a wish in retrospect that we *had* been directive, at least in emphasizing what an affected child meant in terms of daily living; "I understand the statistics in my head, but I don't *feel* it."

Taking Action

The decision reached may demand definitive action. A decision not to have further babies of their own will require further decisions by the parents about the use of contraceptives, sterilization, artificial insemination, adoption, and so on. If contraception fails, the question of abortion may arise. Several recent developments have changed the situation radically with respect to this kind of decision.

The Pill

The advent of contraceptive pills has made it easier for those who wish to avoid having further children to do so. Diaphragms, condoms, and (even more so) the rhythm method were sufficiently unreliable to make the prospect of another—potentially diseased—baby a constant menace. The pill has provided increased security, but on the other hand, for a young woman, the prospect of years "on the pill" and the fear of a single instance of forgetfulness resulting in a high-risk pregnancy can be formidable.

Changing Attitudes toward Abortion

There has been a radical change in social and legal attitudes toward abortion in recent years, from a generally proscriptive to a generally permissive one. In some groups the mere fact that a pregnancy is unwanted is sufficient grounds for termination; in others, there must be danger to the life of the mother to justify killing an unborn human being. The counselor must respect the religious and moral attitudes of the parents, but in many cases these may be in conflict with the law or church. That is, the parents in good conscience may wish to have the pregnancy terminated, but may find it difficult to do so because of legal or religious restrictions. This situation has improved greatly in many areas, but there are still many regions where the law does not allow parents who wish to prevent the birth of a child with a high risk of severe disease from taking the necessary steps.

Prenatal Diagnosis

There have been recent improvements in techniques for prenatal diagnosis. Certain gross malformations can be recognized before birth by special methods of X-ray or ultrasound examination, and methods are being developed whereby the embryo can be viewed directly by optical systems inserted into the uterus. Probably the most valuable approach is amniocentesis. Examination of the amniotic fluid directly or of the fetal cells either directly or after culture can establish the diagnosis of some genetic disorders prenatally and convert an estimate of probability into virtual certainty that the baby will be affected or unaffected (Chapter 22). Table 23-1 lists the kinds of problems that can be dealt with and the North American experience.[4]

TABLE 23-1. Types of Problems Amenable to Prenatal Diagnosis—North American Experience[a]

Indication	*Cases Studied*	*"Affected" Fetuses*	
		No.	*%*
Chromosomal			
Translocation carrier	93	17	18
Maternal age, 40 and +	347	9	3
Maternal age 35–39	255	4	2
Previous Down syndrome	485	5	1
Miscellaneous	188	1	
X-linked disorders	115	54	47
Inborn errors of metabolism	180	37	21
	1663	127	

[a] Data from Milunsky and Littlefield.[4] Since this article was written there has been a sharp increase in techniques performed for neural tube defects (p. 227).

X-Linked Conditions. In the case of X-linked conditions such as hemophilia, it is possible to examine the chromosomes of cultured cells and to determine whether they are XX or XY. If the mother carries the hemophilia gene, each son has a 50:50 chance of being a hemophiliac, whereas the daughters will be unaffected (although each has a 50 : 50 chance of being a carrier). Amniocentesis can establish whether the embryo is a female or a male, and the parents have the opportunity to seek termination of the pregnancy if it is male.

Chromosomal Aberrations. Culture of cells from the embryo and examination of their chromosomes allow diagnosis of chromosomal anomalies. Thus, in high-risk pregnancies, as when a parent has a chromosomal rearrangement or in mothers near the end of their reproductive period (when the risk of a chromosomal anomaly may be more than 1%), amniocentesis and prenatal diagnosis allow termination of a pregnancy that would inevitably result in a malformed and/or mentally retarded child and allow high-risk parents to have only unaffected children.

Enzymatic Defects. Many genetic diseases are due to abnormal enzymes, and, in an increasing number of cases, the enzyme defect can be detected in fibroblasts cultured from the amniotic fluid. For example, parents who have had a child with infantile Tay-Sachs disease may be advised that there is a 1-in-4 risk of recurrence, but that amniocentesis and cell culture will allow diagnosis of the disease if the child falls into the affected class. Termination of the pregnancy would then prevent birth of a child destined to tragic death and allow the parents to try again until they had an unaffected baby. This procedure has already been a great boon to a number of families, and the number of diseases amenable to this approach is continually increasing as research progresses. However, although the procedure carries very little risk to mother or baby in expert hands, some of the biochemical techniques for the diagnosis of rare diseases by enzymatic studies on cultures are reliable only in highly experienced hands. For this reason, steps are being taken to develop centers where samples from surrounding areas can be sent for the necessary tests. With modern methods of transport, material can be sent long distances, and a network of centers now exists that could provide the necessary services for a large country.

The Follow-up

One or more follow-up interviews are desirable for several reasons. First, they may reinforce the parents' understanding of the information given and correct any misapprehensions resulting from reinterpretation of the information. Not infrequently, follow-up interviews reveal that the figures given have been modified upward or downward, either through reinterpretation with the "aid" of friends and relations, or perhaps in response to the parents' own wishes, subconscious or otherwise. For instance, when a 1-in-4 recurrence is translated into a 3-to-1 risk, one wonders whether this reflects the parents' desire not to have more children for any reason.

Finally, there are situations in which the mutant gene or chromosomal rearrangement segregating in the family places certain relatives at high risk. In such diseases as Huntington's chorea and multiple polyposis of the colon (dominant), hemophilia and Duchenne muscular dystrophy (sex-linked recessive), and chromosomal translocation, the study may reveal that certain family members other than the parents are at risk for developing the disease or having affected children. Seeing that these individuals receive counseling may be troublesome, involving questions of breach of confidence or invasion of privacy, but the counselor may also be criticized for failing to provide information that could prevent a tragedy. With the help of the proband's parents and the appropriate family physicians, the persons at risk can usually be notified.

Our family follow-ups have also reassured us that the opinion is false that "people are going to go ahead and have children no matter what you tell them." Some parents do seem willing to take what seems to us a rather inordinate risk, and some ignore the genetic hazards to the point of irresponsibility. However, the majority do heed the risks, and several surveys indicate that, when the risk is low (less than 10%), most parents are prepared to take a chance even with a severe disease, but when the risk is high, the majority take steps (however inadequate) to stop having children. Also, a crippling disease with long-term survival is regarded as more formidable than one that results in early death. It is important, however, to distinguish between parents who have been counseled routinely (e.g., as part of a clinic procedure) and those who have actively sought counseling. Parents who have not sought counseling may not pay much attention to it. Those who have are usually grateful for the knowledge and heed it, making the counselor's task indeed rewarding.

HOW TO LOCATE A GENETIC COUNSELOR

As awareness of genetic counseling and what it can do becomes more widespread, the demand for it increases. Many medical schools and some large hospitals now have departments or divisions of medical genetics or have affiliations with a university genetics department through which referral to an experience counselor can be arranged. Furthermore, an increasing amount of counseling may be done by specialized clinics—diabetes clinics may provide counseling for diabetics, cystic fibrosis clinics for pancreatic cystic fibrosis, and so on.

Both the National Foundation–March of Dimes, 1275 Mamaroneck Avenue, White Plains, New York 10605, and the National Genetics Foundation, 250 West 57th Street, New York City, can direct those in need of counseling to an appropriate source. The National Genetics Foundation sponsors Referral Centers that provide a variety of sophisticated diagnostic tests not routinely available. We also would be happy to answer letters of inquiry. The time is coming, one hopes, when all those who are concerned about the genetic implications of a disease in themselves or their families will have access to expert counsel.

SUMMARY

Genetic counseling depends on accurate diagnosis and definition of etiology where possible. On the basis of the family history, appropriate tests, and a knowledge of the literature, an estimate of the recurrence risk is made. The counselor may then participate in the process of reaching a decision and taking appropriate action, as desired by the family. Decisions include whether to marry, have another baby, use contraceptive measures, seek sterilization, adopt, have antenatal diagnosis, or have a therapeutic abortion. The counseling process may extend to other members of the family, who are (or think they are) at risk for developing the disorder in question, or having affected children. The genetic counseling center is preferably connected with a university medical center with its extensive diagnostic and consultative resources. The counselor should be prepared to provide the complete counseling service, or to support the primary physician who wishes to handle the case himself. Follow-up studies suggest that most families react responsibly to the information given. A more extensive review will be found in a recent report by Fraser.[3]

REFERENCES

1. Carter, C. O.: Genetic clinic. A follow-up. Lancet i:281, 1971.
2. Fraser, F. C.: Counseling in genetics: Its intent and scope. Birth Defects VI:7, 1970.
3. Fraser, F. C.: Genetic counseling. Am. J. Hum. Genet. 26:636, 1974.
4. Milunsky, A., and Littlefield, J. W.: The prenatal diagnosis of inborn errors of metabolism. Ann. Rev. Med. 23:57, 1972.
5. Murray, R.: Screening: A practitioner's view. *In* B. Hilton, D. Callahan, M. Harris, P. Condliffe, and B. Berkely, Ethical Issues in Human Genetics. New York, Plenum Press, 1973.
6. Smith, D. W., and Wilson, A. A.: The Child with Down's Syndrome. Philadelphia, W. B. Saunders, 1973.
7. Stevenson, A. C., and Davison, D. C. C.: Genetic Counselling. London, Heinemann, 1970.

GLOSSARY

Abiotrophy. A disease resulting from a genetic defect that causes progressive failure of some previously normal state or process, and therefore has a postnatal onset—e.g., muscular dystrophy, Huntington's chorea.

Acrocentric. Refers to a chromosome with the centromere near one end, so that one arm is very short.

Actinomycin D. Antibiotic that blocks elongation of RNA chains.

Active site. A region of a protein (particularly enzyme) directly involved in interaction with another molecule.

Adaptor molecules. See *Transfer RNA.*

Adenylcyclase. Enzyme that catalyzes production of cyclic AMP from ATP.

Affinity, cellular. Tendency of cells to adhere specifically to cells of same type, but not of different types. This property is lost in cancer cells.

Alleles, allelomorphs. Alternative forms of a gene. If more than two alleles exist for a given locus, they are called multiple alleles—for example, all the mutant genes at the hemoglobin beta chain locus.

Allograft. A tissue graft from a donor of one genotype to a host of the same species but another genotype. Contrast *Isograft.*

Allosteric. Refers to a protein in which the activity of active site is changed by the binding of a specific small molecule (allosteric effector) at another site.

Amino acids. The building blocks of proteins. Each has an amino group on one end and a carboxyl group on the other and a side group (R) that gives it its specificity.

Amino acids, acidic. Amino acids having a net negative charge at neutral pH (aspartic acid, glutamic acid).

Amino acids, aromatic. Amino acids whose side chains include a derivative of a phenyl group. The aromatic amino acids found in protein are phenylalanine, tyrosine, and tryptophan.

Amino acids, basic. Amino acids having a net positive charge at neutral pH (arginine, lysine, hydroxylysine, histidine).

Anaphase. The phase of mitosis or meiosis at which the chromosomes are drawn by their centromeres from the equatorial plate and pass to the poles of the cell.

Androgen(s). A group of male-determining hormones produced mainly by the testis and adrenal cortex.

Aneuploid. A chromosome number that is not an exact multiple of the haploid number.

Antibody. A gamma globulin formed by immune-competent cells in response to an antigenic stimulus, and reacting specifically with that antigen.

Anticipation. The term used to describe the apparent tendency of certain diseases to appear at earlier onset ages and with increasing severity in successive generations. It usually, if not always, appears to be a statistical artifact.

Antigen. A substance having the power to elicit antibody formation by immune-competent cells and to react specifically with the antibody so produced.

Antigenic determinant. Chemical structure (small compared to macromolecule) recognized by the active site of an antibody. Determines specificity of antibody–antigen interaction.

Ascertainment. The selection through an individual (the proband) of families for inclusion in a genetic study.

Association. The occurrence together, in a population, of two characteristics with a frequency greater than would be predicted on the basis of chance; that is, with a frequency that is greater than the product of the frequency of each. Not to be confused with linkage, where the association occurs only within families where the relevant genes are in coupling.

Assortative mating. Nonrandom mating, resulting from a tendency of parents with a particular characteristic to select mates with that characteristic (positive assortative mating) or shun such mates (negative assortative mating).

Autoimmunity. The formation of antibodies to an individual's own proteins, leading to autoimmune disease.

Autoradiography. A technique whereby the precise location of a radioactively labeled molecule in a cell or tissue can be demonstrated by applying a photographic emulsion to the histological section or cytological slide; the film will be sensitized wherever the label is present. Applied in cytogenetics particularly to delineating DNA synthesis by the chromosome by adding tritium-labeled thymidine to the culture—the label will be incorporated wherever DNA synthesis is proceeding.

Autosome. Any chromosome other than the sex chromosomes.

Backcross. Term from experimental genetics to indicate mating between F_1 hybrid and one of the two parental strains.

Barr body. See *Sex chromatin.*

Base pair. The guanine–cytosine and adenine–thymine pairs of purine (guanine, adenine) and pyrimidine (cytosine, thymine) bases that make up DNA. In RNA, uracil substitutes for thymine. One of the pair is on one chain, the other on the complementary chain.

Base-pairing rules. The requirement that adenine must always pair with thymine (or uracil) and guanine with cytosine, in a nucleic acid double helix.

Bence Jones protein. Light chains of a single antibody species produced by myeloma cells. Commonly detected in urine of human multiple myeloma patients.

β-galactosidase. An enzyme catalyzing the hydrolysis of lactose into glucose and galactose; in E. coli, the classic example of an inducible enzyme.

Bivalent. A pair of homologous chromosomes associated in meiotic pachytene.

Breakage and reunion. The classical model of crossing over between chromatids by physical breakage and crossways reunion of complete chromatids during meiosis. This model has recently been shown to be applicable in at least one case on the molecular level—crossing over between phage-DNA molecules proceeds by breakage and reunion.

Cancer. Strictly refers to carcinomas, but loosely used for diseases characterized by uncontrolled invasive cellular growth. See *Neoplasm.*

Carcinogen. An agent that induces cancer.

Carrier. An individual who carries a gene but does not manifest it—i.e., either an autosomal or X-linked recessive gene or a dominant mutant gene that has not yet resulted in overt disease.

Catalyst. A substance that can increase the rate of a chemical reaction without being consumed in the reaction (e.g., enzymes catalyze biological reactions).

Cell cycle. The cycle undergone by the nuclear DNA from one cell division to the next. It consists of: G1, a period of growth; S, a period of chromosomal DNA replication; G2, a period of further growth; and mitosis. (G stands for "gap" in DNA replication activity, and S stands for "synthesis.")

Centriole. One of the pair of small organelles that form the points of focus of the spindle during cell division in animal cells. The centrioles lie together outside the nuclear membrane at prophase and migrate during cell division to opposite poles of the cell.

Centromere (kinetochore, primary constriction). The constricted portion of the chromosome, separating it into its two arms. It is situated in a heterochromatic region, is the last part of the chromosome to divide, and is attached to the spindle fibers at mitosis and meiosis.

Chiasma. Refers to the X-like crossing of chromatid strands of homologous chromosomes, seen at diplotene of the first meiotic prophase. Chiasmata are evidence of interchanges of chromosomal material (crossovers) between members of a chromosome pair.

Chimera. An individual composed of cells derived from different zygotes; in human genetics, especially used with reference to blood group chimerism, a phenomenon in which dizygotic twins exchange hematopoietic stem cells *in utero* and continue to form blood cells of both types. Distinguish from mosaicism, in which the two genetically different cell lines arise after fertilization.

Chromatid. After the chromosome has made a replica of itself at the beginning of mitosis or meiosis, it consists of two strands, called chromatids, held together at the centromere. Each will become a separate chromosome when the centromere divides.

Chromatin. The material of the chromosomes that stains with nuclear (basic) stains—more or less synonymous with DNA.

Chromatography. A technique for separating compounds from a mixture, by their rate of migration through a medium, followed by appropriate staining.

Chromomeres. Areas of increased optical density and/or increased diameters along the length of a chromosome, especially clearly discernible in prophase of meiosis.

Chromosomes. The carriers of the genes, consisting of long strands of DNA in a protein framework. The exact structure of mammalian chromosomes is still not known. In nondividing cells they are not individually distinguishable in the nucleus, but at mitosis or meiosis they become condensed into visible strands that stain deeply with basic stains.

Cis configuration. See *Linkage*.

Cistron. An operational definition of a gene, coming from microbial genetics, based on complementation tests; its use is largely outmoded and is inappropriate in medical genetics, where the necessary data are lacking.

Cleavage division. Mitotic divisions of the fertilized egg that divide it into smaller and smaller units, until the stage when the original regions of the egg begin to shift relative to one another.

Clone. A group of cells all derived from a single cell by repeated mitosis, and all having the same genetic constitution.

Codominance. See *Dominance*.

Codon. A triplet of three nucleotide bases in a DNA or messenger RNA molecule that codes for a specific amino acid, or the initiation or termination of transcription.

Coefficient of inbreeding. The probability that an individual has received both alleles of a pair from an identical ancestral source; or the proportion of loci at which he is homozygous for such alleles.

Coefficient of relationship. The probability that two persons have inherited a certain gene from a common ancestor; or the proportion of all their genes that have been inherited from common ancestors.

Colinearity. The relationship between two macromolecules (DNA and protein) in which the sequence of components (bases) of the former specifies the sequence of components (amino acids) of the latter.

Complementation test. The bringing together of two mutant genes, either by crossing, coculturing of cell lines, or cell hybridization, to see if together they can produce a normal phenotype. From this, tentative conclusions can be drawn as to whether they are alleles.

Compound. See *Heterozygote*.

Concordant. If both members of a twin pair exhibit a certain trait, they are said to be concordant for that trait. Contrast *Discordant*.

Conditional lethal mutations. A class of mutants whose viability is dependent on growth conditions (e.g., temperature-sensitive lethals).

Congenital. Present at birth. Does not imply either genetic or nongenetic causation.

Consanguinity. Relationship by descent from a common ancestor.

Constitutive enzymes. Enzymes that are synthesized independently of an inducer.

Consultand. The person, in a genetic counseling situation, whose genotype is being evaluated—often the parents of an affected child.

Contact inhibition. The cessation of cell membrane movement that may occur when freely growing cells come into physical contact with each other.

Corepressors. Metabolites that, by their combination with repressors, specifically inhibit the formation of the enzyme(s) involved in their metabolism.

Coupling. See *Linkage*.

Crossing over. The process of exchange of genetic material between homologous chromosomes. The chiasmata seen at the diplotene of meiosis are the physical basis of a previous crossover.

Cyclic AMP. Adenosine monophosphate with phosphate group bonded internally (phosphodiester bond between 3′ and 5′ carbon atoms) to form cyclic molecule. Plays an important role in the mechanism by which hormones (and other compounds) regulate the activity of specific genes. See *Hormone.*

Cytogenetics. The branch of genetics concerned mainly with the appearance and segregation of the chromosomes and their relation to phenotype.

Degenerate code (genetic). One in which two or more codons code for the same amino acid.

Deletion. A chromosomal aberration in which a portion of a chromosome is missing.

Deme. A group defined by the population from which members select their mates—an effective breeding population.

Deoxynucleoside. The condensation product of a purine or pyrimidine with the five-carbon sugar, 2-deoxyribose.

Deoxyribonucleotide. A compound that consists of a purine or pyrimidine base bonded to the sugar, 2-deoxyribose, which in turn is bound to a phosphate group.

Dermatoglyphics. The patterns formed by the ridges of the skin of the palms, fingers, soles, and toes.

Diakinesis. The final stage of prophase of the first meiotic division. During diakinesis the chromosomes become tightly coiled and darkly staining.

Dictyotene. The interphase-like stage in which the oocyte persists from late fetal life until ovulation. The oocyte has not yet completed the first stage of meiosis.

Differentiation. The process whereby the developmental and functional abilities of a cell become restricted to a specific structure and major function.

Dimer. Structure resulting from association of two subunits.

Diploid. Having two complete sets of chromosomes; double the number found in the gametes. In man, the diploid chromosome number is 46. Contrast *Haploid, Triploid.*

Diplotene. The stage of first meiotic prophase during which the paired centromeres begin to repel one another and the chromosomes begin to separate, exhibiting chiasmata.

Discordant. If one member of a twin pair shows a certain trait and the other does not, the twins are said to be discordant for that trait. Contrast *Concordant.*

Disulfide bond. Covalent bond between two sulfur atoms in different amino acids of a protein. Important in determining secondary and tertiary structure.

Dizygotic (dizygous, fraternal). Type of twins produced by two separate ova, fertilized by separate sperms.

DNA (deoxyribonucleic acid). A polymer of deoxyribonucleotides that is the genetic material of all eukaryote cells.

DNA polymerase. An enzyme that catalyzes the formation of new polydeoxyribonucleotide (DNA) strands from deoxyribonucleoside triphosphates, using DNA as a template.

DNA–RNA hybrid. A double helix that consists of one chain of DNA hydrogen bonded to a chain of RNA by means of complementary base pairs.

Dominant. A gene is said to be dominant if the phenotype of the heterozygote is the same as that of the homozygote for that gene. In human genetics the term is used, more loosely, for a mutant gene that is expressed in the heterozygote. If the mutant homozygote is more severely affected than the heterozygote, there is "intermediate" dominance, and if both genes are expressed independently, there is "codominance." Traditionally, the term refers to traits, but is now commonly applied to genes as well.

Dosage compensation. A term, usually used in relation to sex determination, where the effects of structural genes on the X chromosome are the same, whether the X chromosome is represented once, or twice. In many species, including man, the Lyon hypothesis provides a mechanism.

Drift, genetic. Chance variation in gene frequency from one generation to another. The smaller the population, the greater are the random variations.

Drumstick. A small protrusion from the nucleus of a polymorphonuclear leukocyte, found in 3 to 5% of these cells in females but not in males.

Duplication. The recurrence of a segment of chromosome in tandem sequence.

Electrophoresis. A method of separating large molecules by their rate of migration through a medium (e.g., filter paper, starch gel) in an electrical field.

Endogamy. A breeding pattern where mating occurs only between members of a group.

Endonuclease. An enzyme that makes internal cuts in DNA backbone chains.
Endoreduplication. A process in which the chromosomes replicate without cell division.
Enzymes. Protein molecules capable of catalyzing specific chemical reactions.
Epistasis. The masking of the effects of one gene or set of genes by the action of a gene at another locus—e.g., the albino gene is epistatic to the genes determining the normal color of the iris.
Erythroblast. Nucleated cell in bone marrow that differentiates into red blood cell.
Estrogen. Female sex hormone produced mainly by the ovary.
Eukaryote. An organism in which the cells have a nuclear envelope, as contrasted to prokaryotes (e.g., bacteria, blue-green algae). Eukaryotes also have larger cells than prokaryotes and have cytoplasmic organelles.
Euchromatin. Most of the chromosomal material, which stains uniformly. Contrast *Heterochromatin.*
Exonuclease. An enzyme that digests DNA from the ends of strands.
Expressivity. The variability in the degree to which a mutant gene expresses itself in different mutant individuals.

Familial trait. A trait that occurs with a higher frequency in the near relatives of individuals with the trait than in unrelated individuals from the same population.
Feedback (end-product) inhibition. Inhibition of the enzymatic activity of the first enzyme in a metabolic pathway by the end product of that pathway.
Fertilization. Fusion of gametes of opposite sex to produce a diploid zygote.
Fingerprint. (1) The pattern of the ridged skin of the distal phalanx of a finger. (2) The pattern of spots on a two-dimensional chromatogram produced by the peptides of a hydrolyzed polypeptide.
Fitness. The probability that an individual of a given phenotype will transmit his (her) genes to the next generation, relative to the average for the population.
Forme fruste (French: defaced form). An incomplete, partial, or mild form of trait or syndrome.
Fraternal twins. See *Dizygotic.*

G1, G2. See *Cell cycle.*
Gamete. A mature sperm or egg cell with haploid chromosome number.
Gene. A portion of a DNA molecule that is the code for the amino-acid sequence of a particular polypeptide chain.
Gene flow. Transfer of genes from one population to another, by migration of individuals from one population to the other and mating between individuals of the two populations.
Gene redundancy. Presence in cell of many copies of a single gene. Multiple copies may be inherited or result from selective gene duplication during development.
Genetic. Determined by differences between genes.
Genetic code. The relation between the nucleotide triplets in the DNA or RNA and the amino acids in the corresponding polypeptides.
Genetic death. The failure of a mutant gene to be passed on to the next generation because of the phenotypic effects of that gene on an individual.
Genetic heterogeneity. The production of the same phenotype by more than one genotype.
Genetic marker. A readily recognizable genetic difference that can be used in family and population studies.
Genocopy. A trait genetically different from a phenotypically similar one. See also *Genetic heterogeneity.* Contrast *Phenocopy.*
Genome. The complement of genes found in a set of chromosomes.
Genotype. The genetic constitution of an individual, with respect either (a) to his/her complete complement of genes or (b) to a particular locus. Contrast *Phenotype.*
Glycoprotein. Protein in which a carbohydrate is covalently bonded to the peptide portion of the molecule.
Gonosome. Term now rarely used, referring to sex chromosome. Contrast *Autosome.*

Haploid. Having only one complete set of chromosomes. Contrast, e.g., *Diploid.* In man, the haploid number is 23.
Haptoglobin. A serum protein that binds hemoglobin; it exists in several polymorphic variants.

Hardy-Weinberg law. If two alleles (A and a) occur in a randomly mating population with the frequency of p and q, respectively, where $p + q = 1$, then the expected proportions of the three genotypes are AA $= p^2$, Aa $= 2pq$, and aa $= q^2$. These remain constant from one generation to the next. Mutation, selection, migration, and genetic drift can disturb the Hardy-Weinberg equilibrium.

HeLa cells. An established line of human cervical carcinoma (cancer) cells from a patient named Henrietta Lacks; used extensively in the study of biochemistry and growth of cultured human cells.

Hemizygous. Having only one member of a gene pair or group of genes in an otherwise diploid individual. Since males have only one X, they are said to be hemizygous with respect to X-linked genes.

Hemoglobin. The protein carrier of oxygen in red blood cells. A tetramere of two pairs of polypeptide chains, each with an iron-containing heme group.

Hepatoma. A form of liver cancer.

Hereditary, heritable, heredofamilial. Essentially synonymous terms for genetic traits. Formerly "hereditary" was sometimes used in the sense of "dominant." "Heredofamilial" is archaic.

Hermaphrodite. An individual with both ovarian and testicular tissue (not necessarily functional).

Heterochromatin. Chromosomal material with variable staining properties different from that of the majority of chromosomal material, the euchromatin.

Heterogametic. Producing gametes of two types with respect to sex determination. In man the heterogametic sex is the male, who produces sperm bearing X and Y chromosomes, respectively. The female is the homogametic sex, producing only X-bearing ova.

Heterograft. A tissue graft from a donor of one species to a host of a different species—called a xenograft in the modern terminology.

Heteropyknosis. A state in which a region of chromosome is heavily condensed and darkly staining.

Heterozygous (heterozygote). Possessing different alleles at a given locus. Double heterozygote refers to heterozygous state at two separate loci. An individual heterozygous for two mutant alleles, such as those for hemoglobin S and C, may be called a compound heterozygote.

Histocompatibility genes. Genes for antigens that determine the acceptance or rejection of tissue grafts.

Histones. Proteins rich in basic amino acids (e.g., lysine) found in chromosomes except in sperm, where the DNA is complexed with another group of basic proteins, the protamines.

Holandric. The pattern of inheritance of genes on the Y chromosome; transmission from father to all his sons but none of his daughters.

Homogametic. See *Heterogametic.*

Homograft. A graft of tissue between two genetically dissimilar members of the same species.

Homologous chromosomes. Chromosomes that pair during meiosis, have essentially the same morphology, and contain genes governing the same characteristics.

Homozygous (homozygote). Possessing identical alleles at a given locus. Contrast *Heterozygous.*

Hormone. Chemical substance (often small polypeptide) synthesized in one organ of body that stimulates functional activity in cells of other tissues and organs. Many hormones act by stimulating adenylcyclase in the cell membrane to produce cyclic AMP.

^{3}H (tritium). A radioactive isotope of hydrogen, a weak β-emitter, with a half-life of 12.5 years, useful in radioautography.

Hydrogen bond. A weak attractive force between one electronegative atom and a hydrogen atom that is covalently linked to a second electronegative atom.

Hydrolysis. The breaking of a molecule into two or more smaller molecules by the addition of a water molecule.

Hydrophilic (polar). Pertaining to molecules or groups that readily associate with water.

Hydrophobic (nonpolar). Literally, water hater. Describes molecules or certain functional groups in molecules that are insoluble or only poorly soluble in water.

Hydrophobic bonding. The association of nonpolar groups with each other in aqueous solution, arising because of the tendency of water molecules to exclude nonpolar molecules.

Idiogram. A diagram of a chromosome complement.

Immune competent. Capable of producing antibody in response to an antigenic stimulus.

Immunoglobin. Protein molecule, produced by plasma cell, that recognizes and binds a specific antigen. Also called antibody.

Immunologic tolerance. Absence of immune response to antigens.

Immunosuppressive drug. Drug that blocks normal response of antibody-producing cells to antigen.

Inborn error. A genetically determined biochemical disorder in which a specific defect produces a metabolic abnormality that may have pathological consequences.

Incompatibility, immunologic. Donor and host are incompatible if, because of genetic difference, the host rejects cells from the donor.

Incomplete dominance. A term used sometimes as a synonym for intermediate dominance (see dominant), and sometimes as referring to a mutant gene that is expressed in all homozygotes.

Index case. See *Proband.*

Inducer. A small molecule that increases the production of the enzymes involved in its metabolism.

Inducible enzyme. An enzyme whose rate of production is increased by the presence of an inducer.

Interchange. See *Translocation.*

Intermediary metabolism. The chemical reactions in a cell that transform food molecules into molecules needed for the structure and growth of the cell.

Interphase. The stage of the cell cycle between two successive divisions during which the normal metabolic processes of the cell proceed.

Intersex. An individual whose genitalia or gonads show characteristics of both sexes or are ambiguous.

Inversion. End-to-end reversal of a segment within a chromosome; *pericentric* if it includes the centromere and *paracentric* if it does not.

In vitro (Latin: in glass). Refers to experiments done on biological systems outside the intact organism. Contrast *in vivo.*

In vitro protein synthesis. The incorporation of amino acids into polypeptide chains in a cell-free system.

In vivo (Latin: in life). Refers to experiments done in a system such that the organism remains intact. Contrast *in vitro.*

Isochromosome. An abnormal chromosome with two arms of equal length and bearing the same loci in reverse sequence, formed by crosswise rather than longitudinal division of the centromere.

Isogenic. Refers to grafts with identical histocompatibility antigens.

Isograft. A tissue graft in which donor and host have identical genotypes. Contrast *Allograft.*

Isolate. A population in which mating does not occur outside the group. See also *Deme.*

Karyotype. The chromosome set of an individual. The term also refers to photomicrographs of a set of chromosomes arranged in a standard classification.

Kindred. Family in the larger sense, as contrasted to the nuclear family (parents and children).

Lampbrush chromosome. Giant diplotene chromosome found in some species in the oocyte nucleus with loops projecting in pairs from certain regions. Loops are sites of active messenger RNA synthesis.

Leptotene. The first stage of prophase of the first meiotic division, in which individual chromosomes appear as unpaired threads.

Lethal equivalent. A gene that, if homozygous, would be lethal; or a combination of two genes, each of which, if homozygous, would have a 50% chance of causing death; or any equivalent combination.

Leukemia. Form of neoplasm characterized by extensive proliferation of nonfunctional immature white blood cells (leukocytes).

Ligase, polynucleotide. Enzyme that covalently links DNA backbone chains.

Linkage. Gene loci are linked if they are close enough to each other on the same chromosome that they do not segregate independently, but tend to be transmitted together. Genes are linked in coupling if they are on the same chromosome (the *cis* configuration) and in repulsion if they are on homologous chromosomes (the *trans* configuration).

Linked genes. Genes that are located on the same chromosome and that therefore tend to be transmitted together.

Load, genetic. The sum total of death and disease caused by mutant genes.

Lymphocyte. A type of white blood cell important in the immunologic system.

Lymphoma. Neoplasm of lymphatic tissue.

Lyonization (Lyon hypothesis). The process by which all X chromosomes in excess of one are genetically inactive and heterochromatic. In the female, the decision as to which X (maternal or paternal) is inactivated is taken independently for each cell, early in embryogeny, and is permanent for all descendants of that cell.

Map unit (centimorgan); map distance. The measure of distance between two loci on a chromosome as inferred from the frequency (%) of crossing over (recombination) between them. Accurate only for small distances, as double crossovers will not appear as recombinations. Fifty percent recombination is the maximum, corresponding to independent segregation.

Meiosis. The special type of cell division by which gametes, containing the haploid number of chromosomes, are produced from diploid cells. Two meiotic divisions occur. Reduction in number takes place during meiosis I.

Messenger RNA (mRNA). RNA that serves as a template for protein synthesis.

Metacentric. Refers to chromosomes with the centromere near the middle.

Metaphase. The stage of mitosis or meiosis when the centromeres of the contracted chromosomes are arranged on the equatorial plate.

Micron (μ). A unit of length convenient for describing cellular dimensions; it is equal to 10^{-3} cm or 10^5 Å.

Mitogen. A substance which stimulates cells to undergo mitosis.

Mitosis. Somatic cell division resulting in the formation of two cells, each with the same chromosome complement as the parent cell.

Monomer. The basic subunit from which, by repetition of a single reaction, polymers are made. For example, amino acids (monomers) condense to yield polypeptides or proteins (polymers).

Monomeric. Refers to differences determined by genes at a single locus.

Monosomy. A condition in which one chromosome of a pair is missing.

Monozygotic (monozygous, identical). Refers to twins derived from one egg and thus genetically identical.

Mosaic. An individual or tissue with two or more cell lines differing in genotype or karyotype, derived from a single zygote.

Multifactorial Determined by multiple genetic and nongenetic factors, each making a relatively small contribution to the phenotype. See also *Polygenic.*

Multiple allele. See *Alleles.*

Mutagen. Any agent that increases the mutation rate.

Mutant. (1) A gene altered by mutation. (2) An individual bearing such a gene.

Mutation. A permanent change in the genetic material. Usually refers to point mutation, that is, change in a single gene, but in a more general sense includes the occurrence of chromosomal aberrations. In connection with inherited diseases, mutation in the germ cells is most relevant, but somatic mutation also occurs and may be important in relation to neoplasia and aging.

Mutation rate. The rate at which mutations occur at a given locus; expressed as mutations per gamete per locus per generation.

Myeloma. Cancer arising from a clone of plasma cells, and producing a pure immunoglobulin.

Neoplasm. Literally "new growth," a general term for cancers and other tumors in which there has been loss of the normal regulation of mitotic activity.

Nondisjunction. The failure of two members of a chromosome pair to disjoin during anaphase of cell division, so that both pass to the same daughter cell.

Nonpenetrance. See *Penetrance.*

Nonpolar. See *Hydrophobic bonding.*

Nucleic acid. A nucleotide polymer. See also *DNA* and *RNA.*

Nucleolus. Round granular structure found in the nucleus of eukaryotic cells, usually associated with a specific chromosomal site, involved in rRNA synthesis and ribosome formation.

Nucleolus organizer. Secondary constrictions of chromosomes, particularly those related to satellites, seem to have this function.

Nucleoside. The combination of a purine or pyrimidine base and a sugar.

Nucleotide. The combination of a purine or pyrimidine base, a sugar, and a phosphate group. The monomers from which DNA and RNA are polymerized.

Oocyte. Unfertilized egg cell.
Oogenesis. The process of formation of the female gametes.
Operator. A chromosomal region capable of interacting with a specific repressor, thereby controlling the function of an adjacent series of genes (operon).
Organelle. Membrane-bound structure found in eukaryotic cells, containing enzymes for specialized function. Some organelles, including mitochondria and chloroplasts, have DNA and can replicate autonomously.

Pachytene. A stage of first meiotic prophase during which the bivalents (paired chromosomes) shorten and thicken and may be seen to consist of two chromatids per chromosome.
Panmixis. Random mating.
Paracentric. See *Inversion.*
Penetrance. The per cent frequency with which a heterozygous dominant, or homozygous recessive, mutant gene produces the mutant phenotype. Failure to do so is called "nonpenetrance," and penetrance less than 100% is "reduced penetrance."
Peptide bond. A covalent bond between two amino acids in which the a-amino group of one amino acid is bonded to the a-carboxyl group of the other with the elimination of H_2O.
Pericentric. See *Inversion.*
PHA. See *Phytohemagglutinin.*
Pharmacogenetics. The area of biochemical genetics dealing with drug responses and their genetically controlled variations.
Phenocopy. An environmentally induced mimic of a genetic disorder, with no change in the corresponding gene.
Phenotype. (1) The observable characteristics of an individual as determined by his genotype and the environment in which he develops. (2) In a more limited sense the outward expression of some particular gene or genes. Thus a heterozygote and homozygote for a fully dominant gene will have the same phenotype, but different genotypes.
Phytohemagglutinin (PHA). A compound, extracted from beans, that stimulates circulating lymphocytes to enter mitosis. Used in the standard techniques for cytogenetic study of human chromosomes from peripheral blood.
Plasma cell. An antibody-producing cell derived from a lymphocyte (a kind of white blood cell).
Pleiotropy. A mutant gene or gene pair that produces multiple effects is said to exhibit pleiotropy (as seen in hereditary syndromes).
Polar. See *Hydrophilic.*
Polygenic. Refers to determination by many genes, with small additive effects.
Polymer. A regular, covalently bonded arrangement of basic subunits (monomers) produced by repetitive application of one or a few chemical reactions.
Polymorphism. The occurrence of two or more genetically determined alternative phenotypes in a population, in relatively common frequencies. When maintained by heterozygote advantage, it is referred to as a balanced polymorphism.
Polynucleotide. A linear sequence of nucleotides in which the 3′ position of the sugar of one nucleotide is linked through a phosphate group to the 5′ position on the sugar of the adjacent nucleotide.
Polyoma virus. An RNA virus that will transform cells into a neoplastic state in culture.
Polypeptide. A chain of amino acids, held together by peptide bonds between the amino group of one and the carboxyl group of an adjoining one. A protein molecule may be composed of a single polypeptide chain, or of two or more identical or different polypeptides.
Polyploid. Any multiple of the basic haploid chromosome number, other than the diploid number.
Polyribosome. Complex of a messenger-RNA molecule and ribosomes actively engaged in polypeptide synthesis.
Proband (propositus). The affected family member through which the family is ascertained—the index case. Originally a proband was not necessarily affected and a propositus was, but by current usage the terms are synonymous.

Prophase. The first stage of cell division, during which the chromosomes become visible as discrete structures and subsequently thicken and shorten. Prophase of the first meiotic division is further characterized by pairing (synapsis) of homologous chromosomes.

Propositus (female, *proposita;* plurals, *propositi* and *propositae*). Synonyms are *proband* or *index case.* (Proband is preferred.)

Protamines. Proteins rich in basic amino acids found in the chromosomes of sperm.

Pseudohermaphrodite. An individual who has gonadal tissue of only one sex, but who has anomalous development of the genitalia such that the true sex may not be readily apparent. Pseudohermaphrodites are designated as male or female with reference to the sex chromosome constitution and the type of gonadal tissue present.

Quasicontinuous variation. A term applied to discrete traits classified as present or absent (i.e., discontinuous) that are determined by an underlying continuous distribution, multifactorially determined, separated into two parts by a developmental or other threshold.

Radioactive isotope. An isotope with an unstable nucleus that stabilizes itself by emitting ionizing radiation.

Random mating. Selection of a mate without regard to the genotype of the mate (except for sex, of course).

Recessive. Refers to a trait that is expressed only in individuals homozygous for the gene concerned. Usage now justifies applying the term to the gene as well. The definition is an operational one—whether a "recessive" gene is expressed in the heterozygote may depend on the means used to detect it.

Recombination. The formation of new combinations of linked genes by the occurrence of a crossover at some point between their loci.

Reduction division. The first meiotic division, so called because at this stage the chromosome number per cell is reduced from diploid to haploid.

Regulator gene. According to the operon theory of gene regulation of Jacob and Monod, a regulator gene synthesizes a repressor substance that inhibits the action of a specific operator gene, thus preventing the synthesis of mRNA by that operon.

Regulatory genes. Genes whose primary function is to control the rate of synthesis of the products of other genes.

Repressible enzymes. Enzymes whose rates of production are decreased when the intracellular concentration of certain metabolites increases.

Repressor. In the operon model, the product of a regulatory gene, now thought to be a protein and to be capable of combining both with an inducer (or corepressor) and with an operator.

Repulsion. See *Linkage.*

Reticulocyte. Immature red blood cell that has lost its nucleus but is actively synthesizing hemoglobin.

Reverse (back) mutation. A heritable change in a mutant gene that restores the original nucleotide sequence.

Ribonucleotide. A compound that consists of a purine or pyrimidine base bonded to ribose, which in turn is esterified with a phosphate group.

Ribosome proteins. A group of proteins that bind to rRNA by noncovalent bonds to give the ribosome its three-dimensional structure.

Ribosomes. Small cellular particles (200 Å in diameter) made up of rRNA and protein. Ribosomes are the site of transcription of polypeptide chains from the mRNA.

RNA (ribonucleic acid). A nucleic acid formed upon a DNA template and taking part in the synthesis of polypeptides. Instead of thymine, RNA contains uracil. Three forms are recognized: (1) messenger RNA (mRNA), which is the template upon which polypeptides are synthesized; (2) transfer RNA (sRNA, soluble RNA), which in cooperation with the ribosomes brings activated amino acids into position along the mRNA template; (3) ribosomal RNA (rRNA), a component of the ribosomes, which function as nonspecific sites of polypeptide synthesis.

RNA polymerase. An enzyme that catalyzes the formation of RNA from ribonucleoside triphosphates, using DNA as a template.

Robertsonian translocation. A translocation between two acrocentric chromosomes by fusion at the centromeres and loss of the respective short arms.
rRNA. Ribosomal RNA (See *RNA.*)

S (svedberg). The unit of sedimentation (S). S is proportional to the rate of sedimentation of a molecule in a given centrifugal field and is thus related to the molecular weight and shape of the molecule.
Satellite, chromosomal. A small mass of chromatin attached to the short arm of each chromatid of a human acrocentric chromosome by a relatively uncondensed stalk (secondary constriction).
Second set response. The rapid rejection of implanted tissue by a host already sensitized to tissue of that genotype.
Secondary constriction. Narrowed, heterochromatic area in a chromosome. A secondary constriction separates the satellite from the rest of the chromosome. Probably associated with nucleolus formation. See *Nucleolus organizer.* The centromere is the primary constriction.
Secretor. (1) A trait characterized by the presence of the appropriate ABO blood group substance in saliva and other body fluids. (2) The gene responsible for this trait.
Segregation. In genetics, the separation of allelic genes by meiosis, into different gametes.
Selection. In population genetics, the effect of the relative fitness of a genotype in a population on the frequency of the genes concerned.
Sendai virus. A parainfluenza virus isolated in Sendai, Japan, which, in killed suspension, increases cell fusion in somatic cell cultures.
Serum proteins. Proteins found in serum (cell-free) component of blood. Includes immunoglobulins, albumin, haptoglobins, clotting factors, and enzymes.
Sex chromatin. A chromatin mass in the nucleus of interphase cells of females of most mammalian species, including man. It represents a single, condensed X chromosome inactive in the metabolism of the cell. Normal females have sex chromatin, thus are chromatin positive; normal males lack it, thus are chromatin negative. Synonym: Barr body.
Sex chromosomes. Chromosomes responsible for sex determination. In man, the X and Y chromosomes.
Sex-influenced. Refers to a genetically determined trait in which the degree of manifestation of the responsible gene is different in males and females.
Sex-limited. Refers to autosomal traits that occur only in either males or females.
Sex-linked. Determined by a gene located on the X or Y chromosome. Since most sex-linked traits are determined by genes on the X chromosome, the term is often assumed to refer to these; X-linked is preferable term in such cases.
Sex ratio. The ratio of males to females. The primary sex ratio refers to that at fertilization; the secondary sex ratio to that at birth.
Sibs, siblings. Brothers and sisters. Brevity makes *sib* the preferred term.
Sibship. Group of brothers and/or sisters.
Silent allele. An allele that has no detectable product.
Soluble RNA. *Transfer RNA.*
Somatic mutation. A mutation occurring in a somatic cell.
Spermatogenesis. The process of formation of spermatozoa.
Spermiogenesis. That part of spermatogenesis in which spermatids develop into spermatozoa.
S phase. See *Cell cycle.*
Structural gene. One that specifies the amino acid sequence of a polypeptide chain, as opposed to regulatory genes, which may not.
Substrate. A compound acted on by an enzyme in a metabolic pathway.
Synapsis. The process by which homologous chromosomes come to pair side-by-side early in meiosis.
Syndrome. A characteristic association of several anomalies in the same individual, implying that they are causally related.

"T" antigen. Antigen found in nuclei of cells infected or transformed by certain tumor viruses (e.g., polyoma virus and SV_{40}). May be an early viral-specific protein.
Target tissue. In immunogenetics, the tissue against which antibodies are formed.
Telocentric. Refers to chromosome with its centromere at the end.
Telophase. The last stage of cell division, from the time the centromeres of the daughter chromosomes reach the poles of the dividing cell until cell division is complete.

Temperature-sensitive mutant. Mutant that is functional at one temperature but inactivated at another.
Teratogen. An agent that causes congenital malformations.
Tetramer. Structure resulting from association of four subunits.
Trait. Any specific, classifiable characteristic.
Trans configuration. See *Linkage.*
Transcription. The process whereby the genetic information contained in DNA is transferred by the ordering of a complementary sequence of bases to the messenger RNA as it is being synthesized.
Transduction. The transfer of bacterial genes from one bacterium to another by a bacteriophage particle.
Transfer RNA (tRNA, sRNA). Any of at least 20 structurally similar species of RNA, all of which have a MW 25,000. Each species of sRNA molecule is able to combine covalently with a specific amino acid and to hydrogen-bond with at least one mRNA nucleotide triplet. Also called adpater RNA or soluble RNA (sRNA).
Transferases. Enzymes that catalyze the exchange of functional chemical groups between substrates.
Transformation, cell. A permanent change in cell phenotype occurring in somatic cell cultures, in which the resulting cell strain manifests neoplastic features, including many, but not necessarily all, of the following: loss of contact inhibition, and thus change in cell and colony morphology; progressive changes in karyotype; formation of neoplasms on transplantation to host.
Transformation, DNA. The genetic modification induced by the incorporation into a cell of DNA from a genetically different source.
Translation. The process whereby the genetic information present in an mRNA molecule directs the order of the specific amino acids during protein synthesis.
Translational control. Regulation of gene expression by control of the rate at which specific mRNA molecules are translated.
Translocation. (1) The transfer of a piece of one chromosome to another. (2) The resultant chromosome. If two chromosomes exchange pieces, the translocation is reciprocal. See also *Robertsonian translocation.*
Triplet. In molecular genetics, a unit of three successive bases in DNA or RNA, coding for a specific amino acid.
Triploid. Having three sets of the normal haploid chromosome complement.
Triradius. In dermatoglyphics, a point from which the dermal ridges course in three directions at angles of approximately 120°.
Tritium. See ^{3}H.
Truncate selection. The selection of families in a population in such a way that one or more kinds of sibships are not ascertained. Usually the unascertained sibships are those in which no member is affected.
Tumor virus. A virus that induces the formation of a tumor.

Ultracentrifuge. A high-speed centrifuge that can attain speeds up to 60,000 rpm and centrifugal fields up to 500,000 times gravity and thus is capable of rapidly sedimenting macromolecules and separating them by differences in migration along a gradient.
Ultraviolet (uv) radiation. Electromagnetic radiation with wavelength shorter than that of visible light (3900-20000 Å). Causes DNA base-pair mutations and chromosome breaks.

Viruses. Infectious disease-causing agents, smaller than bacteria, possessing a DNA or RNA genome and a protein coat; they require intact host cells for replication.

Wild type. The normal allele of a rare mutant gene, sometimes symbolized by +.

X-ray crystallography. The use of diffraction patterns produced by x-ray scattering from crystals to determine the 3-D structure of molecules.

Zygote. The fertilized ovum or (more loosely) the organism developing from it.
Zygotene. The stage of prophase of the first meiotic division in which pairing (synapsis) of homologous chromosomes occurs.

Appendix A

ANALYSIS OF CHROMOSOMES

FLUORESCENT BANDING TECHNIQUES

A standardized system of identification of the human somatic karyotype based on the fluorescent pattern is described.[3] Because most laboratories do not have available the equipment for densitometric measurements, the decription has been confined to visually recognizable patterns, but confirmed by comparison with the densitometric results of Caspersson *et al.*[1]

General Aspects of Terminology. The previously used definitions by length, centromeric index, autoradiography, or secondary constrictions that exist for chromosomes number 1, 2, 3, 4, 5, 9, 13, 14, 15, 16, 17, 18, and Y have been accepted.[2] (X chromosomes in numbers greater than one can be identified by their late replicating characteristics.) The numbers of the remaining autosomes are based on their fluorescent banding pattern as given by Caspersson *et al.*[1] The chromosome associated with Down syndrome, although smaller than number 22, has been retained as number 21.

Descriptive Terms. In the description that follows, "A" are the diagnostic features that can be seen in a fluorescent metaphase of fair technical quality, "B" are the details that are usually only visible in cells of good quality. When these details are not included in the text they are identical to "A." "C" denotes features that may vary in fluorescent intensity and/or length between individuals and between homologues.

Intensity of Fluorescence. Some mitoses show considerable nonuniformity in that two homologous chromosomes differ greatly in overall fluorescence and relative length. Identification must therefore be based on the fluorescent banding pattern of a chromosome rather than on its overall intensity. However, the latter may serve as a secondary criterion if due allowance is made for nonuniformity. In the descriptions the following terms have been used to indicate the approximate intensity of fluorescence:

Negative—no or almost no fluorescence
Pale —as on distal 1p
Medium —as the two broad bands on 9q
Intense —as the distal half of 13q
Brilliant —as on distal Yq

Proposed System of Identification for Individual Chromosomes. In the following section, only major fluorescent bands will be referred to, even though in some cells these may appear to consist of several smaller bands. Except when they are of special significance, nonfluorescing bands are not referred to and it may be assumed that they separate the major fluorescent bands or are located at the end of the chromosome arms.

No. 1 Long arm is that previously defined as the arm with a proximal secondary constriction.

A *p:* Distal, pale segment grading to a proximal, medium fluorescent segment.
q: Central, intense band. Proximal, negative secondary constriction.

B *p:* Proximal, medium fluorescent segment; divisible into two bands.
q: Five medium fluorescent bands; central one most prominent.

C *q:* Negative secondary constriction variable in length.

No. 2

A Medium fluorescence along the whole length.

B *p:* Four medium fluorescent bands; two central ones often appear as a single segment.
q: Two central bands, sometimes accompanied by another two, all of medium fluorescence. Additional bands can be seen sometimes.

No. 3

A Single pale band in center of each arm separating medium fluorescent segments. Distal, medium fluorescent segment; longer in q than in p.

B Single pale band at end of each arm; longer in p than in q.

C *q:* Proximal band of variable fluorescence.

No. 4

A Medium fluorescence along the whole length.

B *p:* Single central, medium fluorescent band.
q: Proximal, intense band. Distal, pale band.

C Intense centric band.

No. 5

A *q:* Central, long, medium fluorescent segment. Distal, pale segment.

B *p:* Single medium fluorescent band; shorter and brighter than on 4p.
q: Distal, pale segment; divisible into a proximal, pale band and a distal, medium one.

No. 6

A *p:* Central, pale band separating medium fluorescent segments.
q: Medium fluorescence along entire length.

B *q:* Four medium fluorescent bands.

No. 7

A *p:* Distal, short, medium fluorescent band.
q. Two central, intense bands. Distal, medium fluorescent band.

B *p:* Proximal, medium fluorescent band.

No. 8

A Medium fluorescence along the whole length; q brighter than p.

B *p:* Two evenly spaced, medium fluorescent bands.
q: Two medium fluorescent bands in distal half; brighter than those on p.

No. 9

A *q:* Proximal, negative segment corresponding to the secondary constriction. Two evenly spaced, medium fluorescent bands distal to the negative segment.

B *p:* Central, medium fluorescent band.

C *q:* Proximal, negative band (secondary constriction) variable in length.

No. 10

A *p:* Medium fluorescence.
q: Three evenly spaced bands; the most proximal one intense and the others medium in fluorescence.

No. 11
A *p:* Medium fluorescence; longer than 12p.
q: Short medium fluorescent band adjacent to the centromere; separated by a negative band from a more distal, medium fluorescent segment.

No. 12
A *p:* Medium fluorescence; shorter than 11p.
q: Medium fluorescent band adjacent to the centromere; separated by a short, negative band from a more distal, medium fluorescent segment. Distal segment longer than that of 11q.

No. 13
A *q:* Distal half intense.
B *q:* Distal half intense; divisible into two bands.
C *p:* Satellites and/or short arms with variable fluorescence.
q: Proximal, intense band.

No. 14
A *q:* Proximal half intense. Distal half pale; medium fluorescent band close to the distal end.
C *p:* Satellites and/or short arms with variable fluorescence.

No. 15
A *q:* Proximal half medium in fluorescence. Distal half pale; less fluorescent than either 13q or 14q.
C *p:* Satellites and/or short arms with variable fluorescence.

No. 16
A *p:* Medium fluorescence, less fluorescent than q.
q: Proximal, negative segment corresponding to the secondary constriction. Distal to it, a medium fluorescent segment.
C *q:* Negative secondary constriction variable in length.

No. 17
A *p:* Overall pale fluorescence.
q: Two segments of similar length; proximal one pale and distal one medium in fluorescence.
B *q:* Narrow negative band separating proximal and distal segments.

No. 18
A *p:* Overall medium fluorescence.
q: Medium fluorescence; brighter than p.
B *q:* Two bands of medium intensity; proximal one longer and brighter than distal one.

No. 19
A Most weakly fluorescent chromosome in the karyotype. Short, proximal fluorescent bands on both arms; pale when compared to the whole karyotype.
B Fluorescent band longer and brighter on p than on q.

No. 20
A Overall pale fluorescence; p medium and q pale in fluorescence.

No. 21
A *q:* Proximal, intense segment. Distal, pale segment.
C *p:* Satellites and/or short arms with variable fluorescence.

No. 22
A Overall pale fluorescence.
B *q:* Narrow, pale band in center of arm.
C *p:* Satellites and/or short arms with variable fluorescence.

X Fluorescent pattern similar on both X chromosomes of the female.
A *p:* Proximal, pale segment. Central, medium fluorescent band.
q: Proximal, pale segment. Distal to it, a medium fluorescent band.
B *q:* Three evenly spaced, medium fluorescent bands; most proximal one brightest.

Y
A *p:* Overall pale fluorescence.
q: Proximal segment pale. Distal segment brilliant.

C *q:* The brilliant fluorescent segment on the end of q may vary in length and may be subdivided into two or more bands. The normal variation in length of the chromosome is associated with variation in length of the brilliant segment.

GIEMSA BANDING TECHNIQUES

In this section the results obtained by G, R, and C techniques are reported. In all instances, Q-banding is used as the reference method in order to establish the identity of the chromosomes and their characterization.

Banding Patterns Obtained by G-, R- and C-staining Methods

Chromosome No.	*G*	*R*	*C*
1	As for Q *except* C-band stains deeply	Reverse of G	Large, extends from centromere into q
2	As for Q		Small
3-8	As for Q	Reverse of G	Medium
9		Reverse of G except for C-band	Large, extends from centromere into q
10		Reverse of G	Medium
11			Medium, but larger than 10 or 12
12			Medium
13			Medium, but sometimes bipartite
14-15			Medium
16	As for Q *except* C-band stains deeply	Reverse of G	Large, extends from centromere into q
17	As for Q		Medium
18			Medium, but larger than 17
19-22	As for Q	Reverse of G	Medium
X			Medium
Y	Variable	Variable	Very small C-band at the centromere, plus large C-band on distal end of q

Comments on Proposed System of Identification. (1) Comparison of staining results.

There was uniform agreement that the G- and R-staining techniques gave comparable results to those obtained by Q-staining apart from the following exceptions (h = secondary constriction):

The G- and R-methods do not in general clearly demonstrate the variant bands which are visible with the Q-method near the centromeres of chromosome numbers 3, 4, 13, 14, 15, 21 and 22.

The morphologic variability of satellite size and density is reflected in the Q-, G- and C-staining techniques by size and staining intensity.

Chromosome Region	*Q*	*G*	*R*	*C*
1qh	negative	+	—	+
9qh	negative	—	—	+
16qh	negative	+	—	+
distal Yq	brilliant	variable	variable	+

(2) R bands.

The lightly stained bands in the G-staining technique are darkly stained by the R-band technique. The one exception is the secondary constriction region of chromosome 9, which is lightly stained by both techniques. The ends of chromosomes stain darkly with the R-band technique.

(3) Variable features.

The C-bands in chromosomes 1, 9, 16 and Y (distal band in q) are all associated with obvious morphologic variation.

(4) Late or early replicating X chromosomes cannot be distinguished by the Q-, G-, R- and C-staining methods.

REFERENCES

1. Caspersson, T., Lomakka, G., and Zech, L.: 24 fluorescence patterns of human metaphase chromosomes—distinguishing characters and variability. Hereditas, 67:89-102, 1971.
2. Chicago Nomenclature: Birth Defects: Original article Series II: 2, 1966. The National Foundation, New York.
3. Paris Conference: Standardization in Human Cytogenetics. Birth Defects; Original Article Series, VIII, 7, The National Foundation, New York, 1972.

Appendix B

SYMBOLS COMMONLY USED IN GENETIC PEDIGREES

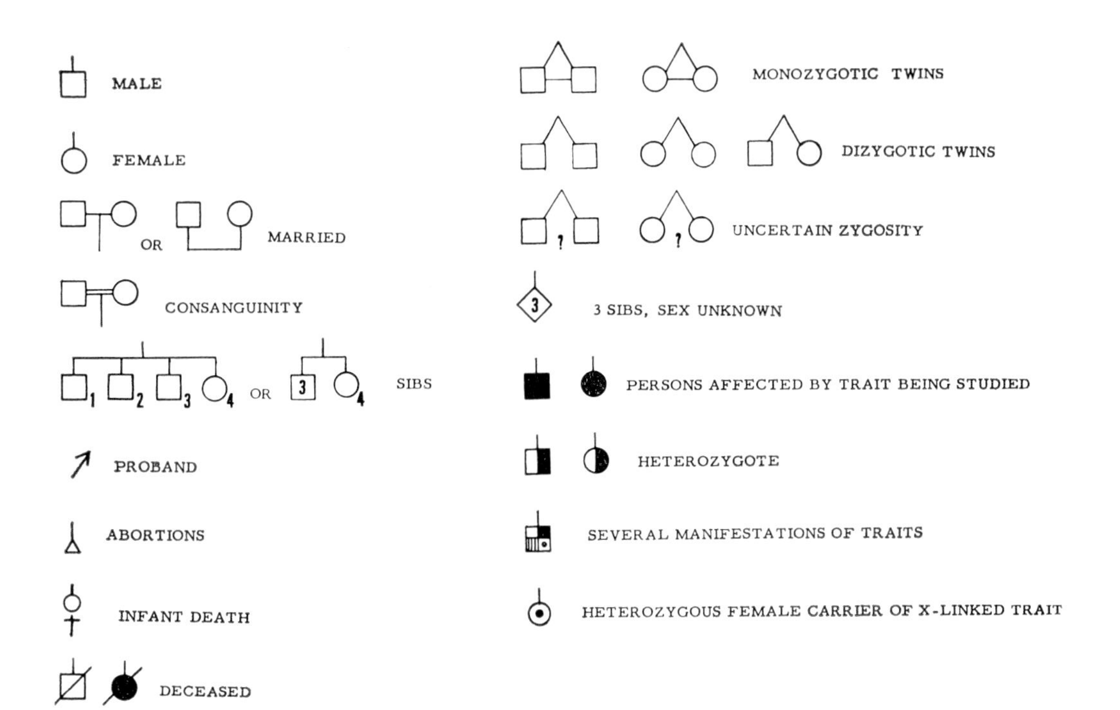

Appendix C

GENERAL SOURCES OF INFORMATION ON MEDICAL GENETICS

Review Article Series

1. Harris, H., and Hirschhorn, K. (eds.): Advances in Human Genetics. New York, Plenum Press.
2. Steinberg, A. G., and Bearn, A. G. (eds.): Progress in Medical Genetics. New York, Grune & Stratton.

Mendelian Traits

3. McKusick, V. A.: Mendelian Inheritance in Man. A catalogue of autosomal dominant, autosomal recessive and X-linked phenotypes. 3rd ed. Baltimore, John Hopkins Press, 1971. (The most complete catalogue of mendelian traits, with references.)
4. Stanbury, J. B., Wyngaarden, J. B., and Fredrickson, D. S. (eds.): The Metabolic Basis of Inherited Disease. 3rd ed. New York, McGraw-Hill, 1972. (A detailed description of the major inborn errors of metabolism and their underlying biochemistry.)

Syndromes and Birth Defects

5. Geeraets, W. J.: Ocular Syndromes. 2nd ed. Philadelphia, Lea & Febiger, 1969. (A well-catalogued but not illustrated collection of syndromes involving, but by no means limited to, the eye.)
6. Gellis, S. S., and Feingold, M.: Atlas of Mental Retardation Syndromes. Visual diagnosis of facies and physical findings. Washington, D.C., U.S. Government Printing Office, 1968. (Another illustrated atlas devoted to syndromes involving mental retardation.)
7. Gorlin, R. J., and Pindborg, J. J.: Syndromes of the Head and Neck. 2nd ed. New York, McGraw-Hill, 1971. (An atlas overlapping Smith's somewhat, but with more emphasis on adults.)
8. Holmes, L. B., Moser, H. W., Halldórsen, S., Mack, C., Pant, S. S., and Matzilevich, B.: Mental Retardation. An Atlas of Diseases with Associated Physical Abnormalities. New York, Macmillan, 1972.
9. Jablonski, S.: Illustrated Dictionary of Eponymic Syndromes and Diseases and Their Synonyms. Philadelphia, W. B. Saunders, 1969. (A handy guide through the confusion created by the wealth of eponyms attached to syndromes.)
10. V. McKusick (ed.): International Conferences on the Delineation of Syndromes. The National Foundation-March of Dimes. (A series of conference proceedings profusely illustrated; each volume is more or less devoted to one or more organ systems, plus one on chromosome syndromes.)
11. Motulsky, A. G., and Lenz, W. (Eds.): Birth Defects. Amsterdam, Excerpta Medica, 1974.
12. Smith, D. W.: Recognizable Patterns of Human Malformations. Philadelphia, W. B. Saunders, 1970. (An excellent review of the problems of "dysmorphogenesis" and catalogue, well-illustrated and annotated, of syndromes, particularly those of the pediatric group.)
13. Warkany, J.: Congenital Malformations. Chicago, Year Book Medical Publishers, 1971. (An exhaustive source book of information and wisdom.)

Chromosomes

14. Hamerton, J. L.: Human Cytogenetics, Vols. I and II. New York, Academic Press, 1971.
15. Paris Conference (1971): Standardization in Human Cytogenetics. Birth Defects: Original Article Series, VIII: 7, 1972. The National Foundation, New York.

INDEX

Page numbers in *italics* refer to illustrations; those followed by t refer to tables.